Molecular Biology Techniques

A Classroom Laboratory Manual

Fifth Edition

Susan Carson
Department of Plant and Microbial Biology, North Carolina State University, Raleigh, NC, United States

Heather B. Miller
Department of Chemistry, High Point University, High Point, NC, United States

D. Scott Witherow
Department of Chemistry, Biochemistry, and Physics, The University of Tampa, Tampa, FL, United States

Melissa C. Srougi
Department of Molecular Biomedical Sciences and Biotechnology Program, North Carolina State University, Raleigh, NC, United States

ELSEVIER

ACADEMIC PRESS
An Imprint of Elsevier

Academic Press is an imprint of Elsevier
125 London Wall, London EC2Y 5AS, United Kingdom
525 B Street, Suite 1650, San Diego, CA 92101, United States
50 Hampshire Street, 5th Floor, Cambridge, MA 02139, United States

ISBN: 978-0-443-26714-7

For Information on all Academic Press publications
visit our website at https://www.elsevier.com/books-and-journals

Publisher: Peter Linsley
Acquisitions Editor: Emma Hayes
Editorial Project Manager: Ruchi Bhargava
Publishing Services Manager: Deepthi Unni
Production Project Manager: Surya Narayanan Jayachandran
Cover Designer: Mark Rogers

Front image credit: The cover artwork was designed by Electra Scott, using knowledge of molecular cloning gained while performing many of the experiments described in this book during her undergraduate studies in biochemistry at The University of Tampa, Tampa, FL. All rights reserved.

Typeset by MPS Limited, Chennai, India

Printed in India

Last digit is the print number: 9 8 7 6 5 4 3 2 1

Contents

For additional instructional material, such as answer keys, please see the following instructor site for the book https://www.educate.elsevier.com/book/details/9780443267147

About the Authors

Dr. Susan Carson is a Professor of Plant and Microbial Biology at North Carolina State University (Raleigh, NC) where she has served on the faculty since 2001. She directs the Master of Microbial Biotechnology Program and also leads a university-wide faculty development program focused on enhancing students' critical and creative thinking skills across disciplines. Prior to her current role, Dr. Carson spent over a decade leading curriculum development for the North Carolina State Biotechnology Program and two years as a Program Officer at the National Science Foundation Division of Undergraduate Education. Her current work focuses on college-level biology education, enhancing students' higher order thinking across disciplines, and integration of Design Thinking in a Professional Science Masters (PSM) program. She graduated from Rutgers University (New Brunswick, NJ) with a BS in Biotechnology, and from the University of North Carolina (Chapel Hill, NC) with a PhD in Microbiology.

Dr. Heather B. Miller is an Associate Professor of Biochemistry at High Point University (High Point, NC). Her area of scientific expertise is RNA biology. Her research focuses on MRSA and other pathogens' gene expression when challenged with antibiotics and novel antibiotic adjuvants. She has mentored over 30 undergraduate students and is the Principal Investigator of a National Institutes of Health (NIH) Academic Research Enhancement Award (AREA). She was named a Henry Dreyfus Teacher-Scholar for her outstanding accomplishments in research and education in the chemical sciences. She has developed and taught multiple biochemistry and biotechnology courses and has published and presented a number of peer-reviewed papers in the scholarship of teaching and learning. Dr. Miller graduated from Clarion University of Pennsylvania (Clarion, PA) with a BS in Molecular Biology/Biotechnology, and from Duke University (Durham, NC) with a PhD in Molecular Genetics and Microbiology. She completed a teaching postdoctoral position in the Biotechnology Program at North Carolina State University.

Dr. D. Scott Witherow is an Associate Professor of Biochemistry at The University of Tampa (Tampa, FL). His research focuses primarily on the regulation of G-protein-mediated signal transduction processes and biochemical education and pedagogy. In addition to mentoring over 20 undergraduate research students, Dr. Witherow has designed and published biochemistry lab courses and has been recognized by the American Society for Biochemistry and Molecular Biology as an education fellow for his efforts in the areas of instruction and assessment. He graduated from Rollins College (Winter Park, FL) with an AB in Chemistry and from the University of Miami (Miami, FL) with a PhD in Molecular and Cellular Pharmacology. Following traditional postdoctoral research positions at Duke University, he transitioned into pedagogy through a teaching postdoctoral position at North Carolina State University.

Dr. Melissa C. Srougi is an Associate Professor in the Department of Molecular Biomedical Sciences and the Biotechnology Program at North Carolina State University (Raleigh, NC) with over 16 years of research experience in higher education. Her scientific areas of expertise are in experimental cancer chemotherapeutics as well as the scholarship of teaching and learning. She is passionate about expanding opportunities for undergraduate students to participate in research and actively trains undergraduate students in her laboratory. She is co-Principal Investigator of a National Science Foundation (NSF) funded Research Experience for Undergraduates Program and the Director of the Graduate Certificate in Molecular Biotechnology. Dr. Srougi has received funding from various federal and private agencies, including the NSF, BioMade, Department of Defense, Department of Education and the American Cancer Society. She teaches a wide variety of undergraduate/graduate interdisciplinary STEM lecture and lab courses focusing on biochemistry, cancer drug discovery, and biotechnology. Dr. Srougi serves as an Associate Editor for *Frontiers in STEM Education*. She has published numerous peer-reviewed papers on the scholarship of teaching and learning with a focus on collaborative learning, Course-based undergraduate research experiences (CUREs) and metacognition. Dr. Srougi graduated from the University of Toledo (Toledo, OH) with a BS in Biology and from Case Western Reserve University (Cleveland, OH) with a PhD in Pharmacology.

About the authors

Preface

Recombinant DNA technology touches many aspects of our lives. From drought-tolerant crops to biofuels to pharmaceuticals, genetically modified organisms surround us. Throughout its history, biotechnology has relied heavily on DNA cloning to produce protein products more efficiently or to make modified genes or combinations of genes. *Escherichia coli* or other hosts serve as cellular "factories" to produce large amounts of protein. Human insulin, for example, is produced recombinantly using many of the same techniques described in this book.

In the seven years since the fourth edition of *Molecular Biology Techniques* was published, molecular biology research and its tools have undergone major transformations and continue to evolve.

Modulating gene expression has become a mainstay in many laboratories. RNAi and CRISPR, especially, have revolutionized the way scientists study gene function, as well as design selective therapies for various disease states. As these technologies are becoming a cornerstone in all facets of modern medicine and biotechnology, students will benefit from a working knowledge of their theory and methods. Additionally, in silico tools to aid biotechnology research are increasing in number and accessibility. Many of these tools are now integrated into electronic lab notebooks, which greatly accelerate what can be done at the bench by planning and predicting on the computer. Moreover, gene expression experiments have taken a leap to high-throughput platforms that allow scientists to take a snapshot of entire transcriptomes in an experimental setting of their choice. Therefore, the fifth edition of *Molecular Biology Techniques* includes expanded laboratory sessions focused on introducing students to experiments involved with in silico cloning, RNA-sequencing, and molecular docking analyses. These labs are versatile, and geared for use in an existing course or in conjunction with previous lab sessions. The new labs have a common thread, and use experimental tools familiar to students who completed earlier sessions.

What will we learn in the next seven years? Students taking this course today will be at the forefront of groundbreaking advancements, and we can't wait to see what unfolds!

Acknowledgments

We would like to thank the many people who contributed to this edition of *Molecular Biology Techniques*. As this project has grown over the years, so have the number of individuals who made this work possible. First, we would like to thank our editorial team at Elsevier for encouraging us to create a new edition. High Point University biology, biochemistry, and chemistry students were helpful in piloting the transcriptomics laboratory session and providing feedback on the previous edition. Students from The University of Tampa were helpful in providing feedback on the previous edition and piloting some of the advanced experiments. Thanks to North Carolina State University students for their suggestions and comments on the in silico cloning and molecular docking labs. We would also like to thank High Point University Wanek School of Natural Sciences and the Department of Chemistry, and the University of Tampa College of Natural and Health Sciences and the Department of Chemistry and Biochemistry for supporting classes that use this book, as well as the North Carolina State University College of Veterinary Medicine, Department of Molecular Biomedical Sciences and the Biotechnology Program. Finally, the authors express their heartfelt gratitude to their spouses and children, whose unwavering support and patience made this book possible.

Note to Instructors

This lab manual was originally developed in the context of the curriculum offered by the North Carolina State University Biotechnology Program. Over the past 15 years, there have been numerous advances in molecular biotechnology that have changed how scientists test hypotheses. Despite these advances, many common techniques still remain the same. In this fifth edition, we have added a new section of material dedicated to in silico techniques that complement "wet lab" experiments. The laboratory sessions described in this manual are designed to prepare students for more specialized upper-level courses in biochemistry and molecular biology and for independent research projects at both the undergraduate and graduate level.

A typical course using this manual meets at least 4 hours for laboratory per week coupled with several (~30 minutes) interim laboratory sessions throughout the semester. We recommend this schedule for Parts I–IV because, for several of the lab exercises, students must inoculate cultures or perform other short activities prior to their lab day. For classes performing cell culture experiments (Part V: Modulation of Gene Expression), it is also useful to meet more than once a week. Maintaining cells requires careful monitoring and maintenance, dictating the need for interim laboratory sessions. We have alerted instructors in the text that certain steps must be performed at particular times either by the class as an interim laboratory session or by the instructor or teaching assistant, "behind the scenes." Part VI offers flexibility, in that the in silico lab sessions can be integrated into an existing lecture or laboratory course. The sessions in this part of the manual can be completed in three sessions.

The majority of the laboratory exercises do not require the full 4 hours to complete for students who come to lab prepared. Additionally, there are a few labs for which incubation times are simply too long to reasonably include in the exercises. In these cases, the steps are included in the protocols with a note that the instructor will perform that particular part of the experiment for the class (for example, the induction of the fusion protein using IPTG, necessary for several laboratories, takes 2–4 hours). For this reason, we recommend offering the laboratory as an afternoon course so that the instructor can begin the incubations in the morning, rather than the middle of the night.

We are aware that earlier editions of this manual are being used at a wide range of institutions and levels, ranging from prep schools to graduate programs. While our aim is to design the experiments toward introductory molecular biology/biochemistry students, included in this edition are a number of Advanced Alternatives. These alternatives are written primarily for upper-level undergraduates or graduate students who have previous laboratory experience with many of the basic techniques described in the regular lab sessions. These alternatives include whole project ideas, such as cloning your own gene of interest, to more specific alternatives, such as altering the design of a western blot to enable it to be performed semi-quantitatively. These alternatives can be found at the beginning of Parts I–III.

Parts I–III comprise a 10–12 lab session course suitable for an introductory molecular biology or molecular biochemistry lab. These experiments include PCR cloning enhanced green fluorescent protein (*egfp*) cDNA into a bacterial expression vector (Part I: DNA Manipulation); screening for positive clones (Part II: Screening Transformants); and expression, purification, and quantification of the recombinant protein (Part III: Expression, Detection and Purification of Recombinant Proteins from Bacteria). The analysis and quantification of mRNA is described in Part IV, which can be done in three additional lab sessions. For a typical 15-week course, one can fit in all the experiments described in Parts I–IV, although many instructors choose to focus just on Parts I–III, leaving time for other classroom exercises, presentations, etc. Part V, Modulation of Gene Expression, focuses on controlling gene expression in mammalian cells. These lab sessions are designed to provide students with the background theory and working knowledge not only on how to culture mammalian cells, but also on how to alter gene expression in this model system using transient transfection, RNAi, and CRISPR technologies. These experiments are best performed as a separate course from earlier parts of the book, as they can be accomplished in 4–13 lab sessions. Part VI, which is new to this edition, can be integrated into an existing lecture or lab course, as it takes only three lab sessions. The most important thing to note is that all the lab sessions outlined in this manual are not designed to be implemented in a single semester. However, over the course of two semesters, there are a number of ways to perform all the experiments, depending on your specific curricular needs. Please see Fig. 1 in the Introduction for more details.

Throughout the text, we have added both time-saving and money-saving tips. These suggestions are included to help instructors use their resources wisely. In this new edition, we have also added more online resources, where students and instructors can find helpful information. Importantly, all online materials and software are open-access, and freely accessible without added cost.

In preparing students for potential laboratory work after graduation, it may be helpful to introduce a virtual (electronic) lab notebook rather than a typical bound notebook. We have found these to have many advantages. First, the instructor can have access to student notebook entries 24/7 from their computer. For the student, the virtual lab notebook enables one to simply upload gel or blot images, Excel spreadsheets, and more, rather than printing, cutting, and taping into a bound notebook. They can also quickly create tables, checklists, and protocols. These notebooks eliminate the problem of illegible handwriting and lost notebooks. In addition, some have incorporated modern cloning simulations, oligo, sgRNA design, and sequence analysis tools. Entries can still be printed, as well. Benchling (http://www.benchling.com) is one version that is freely available. LabArchives is also popular (http://www.labarchives.com), but there is an associated fee for use. We have integrated Benchling into Lab Session 25 where students can use the CRISPR guide and PCR primer design tools, as well as Lab Session 31, where students can use the Gibson Assembly tool for planning a cloning experiment.

For instructors implementing the experiments described in Part V, Modulation of Gene Expression, these labs require a Biosafety Level 2 (BSL-2) laboratory space for working with HEK293 cells. Before engaging in these experiments, please ensure that you have the requisite approval through your institution's biosafety organization. BSL-2 guidelines can be found through the Centers for Disease Control and Prevention (https://www.cdc.gov/training/quicklearns/biosafety/).

We recommend using commercially prepared competent *Escherichia coli* for the experiments in this course, as they tend to perform consistently from batch to batch. If your budget does not permit this, it is possible to prepare "homemade" competent cells and store them in a −80°C freezer. We have included the protocol for preparation of competent cells in the Appendix.

All antibodies described in this manual are available commercially. The pBIT and pEGFP-N1 plasmids are available at no cost to institutions of higher education for educational purposes from Dr. Scott Witherow at The University of Tampa. Contact Dr. Witherow at switherow@ut.edu and include in the subject heading "pBIT request." The pET-41a, pLentiCRISPR-E, and pLentiCRISPR-EGFP vectors are available commercially, and we are not licensed to distribute them. HEK293 mammalian cells, as well as HEK293 cells stably expressing GFP, can be purchased commercially for use in Part V, Modulation of Gene Expression.

Nomenclature

In the literature, the nomenclature for the abbreviations of the enhanced green fluorescent protein gene and its gene product has been inconsistent, at best, and downright confusing at worst. In this publication, we will use "*egfp*" to refer to the gene (either DNA or mRNA) and "EGFP" to refer to the gene product. Likewise, we will use "*gst*" for the glutathione-*S*-transferase gene and "GST" for its gene product. Bacterial genes discussed in this book will use standard bacterial nomenclature with the gene name lowercase italicized, and the gene product with a capitalized first letter and not italicized. For example, the gene for the *lac* repressor is "*lac*I" and its gene product is "LacI."

Introduction

CONCEPTUAL OUTLINE FOR EXPERIMENTS

Parts I, II, and III of this manual describe the process of making a fusion protein by joining genes from two organisms: one from *Escherichia coli* (*gst*) and an enhanced gene derived from the green fluorescent jellyfish *Aequorea victoria* (*egfp*). Expression of the fused gene will produce a single protein in bacteria, which can be readily purified due to the GST portion of the fusion protein. The purified protein can be visualized by fluorescence and quantified.

Part IV moves beyond the protein level to study mRNA levels using reverse transcription PCR (RT-PCR). Levels of mRNA in the presence and absence of inducing molecules can be quantified and compared.

In Part V, we leave the bacterial system and move into mammalian cell culture. In this part of the manual, EGFP is constitutively expressed in a mammalian cell line and gene expression is modulated using two different methods: RNAi and CRISPR.

In Part VI, students are introduced to several in silico tools to study a bacterial protein binding an antibiotic. This part of the manual includes sessions that use an open access web-based cloning tool, publicly available RNA-sequencing data, and open access molecular docking tools.

In summary, the various parts of this manual are designed to explain the theory, outline the experimental design, and guide in the analysis of the data for common biochemical and molecular biology techniques used in many research laboratories Fig. 1.

GENERAL OPERATING PROCEDURES

- Reagents: Aliquoting reagents and supplies for everyone in the course at the same time is difficult. Your patience and cooperation are greatly appreciated. If you run out of a reagent or enzyme and cannot find it, please ask an instructor or TA for more.
- Enzymes: Enzymes are frequently expensive and can be ruined by prolonged exposure to room temperature or by contamination. Always use a fresh, sterile pipette tip for each enzyme. Keep the enzyme on ice; never put it in a microcentrifuge rack at room temperature. Always add enzymes to your reactions as the last component; addition of enzyme to unbuffered solutions can kill its activity.
- Pipette tips: You will be provided with three boxes of pipette tips: P10, P20/P200, and P1000. When you run out of tips, notify your instructor.
- Equipment: Please make sure that you understand the proper use of this equipment before attempting to operate it. If in doubt, ask an instructor, not another student. In addition to costly repairs, improper use of this equipment can be highly dangerous. If equipment is damaged in any way, please report it so that we can have it repaired.
- Lab Safety: Please abide by all lab safety instructions provided by your instructors. A list of general lab safety rules is provided, but is superseded by rules specific to your institution and course.

LABORATORY SAFETY

Hazards that you may be exposed to during the course of the laboratory exercises include working with toxic compounds and ultraviolet (UV) irradiation. Special precautions must be taken when working with recombinant DNA. To ensure the safety and well-being of students and support staff, the following rules will be strictly enforced.

These rules *must* be observed at all times in the laboratory:

1. No drinking or eating is allowed in the laboratory. No food or drinks should be brought into the laboratory.
2. Safety glasses provided must be worn at all times in the laboratory.

EXPERIMENTAL PROCEDURES

Part I: Manipulation of DNA — 5 Lab Sessions
- Isolate plasmid DNA using cultures of bacteria containing the *E. coli* expression vector pET-41a
- Use restriction enzymes for forced cloning to cut the pET-41a vector
- Use PCR to amplify the insert (*A. victoria egfp* DNA) from pEGFP-N1, and restriction digestion to form sticky ends
- Use DNA ligase to "paste" the vector and insert DNA together
- Introduce the ligated DNA into *E. coli* to create the recombinant *gst::egfp* DNA molecule

Part II : Screening Transformants — 1-3 Lab Sessions
- Isolate DNA from transformants and digest with restriction enzymes to further validate the presence of the *gst::egfp* fusion DNA
- Confirm the EGFP-positive clones by fluorescence
- Verify that single nucleotide errors did not occur in the *egfp* gene during PCR cloning by Sanger sequencing

Choose at least one of the three screening methods

Part III: Expression, Detection, and Purification of Recombinant Proteins From Bacteria — 4 Lab Sessions
- Use SDS-PAGE and western blot analysis to confirm the expression of the fusion protein
- Induce a large-scale culture of the transformed bacteria with IPTG to express high levels of the fusion protein
- Purify the fusion protein on an affinity column
- Perform a Bradford protein assay to quantify the amount of protein in eluted fractions
- Use SDS-PAGE of purification fractions to check for purity and degradation

Part IV: Analysis of mRNA Levels — 3 Lab Sessions
- Purify total RNA from a positive bacterial clone
- Assess purity and concentration
- Perform reverse transcription to create a cDNA template
- Perform quantitative PCR to determine mRNA levels after induction with IPTG and/or lactose

Part V: Modulation of Gene Expression — 4-13 Lab Sessions
- Culture a mammalian cell line, HEK293
- Transiently transfect cells with the expression vector pEGFP-N1
- Create stable, EGFP-expressing cells
- Downregulate *egfp* expression using RNAi and/or CRISPR
- Silence a human gene's expression using CRISPR

Choose to perform RNAi and/or CRISPR experiments

Part VI: In silico Tools — 3 Lab Sessions
- Design a Gibson Assembly cloning experiment to insert coding sequence for a *S. aureus* gene, *pbp2*, into the pET-41a vector using a web-based cloning tool
- Access and analyze RNA-seq data to reveal mRNA level changes transcriptome-wide when *S. aureus* is challenged with an antibiotic
- Use molecular docking tools to assess PBP2 protein binding an antibiotic

FIG. 1 Conceptual outline for experiments. The number of lab sessions for each part of this manual are shown to the right. Suggested plans for laboratory courses are detailed in the Note to Instructors section.

3. Wear a lab coat and long pants when working in the lab.
4. No sandals or other open-toed shoes are allowed in the laboratory at any time.
5. Always wear gloves when working with any hazardous or potentially hazardous substance. Remove your gloves before leaving the laboratory and change your gloves frequently to prevent contamination of equipment, etc., with caustic agents.
6. Long hair must be tied back at all times and avoid loose-fitting clothing to avoid hazards associated with open flames, sterile cultures, and hazardous chemicals.
7. Dispose of microorganisms, including the tubes used for their growth, in bags marked "BIOHAZARD" for autoclaving. Liquid medium containing microorganisms will be collected in specially marked containers containing bleach, frequently referred to as the bacterial graveyard.

8. Dispose of glass *only* in properly marked boxes designated for glass disposal. Do not put glass or any sharp object in the autoclavable bags marked biohazard!
9. Keep your lab bench free of unnecessary clutter. Use cabinets and drawers for storing personal items and extra supplies. At the end of the day, your bench should be clean, and your equipment put away.
10. Wear ear protection when working with the sonicator.
11. Always wear a UV-protective full-face shield when using the transilluminator if it is not in a safety cabinet. Your safety glasses are *not* UV protective. Do not try to analyze your gel on the transilluminator. Observe the gel on the monitor if you are using a digital imager or take a photograph and analyze the photograph if you do not have a digital imager.
12. Wash your hands thoroughly before you leave the laboratory.
13. All spills should be cleaned up immediately. Notify an instructor if you spill a potentially hazardous chemical or liquid containing live microorganisms.
14. Immediately report all accidents such as spills, cuts, burns, or other injuries to an instructor.
15. Know the location of the fire extinguisher, eye wash station, emergency shower, and emergency exits.
16. If you have trouble with a power supply or the leads to a gel, report it to an instructor. If you see someone receiving an electrical shock, use a non-conducting object, such as a plastic beaker, to break the circuit or you may also receive a shock.
17. Leave all laboratory facilities and equipment in good condition at the end of the class. Before leaving the laboratory, check to make sure that all electrical equipment is turned off and that the gas to the Bunsen burner is turned off.
18. No pets are allowed in the laboratory.
19. Dispose of hazardous chemicals only in designated containers. Do not pour them down the sink.
20. An up-to-date immunization against tetanus is strongly recommended.
21. If working with HEK293 cells (Part V), be aware that these cells require Biosafety Level 2 (BSL-2) compliance (https://www.cdc.gov/training/quicklearns/biosafety/). This means that these materials pose a moderate risk to those handling them and the surrounding environment and must be handled in an approved BSL-2 environment.

GENERAL HELPFUL TIPS WHILE WORKING IN THE LAB

1. When using water in the lab, you should always use, at minimum, deionized water. For certain applications, water that has been even more thoroughly treated might be required. Sterile water (deionized water that has been autoclaved) is called for in many experiments throughout this manual to minimize potential bacterial contamination. Additionally, when working with RNA molecules, special RNase-free (nuclease-free) water is often necessary. Under no circumstances should regular tap water be used.
2. When pipetting, be sure to change pipette tips between each sample. Small amounts of liquid remain on the pipette tip after pipetting. If a used tip is introduced into another sample, contamination is possible.
3. Be sure to mix samples well. One of the main reasons experiments fail is due to improper mixing. In many cases, a vortex can be used to mix. In others, the force of the vortex is too strong and can be problematic for your sample. For example, when performing plasmid DNA purification from bacteria, vortexing the sample after cell lysis can cause undesired shearing of the plasmid DNA. In these cases, more gentle mixing is called for (described in the lab manual).
4. When growing bacterial cultures overnight at 37°C, cultures should be grown for approximately 16 hours. Longer incubations can lead to the formation of satellite colonies, which are smaller colonies lacking the plasmid of interest surrounding the main colony. If longer incubation times are needed, growing at a lower temperature for a longer time also works. Typically growth that occurs at 16 hours at 37°C can also be achieved by growing cultures at room temperature for 48–72 hours. Those steps are mentioned in the protocol as time-saving tips.
5. Bacterial plates should be unwrapped and inverted (lid side down) in the bacterial incubator when colonies are originally forming. When being stored in the refrigerator for long periods of time, plates should be stored wrapped in Parafilm and placed inverted (lid side down) to avoid condensation accumulating on the surface of the place.
6. Loading of agarose gels should be done with regular pipette tips (P10, P20/200). For loading of SDS–PAGE protein gels, narrow gel-loading tips are frequently preferred. Gel-loading tips work best when the plunger is pressed/depressed slowly to avoid the formation of air bubbles. Additionally, when pipetting your sample into the wells of the gel, do not press the plunger all the way down. Doing so will cause air to be pushed into the well, causing the sample in the well to be displaced. It is much better to leave a small volume (<1 μL) in your tip than to push your sample out of the well. Regular pipette tips can be used to load SDS–PAGE gels, but care needs to be taken that the tip is rested at the top of the well and not forced between the two plates surrounding the gel, as the tip can cause the plates to separate.
7. Spray your work area with 70% ethanol and wipe the area down before starting to work and when you finish. This will help minimization contamination, particularly when doing bacterial work. This is an absolute requirement if working with mammalian cells in a cell culture hood.

Part I

Manipulation of DNA

The goal of these laboratory exercises is to fuse a jellyfish gene with a bacterial gene and to express a single protein from this hybrid DNA sequence. Why would you want to do this? Molecular shuffling of genetic sequences, or gene cloning, is a powerful tool for understanding biological processes and for biotechnological applications. Using basic tools developed in *Escherichia coli*, we can ask questions about other, more complicated organisms.

Scientists have exploited *E. coli* both as a workhorse for producing DNA and as a source of well-characterized sequences to direct the transcription and translation of foreign DNA into protein. With genetic sequence information being produced at a breathtaking rate, the limiting factor is not in sequencing DNA, but in our understanding of the function of the products of these sequences.

In terms of practical biotechnology applications, it can be a huge advantage to clone the gene encoding a difficult-to-purify protein into *E. coli* so that the purification process can be accomplished less expensively and to a greater degree of purity (and oftentimes more ethically, especially if a human gene is involved!). The first recombinant protein to be produced and marketed was human insulin in the early 1980s, which has been invaluable to countless diabetics. The basic tools you will learn in this class will enable you to clone, express, and purify recombinant proteins. They will enable you to begin to probe the function of any protein for which a gene has been identified, and will give you the conceptual background needed for tackling more advanced techniques.

Other hosts are now commonly used for cloning DNA and expressing recombinant proteins, such as members of the bacterial genus *Bacillus*, as well as eukaryotic hosts including numerous species of yeast and other fungi, plants, insect cell culture, mammalian cell culture, and even whole, live mammals ("pharming"). Many of the recombinant DNA methods used in this course are applicable to cloning in other hosts.

The gene we will be cloning and expressing is the enhanced green fluorescent protein gene, *egfp*. EGFP is a brightness-enhanced variant of the green fluorescent protein from the jellyfish *Aequorea victoria*.[1] The gene encoding the green fluorescent protein (and its variants, including the *egfp* gene) is widely used as a "reporter gene" or "marker." A reporter gene is a gene that is used to track protein expression. It must have phenotypic expression that is easy to monitor and can be used to study promoter activity or protein localization in different environmental conditions, different tissues, or different developmental stages. Recombinant DNA constructs are made in which the reporter gene is fused to a promoter region of interest and the construct is transformed or transfected into a host cell or organism. EGFP can also be used to mark (or tag) other proteins by creating recombinant DNA constructs that express fusion proteins that fluoresce and can be tracked in living cells or organisms. In this project, we are not using *egfp* as a reporter, but rather as a convenient gene to clone and assay for, as we learn the basic techniques of recombinant DNA manipulation and protein expression.

REFERENCE

1. T. Yang, L. Cheng, S.R. Kain, Optimized codon usage and chromophore mutations provide enhanced sensitivity with the green fluorescent protein, Nucleic Acids Res 24 (22) (1996) 4592–4594.

Advanced Alternatives Within Part I

The experiments outlined in this textbook are designed primarily for an introductory level molecular biology or biochemistry course, particularly Part I (Manipulation of DNA), Part II (Screening Transformants), and Part III (Expression, Purification, and Analysis of Recombinant Proteins from Bacteria). Based on feedback from previous editions of the textbook, we have included in this edition suggestions for advanced classes, noted as Advanced Alternatives (AAs), which appear at the beginning of each part. These are designed to offer students unique ways to navigate through a similar project, without following the step-by-step procedures outlined in this book. Students who are familiar with basic techniques (PCR, restriction digests, protein purification) can use these advanced suggestions to create an independent research project. It is important to note that depending on the changes that are implemented, the amount of time to complete experiments might change significantly. Because of that, we encourage instructors to be cautious in their planning and to be mindful of the logistical implementation of any of these changes. Parts IV (Analysis of mRNA Levels) and V (Modulation of Gene Expression) do not have AAs listed, as they are designed to already incorporate advanced techniques as well as somewhat flexible experimental design.

The AAs are broader suggestions that would either change the experimental design completely (i.e., clone a gene other than *egfp*) or alter the sequence of experiments outlined in the book (skip one of the screening experiments). Some suggestions are labeled as additional, while others are adaptations of the specific experiments already described (semiquantitative gel analysis; use of a protease to cleave GST from EGFP).

ADVANCED ALTERNATIVES, PART I—MANIPULATION OF DNA

The goal of this part is to clone *egfp* into an expression vector that can be used to express a fusion protein with a tag to facilitate purification. Advanced students or instructors looking for variety in the class might elect to alter the project, while still keeping the same overall theme of the course (clone, screen, purify). The following are some suggestions for AAs that could be performed by more experienced students or instructors of smaller classes who can interact closely with students in a course-based undergraduate research experience (CURE). In particular, a number of changes suggested for this part—DNA Manipulation—will carry throughout the project. The experimental flow comments are to point out in general terms how the alternative would affect downstream experiments within the course.

AA1.1 Clone *egfp* With Different Primers

Instead of using the primers designed and explained in the text, students can design their own primers. Primers are inexpensive and could be ordered for each student/lab group. Students would gain the experience of designing and ordering primers, and also resuspending and diluting them to their proper concentrations for use. General guidelines for primer design are included in Advanced CRISPR, Lab Session 25. One additional note to be mindful of in designing primers for cloning is that nucleotides introduced for cloning (restriction sites and an extra 3–4 nucleotides on the 5′ end of the restriction sites) do not get factored in to the Tm calculation or the temperature used as the annealing temperature in the PCR reaction. As an instructor, you could dictate which restriction sites were to be used to limit the number of restriction enzymes needed for downstream applications.

Additional materials needed

PCR primers designed by students; restriction enzymes (if different than NcoI and NotI).

Experimental flow

The experiments downstream of this would remain the same (if NcoI and NotI were still used for cloning). Primers should be designed and ordered approximately one week before they are needed.

AA1.2 Clone *egfp* Without PCR

While cloning by PCR is often the fastest, most efficient way to transfer a sequence from one vector to another, in some situations, subcloning can be done easily without PCR using only a restriction digest of both the vector and insert. In the case of pEGFP-N1 and pET-41a, this can be done by digesting both vectors with NcoI and NotI, purifying the fragments from an agarose gel, then proceeding to ligation. Details on gel purification are included in the Advanced CRISPR lab session 25.

Additional materials needed

Sufficient quantity of pEGFP-N1 (1 μg per lab group); DNA gel purification kit (sold by many different companies).

Experimental flow

Both the insert and vector could be purified and digested together in one lab session. The following session could be used for purification of the digested components from an agarose gel using a DNA gel purification kit. The next week, ligation and transformation could be performed. Overall, this procedure should take one week less, and students would still have the opportunity to perform PCR by performing a PCR screen outlined in Part II.

AA1.3 Clone a Gene Other Than *egfp*

The standard experiments in this manual all revolve around the cloning and expression of *egfp*. This gene was chosen because the fluorescence makes for an extremely fast and simple functional assay. Most genes, however, do not come with the same convenience. The entire project outlined in this textbook can be performed with any gene. If an instructor or student is studying a particular gene or protein as a research project, the sequence of experiments outlined here can be performed to create and purify that specific fusion protein. While the experiment would be easiest if the cDNA is present in a different vector, RNA could be purified from cells and cDNA could be synthesized using commercially available kits, which could then be amplified by PCR to clone into pET-41a. Details on those procedures are found in Part IV. Whichever restriction enzymes are used, the student would need to confirm that those recognition sites are not found in the gene of interest.

Downstream experiments dealing with the fusion protein would have to be adjusted. Specific antibodies would need to be purchased for western blot experiments, as functional assays dealing with fluorescence would no longer work. Despite being less convenient, this set of experiments would provide a more realistic research experience to advanced students (including the very real possibility of needing to troubleshoot gene expression conditions to maximize soluble expression of the fusion protein). The benefit of this particular AA is the ability to use the course as a CURE, where the research performed in class links with a student's or instructor's specific research project. A student who clones, expresses, and purifies a protein would be positioned to obtain publishable data by coupling these experiments with independent research to study the protein's function.

Additional materials needed

cDNA of a gene of interest, restriction enzymes (if not using NcoI and/or NotI) PCR primers specific to the gene of interest, antibodies to the protein being studied.

Experimental flow

The experimental flow would remain the same throughout the cloning part of the experiment. The only changes would occur later during the screening and analysis of purification that take advantage of the fluorescence of EGFP. For screening, students would rely on the other experiments (miniprep analysis, PCR screen, sequencing data) to confirm the results of the cloning. Analysis of the purification fractions would only be able to be done by SDS–PAGE.

AA1.4 Semiquantitative Agarose Gel Electrophoresis

Any time an experiment is performed, it makes sense to acquire as much data as possible. The main point of agarose gel electrophoresis is typically to resolve bands of DNA to observe the sizes of the fragments. Additional data can be observed from a gel when precise volumes of the samples are loaded on a gel. Semiquantitative experiments are exactly what they sound like. They are quantitative, but not as quantitative as taking spectrophotometric measurements of a sample. Still, by coming up with a close approximation of the concentration of the DNA in your sample, you can confirm spectrophotometric measurements. In cases where multiple DNA fragments are in a sample, the concentrations of the individual fragments can be determined, while a spectrophotometer will only be able to determine the total nucleic acid concentration.

Many imagers and gel documentation systems have software that can facilitate this quantification process. If you have such an imager, follow the instructions. Even without an imager, a general estimate of the concentration of a DNA sample can be made. First, a DNA ladder must be used where the masses of the individual bands are known. The 1 kb DNA ladder from NEB (catalog number N3232) works for this, but ladders with a wider mass range of standards are also available and preferable for more accurate calculations (such as the High Mass DNA ladder from Invitrogen; catalog number 10496016). By loading a defined amount of the ladder, you will know the mass of each individual band within the ladder. Compare your sample to the bands in the ladder. (It can be helpful to load a few different amounts of your sample to ensure you have a mass of DNA comparable to bands in the ladder.) If your band is the same intensity (brightness) as a band in the ladder, they have the same mass. If your band is twice as bright as a band in the ladder, you can say that your sample has twice the mass. When possible, it is best to compare the intensity of your band to the band in the ladder at a similar size. For instance, if your fragment is 0.9 kb, you would do better to compare it to the 1.0 kb fragment in the ladder as opposed to the 10 kb or a 0.1 kb band, as the size of the band can affect how well it is stained by the GelRed dye.

Once you know the mass of your band, you can calculate the concentration by dividing by the volume of your DNA sample. Keep in mind, the volume of your DNA sample is NOT the volume you loaded on the gel. The sample loaded on the gel also includes DNA loading buffer, which you do not want to include in the calculation for concentration of your DNA sample. If you mixed 10 μL of DNA sample with 2 μL of 6× DNA loading buffer, then loaded the entire sample on the gel, the volume of DNA is 10 μL. If you mixed 10 μL of DNA with 2 μL of 6× DNA loading buffer, then loaded 10 μL of that mixture on the gel, the volume of DNA is 8.3 μL. This general method of semiquantitative electrophoresis can also be applied to other protocols as well, such as SDS–PAGE, which will be discussed later in the book.

Additional materials needed

High Mass DNA ladder (Invitrogen catalog number 10496016 or comparable). The standard 1 kb DNA ladder from NEB (catalog number N3232) can still be used but does not provide as wide a mass range.

Lab Session 1

Getting Oriented; Practicing With Micropipettes

Goal: Starting next week, you will be working on a laboratory project that will build throughout the entire semester. Before embarking on that journey, it is important to familiarize yourself with your lab space and to master the use of the work-horses of the molecular biology/biochemistry lab: the micropipettes. If your instructor has not a given safety orientation yet, they will do so today.

I. EQUIPMENT CHECKLIST

It is important to familiarize yourself with the work environment and laboratory equipment before beginning experiments. If the laboratory space in which you are working is shared by other laboratory sections at different times, much of the equipment can be shared. There are certain items, however, such as buffers and sterile disposable items that should not be shared between lab groups. Take a moment to go through your bench, shelves, and drawers to identify equipment and reagents. Use the checklist below and notify your instructor if anything is missing. Items that are indicated as "per group" should not be shared between different sets of students on different lab days. Label these items with your initials and lab day.

- One power supply box
- One horizontal DNA mini gel apparatus for agarose gels
- Four micropipettes: P10, P20, P200, and P1000
- One box 1000 μL sterile tips *per group*
- One box 200 μL sterile tips *per group*
- One box 10 μL sterile tips *per group*
- One ice bucket (or cooler or styrofoam box for ice)
- One box of Kimwipes (Kimberly-Clark, Roswell, GA)
- One 15 mL and one 50 mL styrofoam test tube rack
- One pack of sterile snap cap tubes (17 × 100 mm) for overnight bacterial cultures
- One test tube rack for snap cap tubes
- One autoclaved container of 1.7 mL microcentrifuge tubes *per group*
- Two microcentrifuge tube racks
- One pack of disposable 10 mL pipettes
- One plastic (or electric) pipette pump
- One 50 mL graduated cylinder
- One 500 mL graduated cylinder
- One 2 L polypropylene beaker
- Two 1 L polypropylene bottles, one for deionized water and one for 1 Tris/Borate/EDTA (TBE) × buffer, *per group*
- One 250 or 500 mL Pyrex orange capped bottle or Erlenmeyer flask for melting agarose *per group*
- One thermal glove for handling microwaved agarose
- Labeling tape
- Permanent ink marker (Sharpie)
- One plastic squeeze bottle for 70% ethanol
- One plastic squeeze bottle for deionized water
- One ring stand with clamp
- One pair of blunt-ended forceps
- Two safety eye glasses or goggles
- One cardboard freezer box *per group*

Molecular Biology Techniques. DOI: https://doi.org/10.1016/B978-0-443-26714-7.00016-8

- Protein polyacrylamide mini gel electrophoresis unit
- Protein mini-transblot assembly
- Vortex mixer
- Microcentrifuge
- Bunsen burner
- Heat block
- Timer
- Parafilm
- Two waste containers; one biohazard and one nonbiohazard

II. MICROPIPETTING

Micropipettes are tools used to measure extremely small volumes of liquid typically necessary when performing molecular manipulations. We will use four different micropipettes in this course. Each micropipette is accurate to measure a defined range of volume, as shown in the Table 1.1. Your brand of micropipette may vary slightly from the ranges shown, particularly with the lower limit of the P1000 micropipette. Please refer to the manufacturer's instructions.

Setting the micropipettes to the desired volume can be a little tricky at first. Also, it is common for beginners to confuse the P20 and P200 since they typically use the same pipette tips: therefore, remember to check which micropipette you are using before drawing in solution. Many students accidentally measure 20 μL instead of 2 μL or vice versa because of such mix-ups.

When using micropipettes, one should always keep in mind significant figures. The correct number of significant figures is determined by all the numbers visible, plus the volume referred to by the tick mark (found on most brands of micropipettes). Since there are variations between brands of pipettes, most protocols will refer to a volume without indicating the exact number of significant figures. You should always set your pipette to the correct volume, including as many significant figures as possible. For instance, if a protocol instructs you to pipette 5 μL, you should set your pipette to 5.00 μL and carry the correct number of significant figures through any required calculations.

Use Fig. 1.1 and the following instructions to help with setting the micropipettes until you are confident to set them on your own.

Follow these instructions when using the micropipettes:

1. Set the desired volume by holding the pipette in one hand and rotating the dials with the other hand. Do *not* dial past the lower limit (000) or the upper limit (shown on the pipette; either 10, 20, 200, or 1000). Familiarize yourself with these settings.
2. Attach a tip to the end of the micropipette. To ensure an adequate seal, press the tip on with a slight twist.
3. Depress the plunger to the first stop. This part of the stroke displaces a volume of air corresponding to that indicated on the dial.
4. Immerse the tip to a depth of 2–5 mm in the liquid to be withdrawn. Immersing the tip deeper will cause liquid to adhere to the outside of the tip, causing errors in measurement.
5. Allow the plunger to slowly return to its original position. If the plunger snaps back, aerosols will form, contaminating the barrel of the micropipette and your solution.
6. Wait 1 second before removing the tip from the solution, to allow the introduced liquid to enter the pipette tip fully. Removing the tip too quickly from the solution may result in air occupying some of the calibrated volume. Check to

TABLE 1.1

Micropipette	Volume Range (μL)
P10	0.5–10
P20	2–20
P200	20–200
P1000	100 (or 200)–1000

No Permission Required

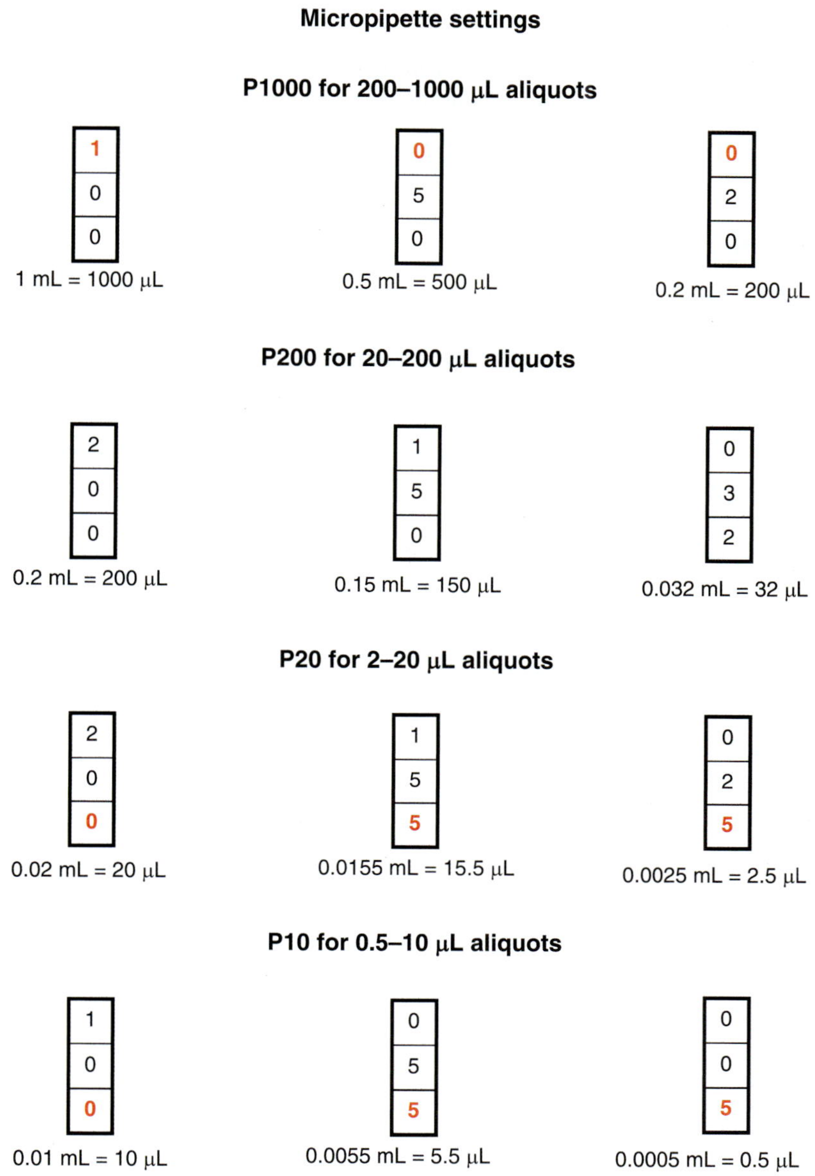

FIGURE 1.1 Micropipette settings cheat sheet. *No Permission Required*

make sure that there are no air bubbles and that the amount of liquid corresponds to the desired amount. Develop an eye for 1 μL volumes, as these are the hardest to pipette.

7. Place the tip against the side wall of the receiving vessel near the liquid interface or the bottom of the vessel. Slowly dispel the contents by depressing the plunger until the first stop. Remaining liquid can be dispelled by depressing the plunger to the second stop. Withdraw the tip from the solution and return the plunger to its original position. Check to make sure that no liquid remains in the tip. If there is a bead of liquid, reintroduce liquid from the receiving vessel to capture the bead and slowly expel the contents.
8. Discard the tip by pressing the ejector button.
9. *Always* use a new pipette tip when pipetting enzymes or the stock solutions may become contaminated. If you accidentally contaminate an enzyme solution, tell an instructor. Always use a new pipette tip for critical volumes, as in a dilution series, because as much as 10% of the volume may stay within the tip after delivery.
10. Working with tiny volumes requires patience and accuracy. The best way to deliver a 1 μL volume is to pick up the receiving tube and make sure that a 1 μL bead is formed on the side of the tube after delivery. In the case of enzymes, schlieren rings should be visible from the glycerol–water interface if the enzyme is dispelled directly into the solution.

III. LABORATORY EXERCISES

Note: While the other laboratory exercises for this course will build on each other, this one will not.

The purpose of this short exercise is to practice using the micropipettes (and test your technique). Each student will perform serial dilutions of the protein bovine serum albumin (BSA) and then compare their results against their lab partner's results using a visualization technique that uses a protein-binding dye.

A. Micropipetting Self-Test

Before proceeding further, each student should do a self-test on their micropipetting skills. Because 1 mL of water weighs approximately 1.00 g (the exact density is 0.99x, with x being temperature-dependent), students can test their micropipetting skills by pipetting into a tared weigh boat on a precision balance. To begin, place a weight boat on a precision balance and tare the balance so it reads 0 g. Next, pipette the appropriate volume of water onto the weigh boat and record the reading. You may tare the scale between volume readings. Continue until you have completed a minimum one set of self-tests.

"Passing" the self-test will ensure that you are selecting the correct micropipette for the given volume, and that your technique is correct. If your self-test does not fall into the right weight ranges, see your instructor for one-on-one feedback about your technique, and to test the calibration of your micropipettes.

Each student should perform at least one set of self-tests; selecting and adjusting their pipettors independently.

Self-Test 1

Volume to Measure (μL)	Weight in g (within 5%)
7.0	0.0070 (0.0067–0.0073)
33.5	0.0335 (0.0318–0.0352)
267.5	0.2675 (0.2541–0.2809)

No Permission Required

Self-Test 2

Volume to Measure (μL)	Weight in g (within 5%)
9.0	0.0090 (0.0086–0.0095)
26.5	0.0265 (0.0252–0.0278)
348.5	0.3485 (0.3311–0.3659)

No Permission Required

Self-Test 3

Volume to Measure (μL)	Weight in g (within 5%)
8.0	0.0080 (0.0076–0.0084)
43.5	0.0435 (0.0413–0.0457)
364.5	0.3645 (0.3463–0.3827)

No Permission Required

Self-Test 4

Volume to Measure (μL)	Weight in g (within 5%)
6.0	0.0060 (0.0057–0.0063)
32.4	0.0324 (0.0308–0.0340)
246.5	0.2465 (0.2342–0.2588)

No Permission Required

B. Preparing Bovine Serum Albumin Dilutions

You will be given a tube with 10 μL of a 10 mg/mL BSA solution. Prepare a dilution series of BSA standards in five tubes (labeled 1–5) according to the following scheme and Fig. 1.2. Attach a new pipette tip to the micropipette each time you make the dilutions. *Each lab partner should do a set of dilutions.*

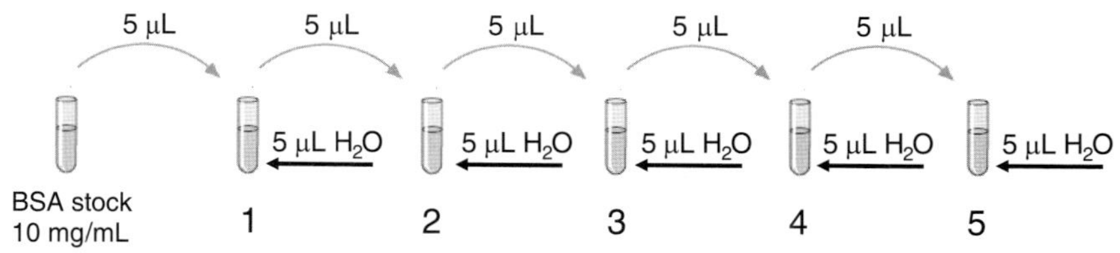

FIGURE 1.2 Serial dilution scheme. *No Permission Required*

Tube	Dilution	Protein Concentration
1	5 μL of stock (10 mg/mL)+5 μL dH_2O *This is a 2-fold dilution*	5.00 mg/mL
2	5 μL from tube 1+5 μL dH_2O *This is a 2-fold dilution*	2.50 mg/mL
3	5 μL from tube 2+5 μL dH_2O *This is a 2-fold dilution*	1.25 mg/mL
4	5 μL from tube 3+5 μL dH_2O *This is a 2-fold dilution*	0.63 mg/mL
5	5 μL from tube 4+5 μL dH_2O *This is a 2-fold dilution*	0.31 mg/mL

No Permission Required

C. Performing a Nitrocellulose Spot Test

Ponceau Red is a stain that quantitatively binds a protein. You will use a micropipette to deliver small amounts of BSA serial dilutions to the nitrocellulose membrane and then stain the nitrocellulose. This will enable you to visualize the relative protein amounts in each sample and provide visual feedback on your pipetting/dilution technique.

1. Obtain a piece of nitrocellulose (always wear gloves when handling nitrocellulose). Place the nitrocellulose on a piece of 3MM paper (Whatman, Clifton, NJ). You will share the piece of nitrocellulose with your partner.
2. Spot 2 μL volumes of deionized H_2O (dH_2O; control) and each of the BSA dilutions. One partner should spot the top row with their samples, and the other partner should spot a row below.
   Spotting of the 2 μL is best done by holding the pipette tip just above the paper. Expel liquid such that a drop forms on the end of the tip. Touch the drop to the paper and the liquid will be drawn into the paper by capillary action.
   CAUTION: Make certain that you leave enough room between each addition so that the spots do not touch each other.
3. After the spots have dried completely, stain the nitrocellulose membrane by placing it in a tray (a square Petri dish works well for this purpose) and adding enough Ponceau Red staining solution to cover the membrane. Allow the stain to incubate for 1 to 2 minutes, with gentle rocking/shaking.
4. Pour off the stain (into a bottle labeled "Used Ponceau Red") and rinse the membrane three times with dH_2O. Add enough water to cover the membrane and rock/shake gently on a platform rocker or an orbital shaker. Change to fresh dH_2O after 5 minutes and rock/shake gently until the background is white. *Note: Ponceau Red can be reused multiple times, but eventually will stop working as it becomes diluted.*
5. Place the nitrocellulose on 3MM paper to dry. Compare the intensities of each spot. Do the intensities of your spots match your lab partner's? Does each spot appear to be half as intense as the last? If not, you need to practice your micropipetting technique.

DISCUSSION QUESTIONS

1. What are some real-life applications of biotechnology? What are some important recombinant proteins and/or recombinant organisms that are used today?
2. What are your goals in taking this class? What are you hoping to learn, and how do you hope it will expand your career or future research?

Lab Session 2

Purification and Digestion of Plasmid (Vector) DNA

Goal: Today you will isolate plasmid DNA. pET-41a is the expression vector that you will clone your *egfp* insert DNA into in a later lab session. You will perform the plasmid purification using the QIAprep Spin Miniprep Kit. This protocol starts with an alkaline lysis procedure to break open the cells and separate the plasmid DNA from the bacterial chromosomal DNA and is then followed by silica adsorption for further purification from soluble cellular proteins and other cellular debris. You will then quantify and digest the DNA.

I. INTRODUCTION

In molecular biology, *Escherichia coli* serves as a factory for the synthesis of large amounts of recombinant DNA. Today you will isolate plasmid DNA from *E. coli* for in vitro manipulation.

When plasmid DNA is transformed into bacteria, identical copies of the plasmid are made. Thus, bacterial cells contain a complex mixture of plasmid DNA, chromosomal host DNA, proteins, membranes, and cell walls. The trick in isolating pure plasmid DNA is to separate it from chromosomal DNA and the rest of the cellular components.

A. Alkaline Lysis

The most common method used for separating plasmid DNA from chromosomal DNA is the alkaline lysis method developed by Birnboim and Doly.[1] They exploited the supercoiled nature and relatively small size of plasmid DNA to separate it from chromosomal DNA.

First, cells are broken open using a detergent sodium dodecyl sulfate (SDS) under alkaline conditions. Under these conditions, both chromosomal and plasmid DNA are released and denatured (hydrogen bonds are broken, and DNA becomes single-stranded). Denatured DNA can reanneal at neutral pH if it is not kept in alkali for too long and if the complementary strands are able to find each other. Since DNA is supercoiled in the bacterial cell, the two halves of the plasmid DNA remain somewhat intertwined during the incubation in alkali and they are in close proximity for reannealing. Because the chromosomal DNA is so large, it remains bound to cellular proteins and lipids, and in the neutralization step, it is precipitated out of the solution along with denatured proteins and lipids with the addition of potassium acetate.

The precipitated chromosomal DNA and other impurities are usually removed by filtration or centrifugation. RNA is also generally degraded during the alkaline lysis step by simply adding RNase to the resuspension buffer. Double-stranded plasmid DNA remaining in the solution can then be precipitated by ethanol or be purified to a higher level either by anion exchange chromatography or by running the sample over a silica membrane. Steps 5–8 of the plasmid DNA purification protocol used today represent the alkaline lysis portion of the purification protocol.

B. Silica Adsorption

Many cellular components are removed during alkaline lysis and the subsequent centrifugation/filtration step, including the chromosomal DNA and insoluble/denatured proteins and lipids. However, many cellular proteins and metabolites still remain. Therefore, for high purity, you must further purify the plasmid DNA. The method you will use to accomplish this is selective adsorption by a silica membrane. Plasmid DNA is selectively absorbed by a silica membrane under optimized high salt conditions. Impurities are washed through, and then pure plasmid DNA is eluted under low salt conditions.

Molecular Biology Techniques. DOI: https://doi.org/10.1016/B978-0-443-26714-7.00015-6

C. DNA Quantification

DNA quantification is accomplished by reading the absorbance of ultraviolet (UV) light in a sample at 260 nm. The average extinction coefficient of pure double-stranded DNA (dsDNA) is 50 μg/mL. This means that one A_{260} unit of dsDNA corresponds to 50 μg of DNA per mL. To assess the purity of a DNA sample, the ratio of the absorbance at 260 nm over the absorbance at 280 nm is calculated. A 260/280 ratio of 1.8 is ideal for dsDNA. A sample with a higher ratio may have RNA contamination, although residual buffers from the procedure can also absorb UV light. A sample with a lower ratio may have protein contamination. The 260/230 ratio, if measured, can also be useful in judging the purity of your sample. A 260/230 ratio in the 2.0–2.2 range is considered optimal, while lower values often indicate residual guanidine (particularly when using column-based purification kits) or other organic contaminants.

II. INTRODUCTION TO EXPRESSION VECTORS

In general, cloning vectors are plasmids used primarily to propagate DNA. They replicate in *E. coli* to high copy numbers and contain a multiple cloning site (also called a polylinker) with restriction sites used for inserting a DNA fragment. A selectable marker, such as an antibiotic resistance gene, is included to select for bacteria containing the plasmid and to ensure the survival of the plasmid. A screenable marker, such as β-galactosidase, is also often included. In the case of β-galactosidase, it enables blue/white screening to distinguish bacteria that acquired the plasmid from those that did not.

An expression vector is a specialized type of cloning vector. Expression vectors are designed to allow transcription of the cloned gene and translation into a protein. They have some features in common with the general cloning vectors that are used only for propagating DNA, such as the multiple cloning site and the selectable marker, but they tend to have a lower copy number within cells and rarely have a screenable marker. They also have some important additional features which allow them to express genes and make proteins, including a promoter, ribosome binding site (RBS), ATG start codon, and often (not always) a fusion tag to aid in the purification steps.

A. Principles of Gene Expression

In order to understand how expression vectors function, it is important to recall the Central Dogma of Molecular Biology (Fig. 2.1). For a gene to be expressed, it must first be transcribed into messenger RNA (mRNA), and then translated into a protein.

In the simplest example, RNA polymerase binds to the promoter of a gene and then proceeds with transcription, producing mRNA. Transcription ends at the terminator sequence.

The ribosome then binds the mRNA at the RBS, and the ribosome moves along the mRNA. As the ribosome moves along the mRNA, transfer RNA is responsible for decoding the mRNA and specifically depositing an amino acid residue on the nascent polypeptide chain. The translational start codon, which is usually encoded by AUG (ATG on the DNA), encodes the first amino acid (usually methionine), and the translation stop codon (TAA, TAG, or TGA) ends translation.

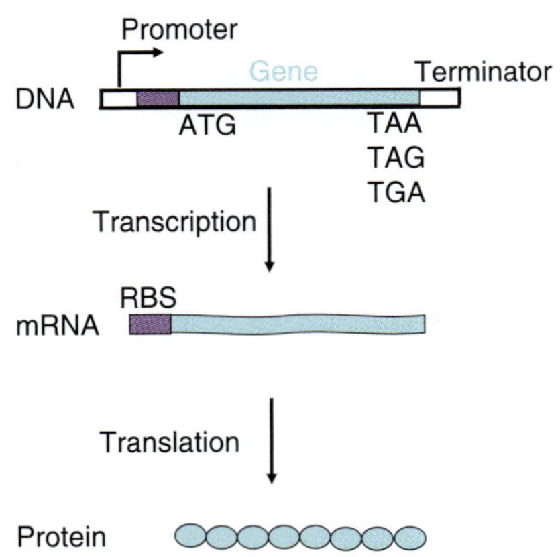

FIGURE 2.1 Central dogma of molecular biology. DNA is transcribed into mRNA, which in turn is translated into protein. RNA polymerase binds to the promoter of a gene on DNA and proceeds with transcription, producing a new mRNA. The ribosome and tRNA work together to translate the nascent mRNA into a protein. *mRNA*, Messenger RNA; *tRNA*, transfer RNA. *No Permission Required.*

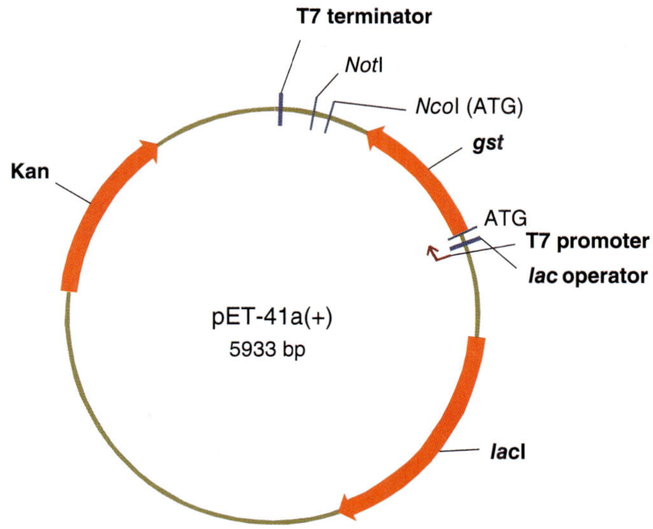

FIGURE 2.2 Salient features of pET-41a. In pET-41a, the T7 promotor are terminator are used to control expression of the *gst* gene. The NcoI and NotI restriction sites are also shown to depict where the *egfp* gene will be inserted to create the *gst::egfp* fusion gene. The *lacI* gene expresses the *lac* repressor protein, which inhibits gene expression by binding to the *lac* operator (see Fig. 2.3 for more detail). The kanamycin resistance gene (Kan; sometimes shown as KanR) is the selectable marker. *No Permission Required.*

B. Expression Vectors

The expression vector you will use for your project is pET-41a (Fig. 2.2). This expression vector utilizes a kanamycin resistance gene as a selectable marker and the glutathione-S-transferase gene (*gst*) as a fusion tag. The multiple cloning site is downstream (3′) of the *gst* gene, and there is no stop codon or termination signal following the *gst* gene. Therefore, when our gene of interest (*egfp*) is cloned into the multiple cloning site, it will be expressed as a fusion protein with *gst*, resulting in the expression of the fusion protein, GST::EGFP. You will learn more about how creating this fusion protein will aid in the purification of the EGFP protein later in the semester.

The expression of the *gst* gene, and consequently the fusion gene in your future construct, is under the control of the T7 promoter and is inducible using isopropyl-β-d-thiogalactopyranoside (IPTG). In nature, the promoter is induced by lactose, and IPTG mimics lactose with regard to the induction properties but is not cleaved by the *E. coli* enzyme β-galactosidase. Inducibility is due to the fact that pET-41a uses two components of the *lac* operon, the *lac* operator and the *lac*I gene, to regulate transcription. In this vector, the *lac* operator is located adjacent to the T7 promoter. *lac*I encodes a repressor and is constitutively expressed, so the repressor protein LacI is always present. LacI binds to the *lac* operator in the absence of an inducer and prohibits RNA polymerase from initiating transcription from the T7 promoter. When the inducer molecule IPTG is added, it interacts with LacI in such a way that LacI will no longer bind to the *lac* operator, and thus transcription by the T7 RNA polymerase proceeds. This process is called derepression of the promoter (Fig. 2.3).

C. Orientation

When cloning the gene of interest into an expression vector, it is critical for the gene to be both in the correct orientation and proper reading frame with respect to the start of translation (the ATG start codon encoded by the vector). To ensure that our gene of interest will be inserted in the proper orientation, you will employ the method of directional cloning, also called "forced" cloning (Fig. 2.4). In forced cloning, the polylinker of the vector is digested with two different restriction endonucleases that leave incompatible cohesive ends, and the small "stuffer fragment" between the two restriction sites is excised. The insert (your gene of interest) is then cut out with the same two restriction endonucleases and ligated into the vector. Cloning in this manner rather than cloning by cutting with only one restriction endonuclease has two advantages:

1. The incompatible cohesive ends will prevent the vector from ligating without incorporating the insert (though if the stuffer fragment is not removed during purification, it can be inserted back into the larger portion of the vector instead of the desired insert).

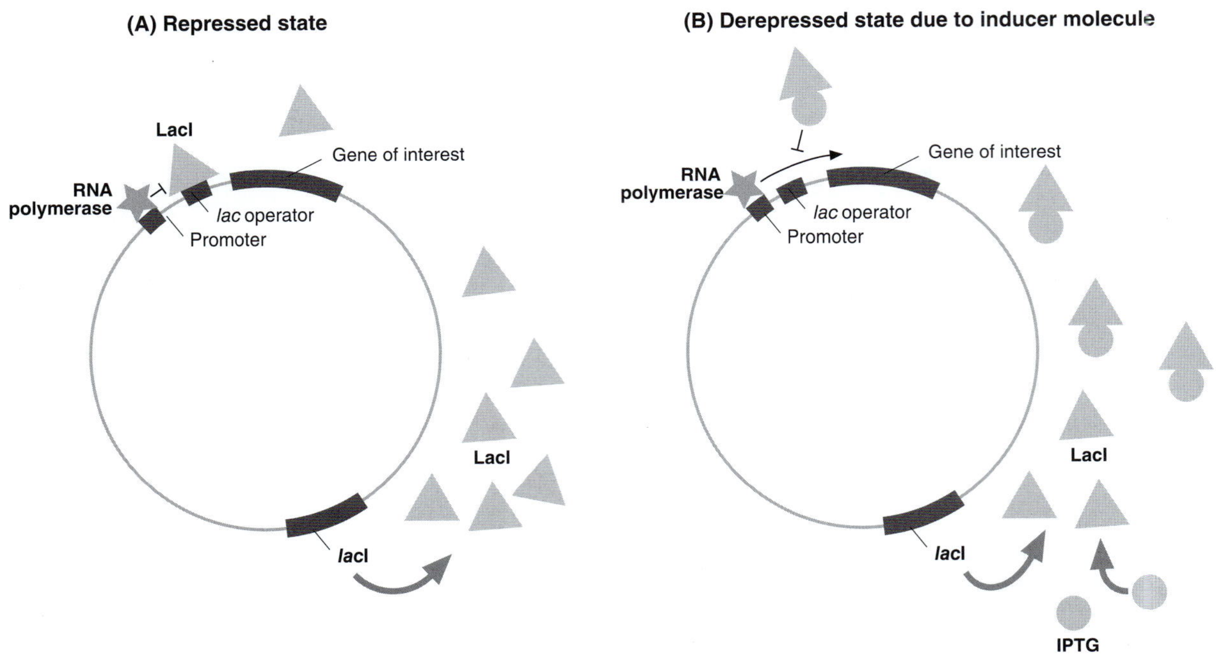

FIGURE 2.3 Promoter repression by LacI and derepression by IPTG. (A) The repressed state of the promoter. (B) The derepressed state of the promoter due to the inducer molecule, IPTG. *IPTG*, Isopropyl-β-d-thiogalactopyranoside. *No Permission Required.*

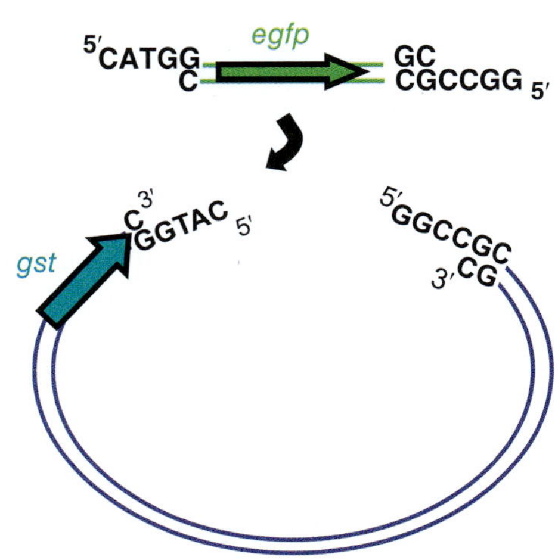

FIGURE 2.4 An example of forced cloning using the NcoI-HF and NotI-HF restriction endonucleases. The insert can only be incorporated in one orientation. *HF*, High-fidelity. *No Permission Required.*

2. The orientation of the insert is forced in a single direction: that is, the 5′ end of the gene can ligate with only one end of the cut vector. Because you know the sequence of both the vector and insert, you know that the fusion protein gene will be transcribed and translated (expressed) correctly.

You will cut the vector with two restriction enzymes that have recognition sequences within the multiple cloning site: NcoI and NotI. Later, the *egfp* PCR product will be digested using the same two restriction enzymes (NcoI at the 5′ and NotI at the 3′ end) and ligated into the expression vector.

Remember that the Watson and Crick strands are anti-parallel. If NcoI was the *only* restriction enzyme used to both cut out the *egfp* insert and the pET-41a vector, the insert would have the ability to be incorporated into the vector in either orientation. In this case, only 50% of your transformants would contain the *egfp* gene in the proper orientation. Only transcription from DNA in the correct orientation will result in the correct mRNA and the correct amino acid sequence being produced. DNA can only be transcribed in a 5′–3′ direction, and the sequence on the bottom strand, 5′–3′, is different from the sequence on the top strand.

D. Reading Frame

The reading frame with respect to the translational start site must be maintained for correct expression. In pET-41a, the junction of foreign DNA with *gst* has to be in the proper reading frame in order to create the desired GST::foreign peptide fusion protein. Most expression vectors are designed in families of three members. Typically, all three expression vectors in the family are identical, except that the reading frames with respect to the multiple cloning sites differ. For example, for a given restriction site, the first vector may put an insert in the +1 reading frame with respect to *gst*, the second in the +2 reading frame, and the third in the +3 reading frame. Only one of the three vectors will maintain the correct reading frame for a given insert. The other two will result in the insert being in the wrong reading frame.

pET-41a and many other expression vectors have an additional feature of the multiple cloning site: a NcoI restriction site. The NcoI recognition sequence is useful in that it contains an ATG sequence, the start codon for most proteins. The complete NcoI recognition sequence is CCATGG. The NcoI recognition sequence in pET-41a is located such that the ATG of the sequence is in-frame with the ATG start codon of the *gst* fusion tag. Therefore, if your gene of interest starts with an ATG that is part of a NcoI site, then the vector and the 5′ end of the insert (your gene of interest) can both be cut with NcoI, and the gene will automatically be in the correct reading frame for translation. Note that the ATG of the gene of interest will not serve as a start codon once ligated into the vector; it will simply encode methionine. The ATG start codon of the *gst* fusion tag found in pET-41a is still used for signaling translation of the fusion protein.

Fortunately, the *egfp* gene that you will clone into pET-41a does contain the NcoI recognition sequence at the ATG start of translation. Therefore, both the vector and the 5′ end of the insert may be cut with NcoI, and then ligated together with confidence that the insert will be in the correct reading frame.

Illustrated is a portion of the multicloning region of the pET-41 family of expression vectors. Note how the NcoI site is in the same reading frame for all of the vectors, but addition or deletion of a single base pair downstream of the NcoI site changes the amino acid sequence, while also setting up the downstream restriction sites in different reading frames. BamHI is highlighted as a reference site. The restriction site for NcoI is C/CATGG and the restriction site for BamHI is G/GATCC.

pET-41a: ATG...*gst*...C**C/C-ATG-G**GA-TAT-CGG-G/GA-TCC-GAA-TTC
Met Pro Met Gly Tyr Arg Gly Ser Glu Phe

pET-41b: ATG...*gst*...C**C/C-ATG-G**AT-ATC-GGG/-GAT-CCG-AAT-TC
Met Pro Met Asp Ile Gly Asp Pro Asn

pET-41c: ATG...*gst*...C**C/C-ATG-G**CG-ATA-TCG-GG/G-ATC-CGA-ATT-C
Met Pro Met Ala Ile Ser Gly Ile Arg Ile

Comparison of the three different reading frames available in the pET-41 family of vectors. *No Permission Required.*

III. LABORATORY EXERCISES

A. Alkaline Lysis and Silica Adsorption Protocol

The following protocol is modified from the QIAGEN QIAprep Miniprep Kit handbook (December 2020). An overview of this protocol is shown in Fig. 2.5.

Note to instructor: Numerous other miniprep kits are available for DNA purification. If you choose to use a different kit, follow the protocol for a low-copy plasmid. Be sure to follow the instructions with the kit to prepare the reagents prior to use.

1. Two afternoons before your laboratory session, your instructor picked a single colony of *E. coli* strain NovaBlue (or another *endA*− K-12 or DH5α strain), containing the pET-41a plasmid, from a freshly streaked Luria-Bertani (LB)/kanamycin plate. The morning before your laboratory session, a starter culture of 2–5 mL LB medium containing kanamycin was inoculated and incubated for ~8 h at 37°C with vigorous shaking (~300 rpm).

   Note: Use a snap cap tube or flask with a volume of at least four times the volume of the culture to provide adequate aeration.

   Note: For smaller classes, individual overnight cultures (3–5 mL) can be grown up for each student. These cultures can be used directly for step 3.

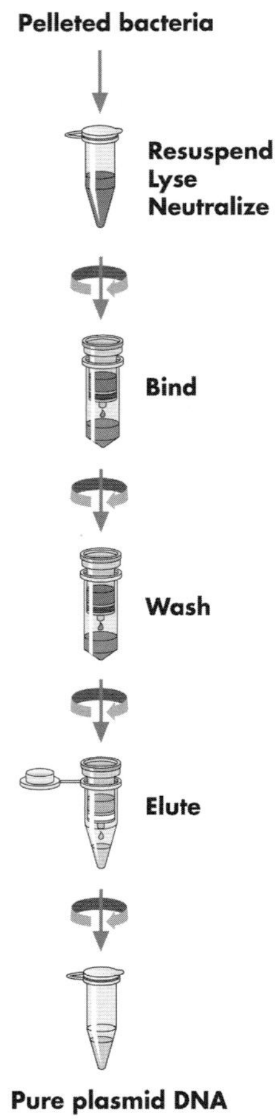

FIGURE 2.5 Overview of QIAprep Miniprep protocol. Brief overview of the purification of DNA using a QIAprep miniprep column. Note that the depiction starts with pelleted bacteria, not the liquid culture. *Modified from QIAGEN Corporation, Copyright 2020. Used with permission.*

2. The evening before your laboratory, your instructor diluted the starter cultures 1/500–1/1000 into 100 mL selective LB/kanamycin medium. They used a flask or vessel with a volume of at least four times the volume of the culture. The culture should reach a cell density of approximately 3–4 × 10^9 cells per mL, which typically corresponds to a pellet wet weight of approximately 3 g/L medium.
3. Obtain 1.5–5 mL of the culture in a microcentrifuge tube. We recommend using approximately 5 mL for this experiment to produce a sufficient amount of purified pET-41a plasmid. You can either (1) split the 5 mL between multiple centrifuge tubes; or (2) fill up one tube, centrifuge as described in step 4, then add more of the culture on top of the pellet left behind after removing the supernatant. Repeat this until all 5 mL has been pelleted.
4. Harvest the bacterial cells by centrifugation at >8000 rpm (6800 × *g*) for 3 minutes.
   Remove all traces of supernatant by decanting or micropipetting the media into a liquid waste container with 10% bleach. (*Note: If you wish to stop the protocol and continue later, freeze the cell pellets at −20°C*).
5. Resuspend the bacterial pellet in 250 µL Buffer P1 (Resuspension Buffer). The bacteria should be resuspended completely until no cell clumps remain. If you have harvested the cells in multiple tubes, add 250 µL to one tube, resuspend, then transfer the resuspended solution to the next tube. The goal is to have all of the pET-41a cells together in one tube and completely resuspended in Buffer P1.
6. Add 250 µL Buffer P2, and mix gently but thoroughly by inverting four to six times.
   Do not vortex, as this will result in the shearing of genomic DNA. If necessary, continue inverting the tube until the solution becomes viscous and slightly clear. Do not allow lysis reaction to proceed for more than 5 minutes. *Buffer P2 contains a detergent (SDS) for cell lysis and sodium hydroxide for denaturation of DNA.*
7. Add 350 µL Buffer N3 to the lysate and mix immediately and thoroughly (but gently) by inverting four to six times. After addition of Buffer N3, a fluffy white precipitate containing genomic DNA, proteins, cell debris, and SDS

becomes visible. The buffers must be mixed completely. If the mixture still appears viscous and brownish, more mixing is required to completely neutralize the solution. *Buffer N3 neutralizes the solution, causing plasmid DNA to reanneal, and acts to precipitate the chromosomal DNA and insoluble proteins.*

8. Centrifuge for 10 minutes at 13,000 rpm (17,900 × *g*) in a microcentrifuge.
9. Apply the supernatant (liquid above pellet) from step 8 on the QIAprep spin column (containing the silica membrane) by pipetting, taking care not to transfer the fluffy white precipitate. Place column in accompanying tube.
10. Centrifuge for 30 seconds. Discard flow-through.
11. Wash the QIAprep spin column by adding 0.5 mL Buffer PB and centrifuge for 30 seconds. Discard the flow-through.
12. Wash the QIAprep spin column by adding 0.75 mL Buffer PE and centrifuge for 30 seconds.
13. Discard flow-through, and centrifuge for an additional 1 minute to remove residual wash buffer.
14. Place the column in a clean 1.5 mL microcentrifuge tube. To elute DNA, add 50 μL Buffer EB (Elution Buffer) to the center of each column, let stand for 1 minute, and centrifuge 1 minute. (If necessary, cut the lid off the 1.5 mL tube with scissors so that it will fit in the microcentrifuge.)
15. Your pure DNA is at the bottom of the tube! Remove and dispose of the column and label the tube with "uncut pET-41a," the date, and your initials and transfer the DNA to the labeled tube.

B. DNA Quantification

Next, you will quantify the plasmid DNA. Depending on the equipment available in your laboratory, you will either use a Nanodrop spectrophotometer or a standard spectrophotometer. Both protocols are given, though your instructor may make modifications to the spectrophotometer protocol depending on the model of equipment and cuvette size.

Option 1: Using the Nanodrop

This method utilizes the Nanodrop spectrophotometer. If you have a Nanodrop available, this is the preferred method, both because it is the simplest and because it uses the smallest volume of your precious DNA. Because there are multiple versions of this instrument available, you will need to check with your instructor on exactly how to use the instrument.

In short, you should use Buffer EB to blank the instrument and measure the concentration of the dsDNA and also record the A_{260}/A_{280} ratio to assess the purity of the sample. Label your tube of purified DNA with "uncut pET-41a," your initials, the date, and the concentration (with the units ng/μL). Keep this sample on ice (and save the remainder in your freezer box at the end of the day).

Option 2: Using a Standard Spectrophotometer

This method for DNA quantification uses a standard UV/Vis spectrophotometer capable of assessing absorptions at 260 and 280 nm, such as the GeneQuant apparatus from Amersham Biosciences. Most of these instruments can use special disposable, UV-transparent cuvettes that accept quantities as small as 50 μL.

Note to instructor: If using this method, you may need to dilute samples prior to measuring to ensure an A_{260} value of less than 1.0.

1. Turn on the spectrophotometer 15 minutes before use.
2. Blank the spectrophotometer using 50 μL of Buffer EB in the UV cuvette. Be especially careful not to get fingerprints on the clear side of the cuvette. If you think you left fingerprints, wipe well with a Kimwipe.
3. In a separate disposable UV cuvette, read the absorbance of 50 μL of your DNA at 260 and 280 nm. Record the readings in your notebook.
4. Calculate the concentration and purity of your original sample.

   To determine the concentration of your DNA, use the equation: (A260) × (50 ng/μL) = DNA (ng/μL). If you diluted your original DNA prior to measuring in the cuvette, factor that in when calculating the concentration of your original sample.

   To determine the purity of your DNA, calculate the ratio: A_{260}/A_{280}
5. Record the DNA concentration and purity ratio in your notebook. If the A_{260}/A_{280} ratio was significantly different from 1.8, see your instructor.
6. Label your original tube of purified DNA with "uncut pET-41a," your initials, the date, and the concentration (with the units ng/μL). Keep this sample on ice (and save the remainder in your freezer box at the end of the day).

C. Restriction Digestion of Expression Vector DNA pET-41a, a GST Fusion Protein Vector

Goal: You will prepare the expression vector plasmid, pET-41a (Novagen), to be able to accept the gene (*egfp*) you are going to clone in the following weeks. To accomplish this, you will digest pET-41a simultaneously with two restriction endonucleases, NcoI and NotI. This will allow you to clone the *egfp* gene into the vector in a single, forced orientation, ensuring correct translation of a GST::EGFP fusion protein.

Restriction Enzyme Digestions

Restriction enzyme activity is defined as the amount of enzyme (measured in units, U) that will cleave 1 µg of DNA (usually lambda DNA) to completion in 1 hour at the optimum temperature for the enzyme, usually 37°C. Buffers are usually supplied with restriction enzymes at a 10× concentration. As a general set of rules, to set up a restriction enzyme digest:

- Determine the amount of DNA to be cleaved.
- Use at least a fivefold excess of enzyme.
- Ensure that the volume of enzyme does not exceed 10% of the final volume.
- 10× buffer should be diluted to a final concentration of 1×.
- Enzymes should be added to the reaction last.

The New England Biolabs (NEB) protocol for a "typical" restriction digest is described in this table:

Incubation time: 5–15 minutes for Time-Saver enzymes; 60 minutes for others. See product info for this designation.

Incubation temperature: Enzyme dependent (most commonly 37°C)

Some enzymes will cleave at a second site under sub-optimal conditions, producing what is referred to as "star activity." However, many enzymes are now available in high fidelity (HF) versions that minimize or eliminate star activity. In addition, HF versions of enzymes tend to work faster, allowing reactions to be carried out in 5–15 minutes. When HF versions of the enzymes are available, they are preferred.

As companies continue to improve their technologies, protocols can become rapidly outdated. Also, different companies will have different reaction conditions for their particular enzymes. The best way to ensure that your reaction will work correctly is to check the product insert that comes with purchasing the enzyme or by looking online at the reaction conditions for the enzyme(s) you are using. Enzyme activity is highly dependent on the buffer. When you are performing a double digest (cutting a plasmid using two enzymes simultaneously in one reaction), an important consideration is to ensure that both enzymes are active in the same buffer. If both enzymes are not active in the same buffer, typically the plasmid is cut with one enzyme, purified on a spin column (see Lab Session 3), then digested with the second enzyme. We recommend the use of enzymes from NEB, but others can be used as well.

Setting Up Digestion of Your pET-41a Vector With NcoI and NotI

Note: Restriction digest protocols are all extremely similar but vary slightly between manufacturers and even as a particular manufacturer updates their reagents. Because of that, we strongly suggest confirming the protocol before beginning. The following protocol is current for NEB enzymes at the time of this publication:

1. Look up the protocol for your restriction enzymes based on the company your enzymes are from (NEB, Fermentas, etc.). This can be done online or with the product sheet that is provided with the enzymes. For instance, going to NEB.com and searching for NcoI-HF, under Product Information for the enzyme it states the following reaction conditions: 1× CutSmart Buffer at 37°C incubation temperature. NotI-HF (and most other enzymes) is also active in 1× CutSmart Buffer at 37°C. Once this is confirmed, you can set up your digest accordingly.
2. In your lab notebook, create a table (similar to Table 2.1) for your digestion that includes all the materials you will be adding. This is a very helpful practice and allows you to check off materials as you add them. If you are performing large numbers of reactions or a different type of procedure that requires adding multiple reagents, checking off the reagents as they are added can prevent experimental mistakes.

Reagent	Final Amount/Concentration
DNA	1 µg
10× Buffer	1×
Water	To bring to a final volume of 50 µL
Restriction enzyme(s)	At least 5 U per µg of DNA

TABLE 2.1 pET-41a Restriction Enzyme Digestion Setup (50 µL reaction)

Reagent	Initial Concentration	Final Amounts/Concentrations	Volume (µL)
pET-41a DNA		2000 ng	
CutSmart Buffer	10×	1×	5
Sterile Water	N/A	N/A	
NcoI-HF		10 U	
NotI-HF		10 U	

No Permission Required

In this case, the reagents are pretty simple: DNA, buffer, water, and NcoI-HF and NotI-HF enzymes. Remember, enzymes are always added last to enzymatic reactions! Your table in your lab notebook should look like Table 2.1, with the blanks filled in with the appropriate values. Let's look briefly at each of the components individually.

DNA: In your digest, use 2 µg of pET-41a as recommended to ensure enough material for downstream applications. If you don't have enough DNA from your miniprep to digest 2 µg, you can use less DNA. As little as 750 ng of digested DNA should provide enough material for subsequent experiments. To figure out the volume of plasmid DNA necessary to add to your digest, use the following equation:

$$\text{Volume to add } \|\mu\text{L}\| = \text{Amount you want (1000 ng)/Concentration (ng/}\mu\text{L)}$$

10× Buffer: There are a number of different 10× Buffers available from NEB. Other manufacturers are similar. The most common NEB buffer currently used is CutSmart Buffer. This buffer is recommended for both NcoI-HF and NotI-HF and is supplied as a 10× stock solution. The final concentration needs to be 1×. This means that you need 1 µL of 10× Buffer for every 10 µL of final solution. Since your restriction digest is going to be performed in a final volume of 50 µL, you will need 5 µL of 10× CutSmart Buffer.

Water: You will add sterile water to bring up your final volume to the desired amount, in this case 50 µL. You can't determine this volume yet though, because you need to calculate the amount of enzymes and DNA that will be added to the reaction.

Restriction enzyme(s): In your table, you should have two different rows, one for each enzyme (NcoI-HF and NotI-HF). It is suggested that you add a minimum of 5 U/µg of DNA. In your reaction today, you should add 10 U of each enzyme to the reaction. The concentration of restriction enzymes is typically given as U/mL or U/µL. Look at the product insert or on the tube of the enzyme to find the concentration of the enzyme you are using. If the concentration of NcoI-HF is 20,000 U/mL, this would require 0.5 µL of enzyme [10 U × (1 mL/20,000 U) × (1000 µL/1 mL)]. Adding too much enzyme, particularly if the final volume of combined enzymes is greater than 10% of the overall reaction, can lead to star activity with many enzymes. Assuming both NcoI-HF and NotI-HF are supplied at 20,000 U/mL, 0.5 µL of each enzyme would result in 1 µL total, which is only 2% of the total volume of the reaction.

3. Now that you have calculated how much enzyme and DNA to add to the reaction, you can go back and calculate the volume of water necessary to bring the reaction to 50 µL. Another reason for having restriction enzymes as the last reagent in the table is as a reminder to always add them last to the reaction!

   Once the calculations in Table 2.1 are completed, you can prepare the digestion reactions. Label a tube with your initials, "pET NcoI/NotI digest," and the date so you can identify it later. Labeling tubes is very important! Pipette the required amounts as described in your table, checking off each component as it is added. Remember to always **add the enzymes last**!
4. Mix your tube well by tapping to ensure the enzymes are well-mixed in the solution, then briefly centrifuge for 2–5 seconds to bring contents to the bottom of tube.
5. Place your tube in a microcentrifuge rack in a 37°C water bath (or incubator) for the required time. For the Time Saver NcoI-HF and NotI-HF enzymes, 15 minutes is recommended. For this step, it is critical for the digestion to go to completion, so always choose the longest of any recommended times. At the completion of the digestion, store your sample in your freezer box and place at −20°C along with the tube containing the remainder of your uncut plasmid. *You will be using both your digested and your uncut vector in the future, so it is critical to save them at −20°C. Also be sure both are labeled so you can distinguish them from each other.*

REFERENCE

1. H.C. Birnboim, J. Doly, A rapid alkaline extraction procedure for screening recombinant plasmid DNA, Nucleic Acids Res 7 (1979) 1513.

DISCUSSION QUESTIONS

1. If you do an absorbance reading after plasmid purification and get a A_{260}/A_{280} of less than 1.8, how could you further purify the sample to get rid of the protein contamination? Is it always necessary to have completely pure DNA? What are some cases where it would or would not be?
2. Why do you increase the pH to denature the plasmid and chromosomal DNA during alkaline lysis rather than using high temperatures, which would also denature DNA?

Lab Session 3

Completion of Vector Preparation and Polymerase Chain Reaction Amplification of *egfp*

I. INTRODUCTION

The gene we will be cloning and expressing is *egfp* (the gene encoding the enhanced green fluorescent protein). Green fluorescent protein (GFP) is a naturally occurring protein found in a species of fluorescent jellyfish called Aequorea victoria. The difference between the fluorescence of the GFP and the enhanced GFP (EGFP) is that EGFP emits 35× the fluorescence of GFP when excited with ultraviolet or blue light; it is much brighter. The increased fluorescence was achieved by making mutations in the nucleic acid sequence which resulted in a small change in the amino acid composition within the chromophore region of the protein: Ser65 → Thr, and Phe64 → Leu.[1] To clone the *egfp* gene, we will use a polymerase chain reaction (PCR)-based strategy.

Cloning by PCR is one of many techniques that utilizes a PCR-based strategy. In this manual alone, we will employ PCR in three different experiments: cloning by PCR, quantitative (real-time) reverse-transcription PCR, and semiquantitative reverse-transcription PCR. There are many other applications that utilize this technique and may be introduced in your lectures, such as site-directed mutagenesis using a PCR-based strategy.

Before proceeding, you should be familiar with both PCR and PCR-based cloning.

What Is the Polymerase Chain Reaction?

In order to understand the PCR cloning started in this laboratory session, a general understanding of this method is required. PCR was developed by Kary Mullis in 1983 and has simplified many procedures in molecular biology and made countless new techniques possible.[2]

PCR is an exponential process in which DNA sequences are amplified. A thermostable DNA polymerase is used in repeated cycles of primer annealing, DNA synthesis, and dissociation of duplex DNA to serve as new templates. The theoretical amplification of template DNA, assuming reagents are not limiting and the enzyme maintains full activity, is $2n$ where n is the number of cycles. After 30 cycles of PCR from a single template, 1×10^9 new DNA molecules could be synthesized.

A typical PCR thermocycling program consists of the steps (exact temperatures and times will vary depending on the specific polymerase):

1. Denature DNA (95°C) ~1 minute.
2. Anneal primers to template (based on the melting temperature of the primers; ~60°C for a typical 20-mer) ~1 minute.
3. Elongate DNA (72°C) ~1 minute/kb to be amplified.
4. Repeat Steps 1–3 ~30 times.

Short nucleic acids are commonly referred to as oligonucleotides, or oligos, for short. In PCR, DNA primers are oligos that are short, single-stranded sequences complementary to the ends of the sequence to be amplified and are oriented in opposite directions. In other words, the two primers must flank the DNA region to be amplified, with one primer annealing to the sense strand and one to the antisense strand, with the primers facing inwards toward each other (Fig. 3.1). Both primers are necessary for exponential amplification to occur.

Molecular Biology Techniques. DOI: https://doi.org/10.1016/B978-0-443-26714-7.00020-X

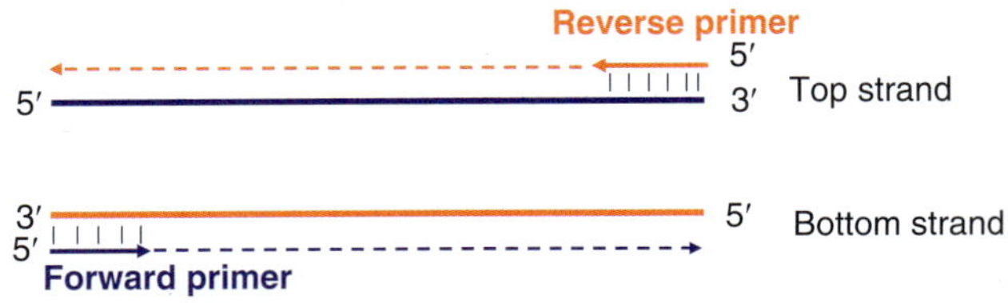

FIGURE 3.1 Orientation of PCR primers in relation to target DNA. The forward primer anneals to the 3′ end of the bottom strand. When the forward primer is extended, a copy of the top strand is created. The reverse primer anneals to the 3′ end of the top strand. When the reverse primer is extended, a copy of the bottom strand is created. If the top strand corresponds to the sense strand, the forward primer creates a copy of the sense strand, even though it binds to the 3′ end of the antisense strand. By convention, the sequence of a gene refers to the mRNA-like strand. The template used by RNA polymerase during transcription is the antisense strand of a gene. This convention makes it easier to conceptualize sequence domains and correlate them with protein motifs. *PCR*, Polymerase chain reaction. *No Permission Required.*

Primers are chemically synthesized on an instrument called an oligonucleotide synthesizer. The annealing temperature of the primers can be estimated from the formula for the melting temperature (T_m) of DNA molecules shorter than 50 bp: $T_m = (4)$ (number of GC pairs) + (2) (number of AT pairs).

The thermostable DNA polymerase most commonly used is *Taq* DNA polymerase, isolated from the thermophilic bacterium *Thermus aquaticus*, which was discovered growing in hot springs at 75°C at Yellowstone National Park. In our experiment, we will be using an alternative thermostable DNA polymerase called Vent from *Thermococcus litoralis*. Vent is often used when a low error rate is required because it has 3′ → 5′ exonuclease (proofreading) activity.

The reagents necessary for the polymerase chain reaction to take place are:

- Template DNA
- Forward primer
- Reverse primer
- Deoxynucleotide triphosphates (dATP, dCTP, dTTP, dGTP)
- Thermostable DNA polymerase
- Buffer
- Magnesium (necessary for enzyme activity and often included in the buffer)

For an excellent review of PCR, visit the website: https://dnalc.cshl.edu/resources/animations/pcr.html.[3]

Why Clone by Polymerase Chain Reaction?

Cloning using a PCR-based strategy has largely replaced traditional cloning for most applications. Traditional cloning is where the DNA to be cloned is cut out of a larger DNA molecule using restriction enzymes, and then the desired DNA fragment is purified from an agarose gel. The purified fragment can then be ligated into a desired vector.

There are numerous advantages of cloning by PCR over traditional cloning, including:

1. There is no reliance on having nature provide restriction sites flanking the region you want to clone. In PCR cloning, restriction sites can be engineered into PCR primers to create flanking restriction sites, or other overhangs suitable for cloning can be created.
2. In traditional cloning, if the DNA molecule you are isolating your DNA from is very large (a genome, for instance), you cannot simply separate your DNA fragment using agarose gel electrophoresis in order to purify it, since there would be too many other fragments of the same or similar size.
3. PCR is fast!

The main disadvantage of PCR cloning versus traditional cloning is that PCR has a higher error rate than replication in vivo; therefore, mistakes are more likely to be incorporated into your cloned gene. To overcome this, we do two things:

1. We use a DNA polymerase that has a proofreading function (such as Vent DNA polymerase).
2. We always sequence putative positive clones to check for sequencing errors that could affect the amino acid sequence of the recombinant protein.

There are multiple methods of cloning PCR products. Three commonly used methods are Gibson Assembly, TA cloning, and cloning by incorporation of restriction sites into the primers.

Gibson Assembly

Gibson Assembly is a powerful DNA cloning method that allows for the scarless joining of multiple DNA fragments in a single, isothermal reaction. It utilizes three key enzymes: exonuclease to create complementary overhangs, polymerase to fill gaps, and ligase to seal the nicks, resulting in a fully ligated DNA molecule. This efficient and versatile technique permits rapid and precise construction of complex DNA constructs regardless of the length or end compatibility of the DNA fragments. Lab Session 31 has more details on this technique, as well as a step-by-step protocol for in silico cloning of a gene into pET-41a.

TA Cloning

One commonly used method, called TA cloning, takes advantage of an unpaired A-overhang that occurs at the 3′ ends of the PCR product in reactions where a non-proofreading DNA polymerase was used. These PCR reactions can be ligated into a vector that has been cut open with an enzyme that leaves blunt ends, and then modified to achieve a single T overhang. The downside of this method is that DNA polymerases which provide a proofreading function (and therefore, a lower error rate) do not create an A overhang.

PCR Cloning by Incorporation of Restriction Sites

The PCR cloning method we will employ involves engineering restriction sites into the PCR primers, so that those restriction sites will be incorporated into the PCR product to be cloned.

Figure 3.2 depicts a PCR reaction where restriction sites are engineered into the primers. The restriction sites are not encoded by the template DNA, so notice that the sites do not anneal to the template. For this reason, when considering the annealing temperature used in the PCR reaction, we do not add in the T_m coming from the non-complementary portion of the primer. Note that we must also add a few additional nucleotides 5′ of the restriction sites on each primer. This is

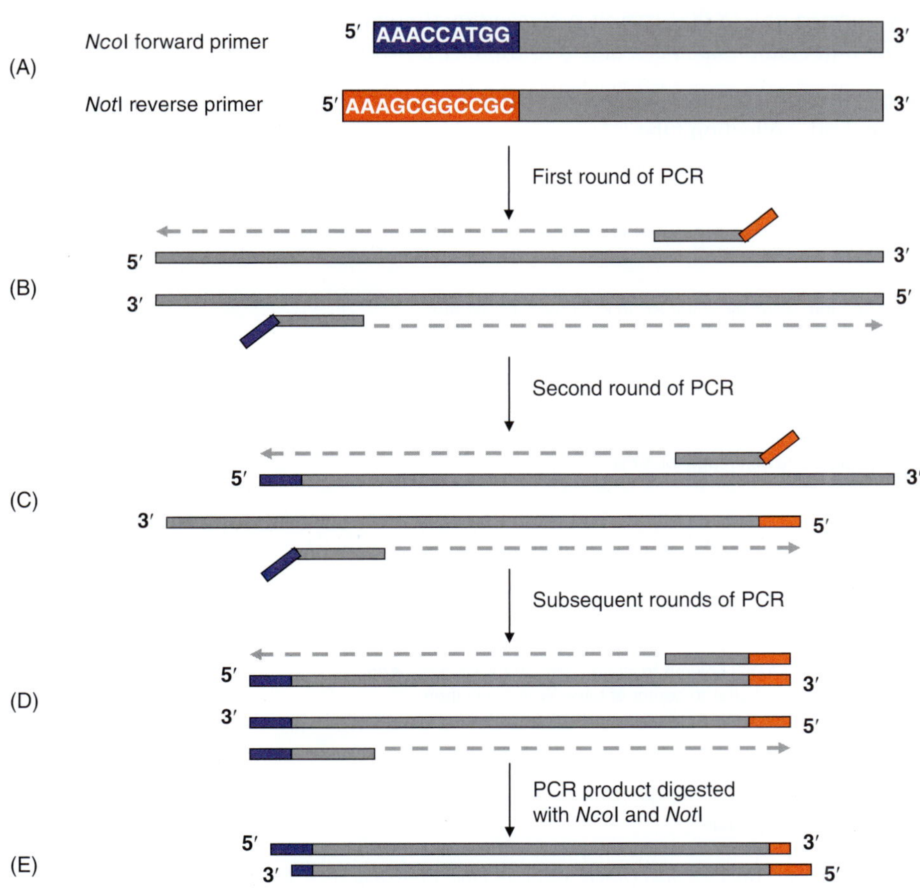

FIGURE 3.2 Incorporation of restriction sites into a PCR product by engineering them into the forward and reverse primers. (A) shows in detail the primers with the engineered restriction sites and additional nucleotides necessary for digestion of the PCR product. (B) and (C) show the first two rounds of PCR. Notice the engineered portions of the primers are not complementary to the template and do not anneal. In (D), the third and subsequent rounds of PCR, the primers do fully anneal to the newly synthesized template. (E) depicts the digested DNA with "sticky" 5 ends. *No Permission Required.*

because our next step will be to digest the PCR product for insertion into the vector, and most restriction endonucleases cannot cut restriction sites that are at the very end of the DNA molecule.

Once we assess that our PCR reaction successfully yielded a single PCR product of the expected size, we will digest the PCR product with the restriction enzymes NcoI-HF and NotI-HF in order to have compatible cohesive ends with the digested pET-41a expression vector and proceed to ligation (in later lab sessions).

Cloning Synthetic Genes

In this method, the entire DNA sequence to be inserted into the vector is created de novo. This has the advantage that the gene sequence can readily be altered to accommodate differences in codon usage between the organism of origin of the gene and the recombinant host. It can also be used to make other site-directed mutations.

II. LABORATORY EXERCISES

A. Clean-Up of Digested pET-41a Vector

Goal: This step removes restriction enzymes, salts, and other impurities from your digested pET-41a vector.

The vector you digested with NcoI-HF and NotI-HF in the previous lab session will be used in a later lab session for ligation with the insert (*egfp*) DNA. Ligations are sensitive to salt concentrations, so it is important to remove the salts present in the restriction digestion buffer.

We will use the QIAGEN QIAquick PCR Purification Kit Protocol (Fig. 3.3) to remove the salts and enzymes. The spin column works on the principle of silica adsorption, just like the column used during the small-scale plasmid purification (miniprep) previously. Alternatively, one could achieve the same goal by performing an ethanol or isopropanol precipitation, but the QIAquick protocol is preferred because it is more rapid and efficient (less sample is lost) than an alcohol precipitation.

Protocol: The following protocol is modified from the manufacturer's handbook (QIAquick Spin Handbook January 2020).

Note: all centrifugation steps are performed at ~13,000 rpm in a conventional microcentrifuge. Balance your tube(s) with that from another lab group. Be sure that ethanol has been added to Buffer PE prior to beginning.

1. Add 250 μL Buffer PB to your NcoI-HF/NotI-HF digested vector from last week and mix.
2. Place a QIAquick spin column in the 2 mL collection tube.
3. To bind the DNA, apply the entire sample to the QIAquick column and centrifuge for 30 seconds.
4. Discard the flow-through. Place the QIAquick column back in same collection tube.
5. To wash, add 0.75 mL Buffer PE to the QIAquick column, and centrifuge for 30 seconds.

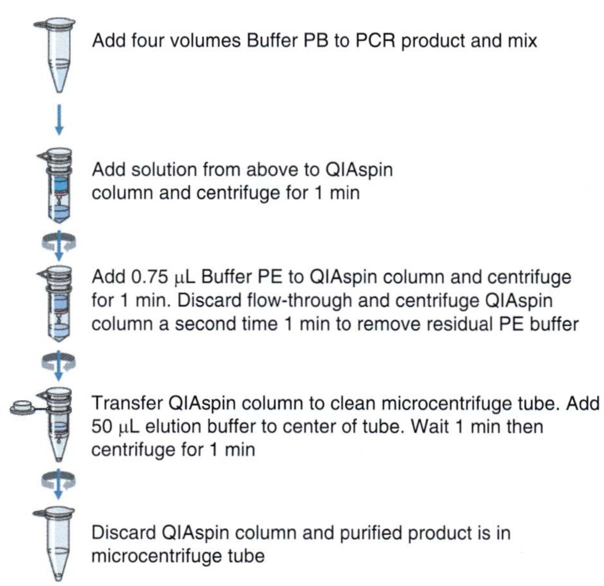

FIGURE 3.3 QIAquick spin flowchart for PCR purification.[4] *Modified from QIAGEN Corporation, Copyright 2020. Used with permission.*

6. Discard flow-through and place the QIAquick column back in the same collection tube. Centrifuge for an additional 1 minute.
7. Place the QIAquick column in a clean, sterile 1.5 mL microcentrifuge tube.
8. To elute the DNA, add 50 µL Buffer EB to the center of the QIAquick membrane, let stand for 1 minute and then centrifuge for 1 minute.
9. Label the tube containing your purified DNA with "purified NcoI/NotI pET41a," your initials, and the date.
10. Determine the concentration of the sample using Nanodrop or spectrophotometer and record (as previously described).

B. Agarose Gel Electrophoresis

Note to instructor: Analysis of the pET-41a DNA fragments using agarose gel electrophoresis can be performed in this session or in the next lab session along with the PCR products.

Agarose gel electrophoresis works on the premise that all DNA is negatively charged. When DNA is added to the negatively charged pole (cathode) of an electrical field, it will move toward the positively charged pole (anode). Because of the pore size of the agarose matrix, large DNA molecules will be retarded in the matrix more than small molecules, so the small molecules will migrate more rapidly through the gel. Distinct bands ordered according to size can be visualized if the electric field is stopped before the bands run off the end of the gel.

You will separate your digested, purified vector DNA on an agarose gel to visualize whether the DNA cut to completion. Linear DNA after restriction digestion migrates as a single band of a predictable size on an agarose gel. Circular plasmid DNA, on the other hand, can appear as multiple bands and does not migrate at the same rate as linear DNA because of its secondary structure. When analyzing circular DNA on an agarose gel, you will frequently see up to four different bands (Fig. 3.4).

- Supercoiled DNA. This is the endogenous, native form of the DNA. This should be the primary band in terms of intensity following plasmid purification. Supercoiled DNA migrates faster than linear DNA, despite having the same number of base pairs, due to its condensed secondary structure having less resistance in moving through the agarose.
- Nicked, circular DNA. This form is typically the highest up on the gel, migrating slower than linear DNA. Nicked DNA results from a single-stranded break in the DNA, which frequently occurs during the purification process. The single-stranded break causes a loss of supercoiling, which causes the DNA to become a larger circle, creating more resistance when migrating through the agarose.
- Linear DNA. This is the product you expect to see following a restriction digest where you use a single enzyme to create breaks in both strands of DNA. However, it can also appear when analyzing undigested plasmid DNA as well. If during the purification process both strands of DNA become cut at the same place, the result is linear DNA. This can result from nuclease contamination of the sample or from harsh treatment during purification. If you see a band in your uncut DNA sample running between nicked, circular and supercoiled DNA with the expected size of your linearized DNA (as determined by your DNA ladder), this is likely the result.
- Single-stranded circular DNA. This form of DNA sometimes results during plasmid purification incorporating alkaline lysis. During the procedure, high pH levels disrupt the hydrogen bonds between the DNA base pairs on opposite strands. Neutralization allows the hydrogen bonds to reform. Sometimes, particularly if the alkaline conditions persist for too

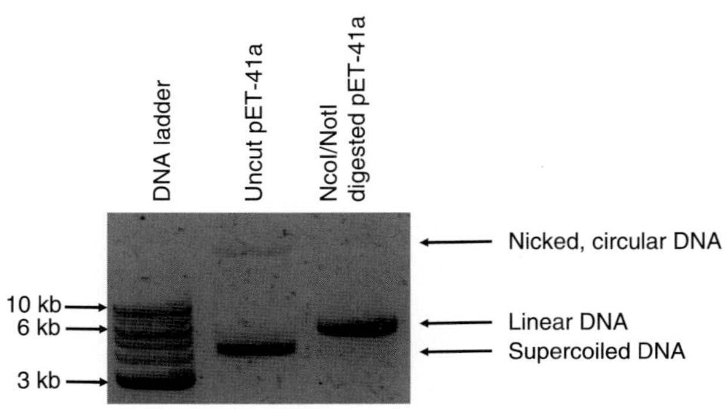

FIGURE 3.4 DNA agarose gel of pET-41a before and after digestion with NcoI-HF and NotI-HF. Note that neither single-stranded, circular nor linear DNA are observed in the uncut pET-41a lane, as neither species is typically present following plasmid purification. As expected, linear DNA is the only band present following NcoI and NotI digestion. In this gel, there also appears to be a slight doublet of supercoiled DNA, likely representing plasmid with a different degree of supercoiling. *No Permission Required.*

long, the DNA becomes permanently denatured and cannot re-anneal. The result is single-stranded circular DNA that runs faster than supercoiled DNA on your agarose gel. In most cases, you will not observe this band on your gel (as is the case in Fig. 3.4).

In this exercise, you will compare the appearance of the linearized (digested and purified) DNA from your reaction to an undigested sample. If your sample is not completely digested, there will be two or more bands instead of one. Make sure that you do not see the circular DNA in the digested sample to ensure that the digested DNA is cut to completion. If you see two or more bands, consult your instructor.

Ethidium bromide (EtBr) was traditionally used to visualize DNA on a gel. EtBr intercalates into double-stranded DNA, and then emits orange/pink fluorescence when excited by UV light. However, EtBr is a carcinogen because if it enters cells, it can intercalate into DNA causing thymidine dimers, which can lead to mutations. GelRed detects DNA in an analogous way but is unable to cross the cell membrane and thus does not pose the same health risk.

You will each receive a stock of 10,000× GelRed Nucleic Acid Stain. *Note: This tube should be saved and stored such that it is protected from light. You will use this stock of GelRed throughout the semester.*

1. To prepare a 1% agarose gel, add 1 g of agarose to 100 mL of 1× TBE Buffer in a 250 mL Erlenmeyer flask or glass bottle. Microwave for 30 seconds with the cap of the bottle on loosely. Watch your flask carefully to avoid your solution bubbling out. Swirl and repeat until agarose is completely in solution. Be sure to use rubber "hot-hands" to handle the hot flask or bottle, and never swirl it close to your face in case the liquid boils over.

   CAUTION: Never leave the agarose solution unattended when using the microwave. The solution must be swirled occasionally during the heating process to prevent superheating of local areas. Always use "hot-hands" or autoclave gloves when heating the agarose.
2. Pour approximately 30 mL of your melted agarose into a disposable 50 mL conical test tube. The volume does not have to be precisely measured.
3. Place the comb in the casting tray and assemble the apparatus as demonstrated by your instructor.
4. Add 3 μL of 10,000× GelRed to the 30 mL agarose and mix thoroughly without creating bubbles. Pour into the casting tray. *Allow the agarose to cool slightly before pouring the gel, because steaming hot agarose can warp the gel apparatus.*

   Allow the remaining agarose to solidify in the bottle with the cap tightly closed. If using a flask, you can cover the opening with Parafilm. This agarose will be re-melted in the microwave for pouring gels during the next several exercises.
5. While the gel is solidifying, begin preparing your samples.

   For the following sample preparation: do not add loading dye to the entire original sample: if you do, it cannot be used for cloning.

   To prepare your samples, you will be mixing together DNA, water, and loading dye. There are numerous formulations of loading dye for agarose gel electrophoresis. These solutions typically come in 2×, 6×, and 10× concentrations. You can calculate the correct volume of loading dye to add to your reaction by using the $C_1V_1 = C_2V_2$ equation you learned in general chemistry,

$$C_1V_1 = C_2V_2$$

   where C_1 is the concentration of loading dye, V_1 is the unknown volume of loading dye, C_2 is 1×, and V_2 is the final volume (keep in mind the final volume includes the volume of concentrated loading dye added).

   More simply, the "×" can be thought of as the strength of the solution. A 2× loading dye solution is twice as concentrated and needs to be diluted 1:1 (equal parts loading dye and DNA/water mixture), 6× loading dye needs to be diluted 1:5 (1 part loading dye, 5 parts DNA/water), and 10× loading dye needs to be diluted 1:9 (1 part loading dye, 9 parts DNA/water).
6. To prepare your digested, purified DNA sample for loading on the gel, mix together in a separate tube:

   __ μL NcoI/NotI digested pET-41a (to equal 50 ng)
   __ μL loading dye
   __ μL sterile distilled water (to bring the total volume to 10 μL)
   Mix, then centrifuge for 2 seconds (or tap contents to the bottom of tube).
7. To prepare uncut DNA for loading on the gel, mix together in a separate tube:

   __ μL uncut pET-41a (to equal 50 ng) (save the rest in your freezer box for a later laboratory session)
   __ μL loading dye
   __ μL sterile distilled water (to bring the total volume to 10 μL if necessary)
   Mix, then centrifuge for 2 seconds (or tap contents to the bottom of tube).

8. Once the gel has completely solidified, pour about 160 mL 1× TBE over the gel (add enough buffer to cover the gel) and remove dams and comb.
9. Load your gel as follows: Lane 1—empty; Lane 2—150 ng NEB 1 kb DNA ladder (Cat# N0552S); Lane 3—10 μL NcoI/NotI digested pET-41a sample described; Lane 4—empty; Lane 5—10 μL uncut pET-41a described.

   The NEB 1 kb DNA ladder is a molecular weight marker for estimating DNA size. The visible bands are approximately 10, 8, 6, 5, 4, 3, 2, 1.5, 1, and 0.5 kb in size (Fig. 3.5).
10. Once you have loaded the samples, close the apparatus and connect the leads to the voltage pack, as demonstrated by your instructor. Run your DNA gel at approximately 85 V, or at an appropriate voltage for your apparatus (10 V/cm). While the gel is running, proceed to part C (PCR of *egfp*). Once the bromophenol blue dye front runs to the bottom of your gel, stop the current and place the gel on a UV transilluminator. Wear UV protective goggles whenever direct exposure to UV is a possibility. Photograph the gel and examine the photograph. Your instructor will help you determine whether your plasmid cut to completion. There should be a single DNA band in your cut DNA sample lane, and it should run at a slightly different size than uncut DNA.

C. Polymerase Chain Reaction Amplification of *egfp*

Goal: pEGFP-N1 contains the *egfp* gene (720 bp). The PCR primers you are using are specially engineered to have restriction sites incorporated into them, so that when your PCR reaction is complete, you will be able to digest the PCR product so that it will have sticky ends for cloning.

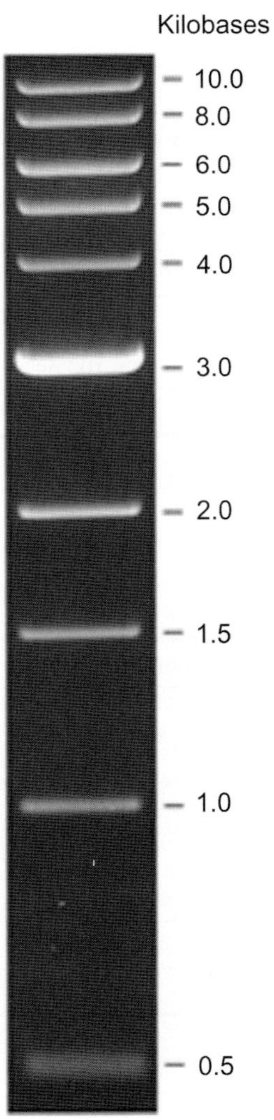

FIGURE 3.5 NEB 1 kb DNA ladder (catalog number N0552S). Note that the 3 kb band is brighter than the adjacent bands. From New England Biolabs, with permission.

The forward primer (egfpNco: 5′-AAACCATGGTGAGCAAGGGCGA-3′) you will use will incorporate a NcoI restriction site on the 5′ end of your PCR product, and the reverse primer (egfpNot: 5′AAAGCGGCCGCTTTACTTGTACA-3′) will incorporate a NotI restriction site on the 3′ end of your PCR product.

PCR Protocol

Before pipetting your PCR reaction, construct a table (similar to Table 3.1) in your lab notebook with the components, their stock concentrations, final concentrations, and the volume to pipette. Note: You will need to complete your table based on the stock concentrations of your specific reagents. Once your table is complete, pipette the reagents in the order listed on the table.

Mix by flicking the tube a few times, then tap the reaction down to the bottom of the tube.

Place the PCR tube in a thermal cycler and run the protocol below (Table 3.2) using a heated lid. Note that the annealing and extension times are dependent on the primers and size of the desired product.

Your instructor will store your PCR reaction in the −20°C freezer for you until next week.

REFERENCES

1. T. Yang, L. Cheng, S.R. Kain, Optimized codon usage and chromophore mutations provide enhanced sensitivity with the green fluorescent protein, Nucleic Acids Res 24 (22) (1996) 4592–4594.

TABLE 3.1 PCR Setup for a 50 μL Reaction

Reagent	Initial Concentration	Final Amounts/Concentrations	Volume (μL)
Nuclease-free H_2O	To bring final volume to 50 μL		
ThermoPol Reaction Buffer	10×	1×	5
dNTPs		200 μM for each dNTP	
egfpNco primer		0.25 μM	
egfpNot primer		0.25 μM	
pEGFP-N1 plasmid		100 ng	
Vent polymerase		1 U	

No Permission Required.

TABLE 3.2 PCR Program

	Temperature	Time
1. Denature	95°C	2 min
2. Denature	95°C	30 s
3. Anneal	60°C	30 s
4. Extend	72°C	1 min
5. Repeat steps 2–4 30 times		
6. Extend	72°C	5 min
7. Hold	4°C	indefinitely

No Permission Required.

2. K.B. Mullis, The unusual origin of the polymerase chain reaction, Sci Am 262 (4) (1990) 56–65.
3. Dolan DNA Learning Center: Biology Animation Library, Cold Spring Harbor Laboratory. <https://dnalc.cshl.edu/resources/animations/pcr.html>.
4. QIAGEN® QIAquick Spin Handbook. <https://www.qiagen.com/us/resources/download.aspx?id=a9de0fd4-e405-4bb7-b3a0-a74b336d613e&lang=en>; January 2020.

DISCUSSION QUESTIONS

1. How efficient would PCR be if we set the annealing temperature higher or lower than the calculated melting temperature? How would a higher or lower temperature affect the annealing capability of the primers and the final quantity or quality of the products? How would a gel of such PCR products look compared to PCR products obtained at the optimum annealing temperature?
2. People are most familiar with using *Taq* DNA polymerase, but in this lab, we are using a different thermostable DNA polymerase. What is it and why are we using it instead?
3. Why can circular plasmid DNA appear as multiple bands on an agarose gel? Why doesn't it run at the same apparent size as linear DNA of the same length on the gel?
4. A different DNA ladder from NEB (Cat #N3200L) comes as 500 μL of a 1000 μg/mL solution and is not pre-mixed with the 6× DNA loading buffer. If you want to load 150 ng of this ladder, how would you prepare the solution and what volume of the solution would you load on your agarose gel?

Lab Session 4

Preparation of Insert DNA (*egfp*) PCR Product

Goal: Today you will confirm the success (hopefully!) of your PCR reaction and digest your *egfp* PCR product to create NcoI/NotI sticky ends to be used for ligation next week. Review Fig. 3.2 from Chapter 3: Completion of Vector Preparation and Polymerase Chain Reaction Amplification of *egfp* prior to this laboratory.

I. INTRODUCTION

The *egfp* gene has been amplified using primers with restriction sites for NcoI and NotI engineered onto the ends (with a few additional nucleotides). Right now, the *egfp* gene still has blunt ends from the PCR reaction. The DNA must be digested with the restriction enzymes to create the sticky ends for cloning. To accomplish this, first the PCR product will be purified to remove salts and other contaminants that might affect the restriction digest. Then, the *egfp* PCR product will be simultaneously digested with NcoI-HF and NotI-HF to create the appropriate compatible ends for ligation within the pET-41a vector. This will allow cloning of the *egfp* gene into the vector in a single orientation, ensuring correct translation of a GST::EGFP fusion protein. Finally, the digested *egfp* PCR product will be purified once more to remove salts and enzymes that could interfere with the ligation that will be performed next week.

II. LABORATORY EXERCISES

A. Spin Column Clean-up of PCR Product

1. Remove 5 µL of your PCR sample and save it in a separate tube labeled "Original PCR."
2. With the remaining PCR product, perform the QIAGEN QIAquick PCR Purification Kit Protocol to remove the salts, enzyme, and unincorporated nucleotides from your PCR product, as previously described in Lab Session 3, Completion of Vector Preparation and Polymerase Chain Reaction Amplification of *egfp* (clean-up of digested pET-41a vector).

B. Quantification of Purified *egfp* PCR Product

Use the Nanodrop to determine the concentration of your undigested *egfp* PCR product following purification. If no Nanodrop is available, proceed directly to the restriction digestion (Part C) and use 25 µL of the purified *egfp* PCR product in your digestion reaction.

C. Restriction Digestion of *egfp* PCR Product

Note: As mentioned in Lab Session 2, Purification and Digestion of Plasmid (Vector) DNA, restriction digest protocols are all similar but vary slightly between manufacturers and even as a manufacturer updates their reagents. Because of that, we strongly suggest confirming the protocol before beginning.

1. Create a table in your lab notebook for your restriction digest, similar to Table 4.1.
2. In a microcentrifuge tube labeled "egfp digest," add the reagents listed in the table in order (enzymes last!). Briefly centrifuge (~2 seconds) tubes that contain small volumes of liquid (enzyme, DNA, buffer, etc.) before removing aliquots. Make sure that buffers that have been stored at −20°C are completely thawed and vortexed before using.
3. Vortex to mix and spin (with another tube used as a balance) for 2–5 seconds to bring the contents to the bottom of the tube and place it in an appropriate rack/holder in a 37°C incubator (or water bath) for a minimum of 15 minutes.
4. While the digest is incubating, prepare your agarose gel and DNA samples for gel analysis (Part E).

Molecular Biology Techniques. DOI: https://doi.org/10.1016/B978-0-443-26714-7.00022-3

TABLE 4.1 PCR Product Restriction Enzyme Digestion Setup for a 50 μL reaction

Reagent	Initial Concentration	Final Amounts/Concentrations	Volume (μL)
Purified *egfp* PCR product		1000 ng	
CutSmart Buffer	10×	1×	5
Sterile dH_2O	N/A	N/A	
NcoI-HF		10 U	
NotI-HF		10 U	

No Permission Required

D. Removing Enzymes and Cleaning Digested DNA Using a Spin Column

Your *egfp* gene now has sticky ends for cloning. Ligations are extremely sensitive to salt concentrations, so it is important to remove the salts used in the restriction digestion buffer. Perform the QIAGEN QIAquick PCR Purification Kit Protocol once again on the digested PCR product, and then check the concentration and estimate purity by using a Nanodrop.

If no Nanodrop is available, run 5–10 μL of your purified digested PCR product alongside the 1 kb ladder on an agarose gel in Part E. Your instructor can help you estimate your DNA concentration comparing against the mass amount in the ladder.

Label this tube as "Purified egfp digest" along with the concentration, and store in your freezer box with your digested, purified pET-41a vector after you start your gel in Part E.

E. Check PCR Reaction and Purification on an Agarose Gel

Analyzing your samples on an agarose gel will allow you to determine whether you successfully amplified and purified the 720 bp *egfp* gene. It is important to confirm this information before proceeding to DNA ligation.

Protocol (Review Lab Session 3b)

1. Pour a 1% agarose gel as described previously in Lab Session 3b. (This should be performed while your sample in Part C is digesting.)
2. Prepare your "Original PCR" sample (from Part A) for loading on the gel by adding 2 μL of Original PCR to 8 μL of water, then adding the appropriate amount of loading dye.
3. Prepare your "Purified egfp digest" sample (from Part D) to load on the gel. You should prepare enough sample to have for 100 ng for a gel this lab session and another 100 ng for a gel next lab session. It is recommended you add 300 ng of the "purified egfp digest" to a tube, the volume of which can be calculated from the DNA concentration determined using the Nanodrop spectrophotometry data. Bring up the volume to 50 μL with dH_2O, then add the appropriate amount of loading dye. With this prepared sample, ⅓ of the volume will be loaded for 100 ng in this lab session, and the rest can be stored at −20°C for the next lab session.
4. Load your gel as:
   Lane 1: 150 ng NEB 1 kb DNA ladder, pre-mixed with loading dye
   Lane 3: 10 μL of the diluted "Original PCR" sample
   Lane 5: 100 ng "Purified egfp digest" sample
5. Run and visualize the gel as previously described.

F. Calculations for Next Week

While your gel is running, perform the following calculations that you'll need for your ligation next week:

1. Calculate the volume (in μL) of your digested, purified pET-41a vector equal to 50 ng.
2. Calculate the volume (in μL) of your purified *egfp* digest (insert) equal to 30 ng.

Make sure to record these values in your lab notebook for next week.

DISCUSSION QUESTIONS

1. What are the potential reasons more than one band may appear when you run your PCR product on a DNA gel?
2. Would the cloning of the *egfp* gene into the pET-41a vector work if we skipped the NcoI/NotI digestion of the PCR amplified *egfp*? How would the results of next week's ligation be affected?
3. Normally, a few additional nucleotides need to be engineered 5′ of the restriction sites on both the forward and reverse primers when cloning by PCR in the method we are using. Why is this necessary? How can you determine how many additional nucleotides should be added to the ends of the PCR primers? Does it matter which nucleotides are added?

Lab Session 5

DNA Ligation and Transformation of *Escherichia coli*

Goal: The goal of this week's laboratory session is to ligate the *egfp* gene (our insert) into the linearized pET-41a expression vector. We will then transform the ligation mixture into *Escherichia coli* host cells.

I. INTRODUCTION

A. Ligation

DNA ligase catalyzes covalent bond formation between the 3′OH and 5′PO_4 on DNA (i.e., phosphodiester bond). It requires two ends of double-stranded DNA and a buffer that includes adenosine triphosphate (ATP). Ligation kinetics are complex, and it is worthwhile to understand some of the parameters that affect both the frequency and products of ligation. Bacteriophage T4 DNA ligase is the preferred enzyme for DNA ligation in most cloning experiments because it can efficiently ligate blunt-ended DNA as well as DNA with compatible cohesive ends (short, complementary 5′ or 3′ extensions of single-stranded DNA created by restriction enzyme digestion).

The optimal ligation temperature for a given reaction has been shown empirically to vary. DNA with compatible cohesive ends are the most amenable to ligation, although blunt ends of DNA can also be ligated together. Different preparations of ligase and ligase buffer call for different ligation times and temperatures. Ligation reaction temperatures are a balance between the activity of the enzyme and the annealing temperature of the compatible ends. For this reason, ligation reactions with compatible cohesive ends are usually performed at room temperature, while we perform blunt-ended ligation reactions at 4°C. Always be sure to check with manufacturers' instructions for recommended conditions for the specific ligase you are using.

When designing ligation reactions, there are several practical rules of thumb that are useful to follow:

- Ligation reactions are generally set up in small volumes of 10–20 µL so that compatible (or blunt) ends will not be too diluted in solution.
- For average sized vectors (2–6 kb), use approximately 50 ng of vector DNA per reaction. Dugaiczyk[1] empirically derived a formula for determining how much vector is ideal in a ligation reaction.
- The ideal amount of insert to use can only be determined empirically, but common vector-to-insert molar ratios used are 1:1, 1:3, 1:5, and 1:10.
- Ligase buffer includes the necessary salts and ATP, and is usually supplied as a 10× solution by the ligase manufacturer. Be sure to include it in your ligation mix or the ligase will not be active.
- 1 µL ligase per reaction from most commercially available sources is sufficient; however, a more concentrated ligase preparation may be useful for blunt-end ligations.

For today's experiment, we will attempt to ligate the pET-41a expression vector with the *egfp* insert using a 1:5 molar ratio of vector-to-insert. The vector-to-insert molar ratio is dependent on the lengths of the individual DNA fragments. The equation used to determine this amount of insert to add (assuming performing a standard ligation with 50 ng of vector) is shown below.

$$\frac{\text{insert size in kb}}{\text{Vector size in kb}} \times \frac{\text{mole of insert}}{\text{mole of vector}} \times 50\text{ ng vector} = \text{x ng insert}$$

In our experiment, we know that *egfp* is 0.72 kb and *pET-41a* is 5.9 kb. Since we desire a 5 to 1 molar ratio of insert to vector using this calculation, you should determine that 30 ng of insert is required. This amount was prepared in Lab Session 4 F last week.

Molecular Biology Techniques. DOI: https://doi.org/10.1016/B978-0-443-26714-7.00011-9

In addition to our experimental ligation and transformation, we will also perform controls to confirm the viability and competence of our competent cells, as well as to confirm that our vector did indeed cut to completion in Lab Session 2. While we previously looked at digested pET-41a on an agarose gel to determine whether the plasmid was linearized, the transformation control is more sensitive. There are two pieces of information that are critical to understanding these controls.

1. DNA strands do not spontaneously ligate, even if the strands have complementary cohesive ends. An enzyme that catalyzes phosphodiester bond formation (such as DNA ligase) *must* be present for ligation to occur.
2. Only circular DNA is transformable into competent *E. coli* cells. Linear DNA can get into the cells, but it cannot replicate once taken up.

B. Transformation

Bacterial cells can be made competent for DNA uptake by treatment with divalent cations, and can be stored in this state indefinitely at −80°C. Once chilled, the cells must not warm to room temperature or they will lose competence. For optimal transformation efficiencies, the cells should be kept on ice.

Because the pET-41a expression vector has a T7 promoter that drives the transcription of the gene of interest, the *E. coli* host strain in our experiment must be a λ lysogen so that it will have a T7 RNA polymerase to bind at the promoter for expression. A suitable strain is BL21(DE3), which can be purchased from many different manufacturers.

Fig. 5.1 shows outcomes of the two controls and the experimental ligation and transformation. The top row shows the control for viability and competence of the competent cells. The second row shows the control for complete restriction digestion of the pET-41a vector, and the bottom row shows the experimental ligation (vector plus insert). The first column shows the type of DNA to be added to the reaction: circular or linearized. The second column indicates whether DNA ligase is added to the reaction. The third column shows whether or not the DNA should transform and be replicated (circular DNA can transform, linear cannot). If the plasmid transforms, then transformed cells should grow on medium containing kanamycin since the vector contains the kanamycin resistance gene. The last column provides some explanations for unexpected results. Importantly, restriction digestions and ligation reactions rarely go to completion. Therefore, it is normal to see a few background colonies on the linearized pET-41a control plate even if the previous results from the agarose gel showed only a single band (as expected), since the sensitivity of transformation is higher. Note that these colonies will not contain insert and should not be selected for screening in later labs.

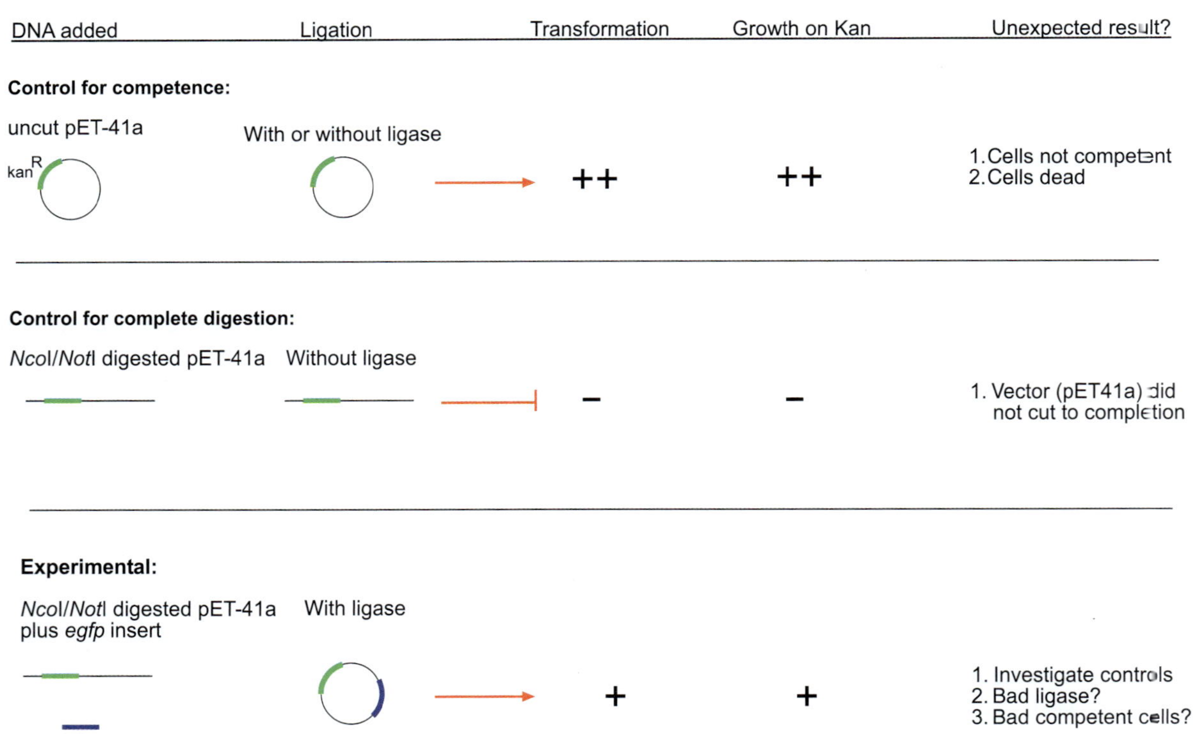

FIGURE 5.1 DNA ligation and transformation controls. *No Permission Required.*

Keep in mind that any DNA fragments with compatible cohesive ends may be ligated together. This means that in the ligation microcentrifuge tube, anything goes! Molecules may exist as one vector plus one insert (your desired clone), recircularized vector alone, vector with concatemerized (multiple, consecutive) inserts, multiple vector combinations, and vector with incorrect inserts (Fig. 5.2).

Although they do occur during the ligation reaction, it would be unlikely to recover insert-only, or concatemerized insert with no vector from a transformant, because such DNA molecules would be unlikely to contain an origin of replication, and therefore could not replicate in vivo. Given the parameters of the protocol, the most likely transformants to be obtained will be either the desired clone or vector alone with no insert. In later exercises, you will learn how to screen for the desired clone.

II. LABORATORY EXERCISES

A. Ligations and Ligation Controls

Note: This protocol uses NEB T4 DNA ligase, catalog number M0202L. If using a different brand of T4 DNA ligase, follow manufacturer's instructions. The incubation time may be longer.

1. Pour a 1% agarose gel as you have done previously. As the gel solidifies, move to Step 2.
2. Set up three reactions in microcentrifuge tubes. Label your tubes clearly [circular (uncut) DNA, linear DNA, and ligation]. Be sure to include tables in your lab notebook that clearly indicate the components of each reaction/control.

   Reaction one (*to control for viability/competency of cells and your technique*): circular (uncut) pET-41a

50 ng pET-41a uncut	____μL
Sterile dH_2O	____μL
	20 μL total

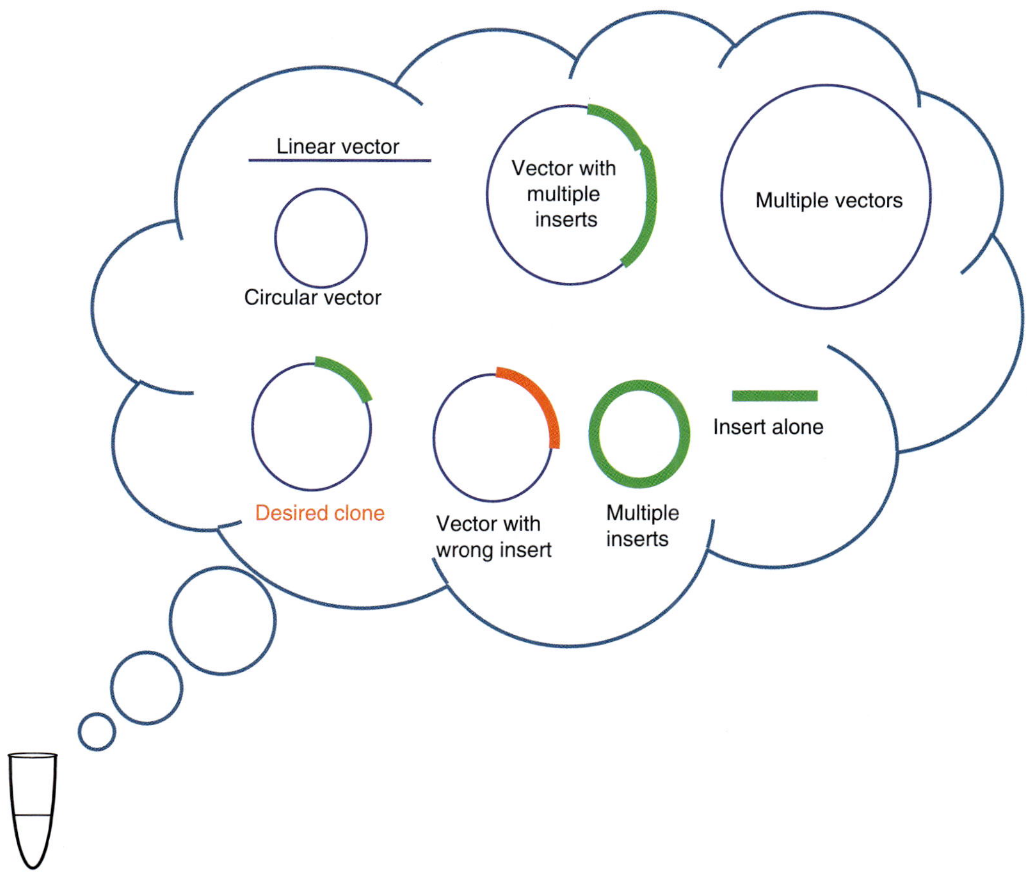

FIGURE 5.2 The multiple and varied products of ligation. While many combinations can form in a ligation mixture, after transformation, we will most likely detect transformants containing vector only and transformants containing our desired clone (vector plus insert). *No Permission Required.*

Reaction two (*to control for the presence of uncut vector*): linear pET-41a

50 ng pET-41a NcoI/NotI	____μL
Sterile dH_2O	____μL 20 μL total

No Permission Required.

Reaction three (*experimental ligation*): 1:5 molar ratio pET-41a: *egfp* insert

50 ng pET-41a NcoI/Not I	____μL
30 ng *egfp* insert Sterile dH_2O 10× ligase buffer Ligase	____μL ____μL 2 μL 1 μL 20 μL total

No Permission Required.

3. Mix gently and bring contents to the bottom of the tube either by a brief (2–5 seconds) centrifugation or tapping. Incubate the reaction for 10 minutes at room temperature. Proceed to transformation (Part B).

B. Divalent Cation-Mediated Transformation

Competent cells are extremely fragile. Never vortex or pipette competent cells vigorously because they will lyse. Never warm competent cells to room temperature because they will start repairing their membranes and will lose their competency. Do not centrifuge competent cells before use, because you will be unable to get them back into suspension.

Transformation protocols will vary between manufacturers and cell types. The protocol below is a general guide for transformation, but for optimal results, follow the protocol for the specific cells you are using. Typically, transformation efficiency is much higher for purchased competent cells (i.e., NEB #C2527) than homemade competent cells (such as those described in Appendix C).

Set up three transformations (one for each reaction in Part A) as per the following description. The protocol described is based on 50 μL of competent cells per transformation. Some protocols may suggest less. In any case, the volume of DNA added to your transformation should never be more than 10% of the total volume of cells. For example, if 20 μL of cells were used, no more than 2 μL of reaction mixture should be added.

Money-Saving Tip: Use only 20 μL of competent cells for each transformation. To each 20 μL aliquot of cells (in pre-chilled tubes), add 2 μL of the reaction mixes. Also adjust the amount of SOC medium added to 80 μL and plate the entire transformation mixture on the LB/kan plates. The rest of the parameters are unchanged from the following protocol.

1. Set the heat block to 42°C and fill holes with water (or use a 42°C water bath). Verify the temperature using a thermometer, as the digital displays are not always accurate.
2. Label three sterile microcentrifuge tubes 1–3 (designations from described reactions). Chill and keep on ice. (Some purchased competent cells come in pre-aliquoted tubes. In this case, you may label those tubes directly and proceed to Step 5.)
3. Obtain 175 μL of competent cells, keeping on ice. *Keep competent cells on ice at all times.*
4. "Finger flick" to be sure cells are in suspension, and aliquot 50 μL into each of the three chilled microcentrifuge tubes. Immediately place cells back on ice after aliquoting.
5. Add 5 μL of each of the reaction mixes into the appropriate tube, being sure to add the DNA directly into the bacterial suspension. Mix the tube gently by "finger flicking."
6. Incubate on ice for 5 minutes.
7. Heat shock the cells for 30 seconds at 42°C. The duration and temperature of the heat pulse is critical.
8. Incubate tubes on ice for 2 minutes.
9. Add 950 μL of room temperature SOC broth or LB broth per tube and shake for aeration for 60 minutes at 37°C. *(Start your agarose gel during this down-time.) Note: some protocols will mention a quick transformation procedure that shortens this step to 5 minutes or removes it entirely. Eliminating this step does not work with kanamycin-resistant plasmids, such as pET-41a. Be sure to incubate tubes for 60 minutes.*

10. Label three, pre-warmed to 37°C, LB/kan plates with your initials and the following reaction conditions: (1) circular positive control; (2) linear negative control; and (3) experimental ligation; *always label the backs of Petri dishes rather than the lids to avoid mix-ups*.
11. Mix the transformed cells in the tube gently, then plate 100–200 µL from each transformation on the appropriately labeled LB/kan plates.
12. Place the plates inverted (lid down) in the 37°C incubator overnight (or RT for 24–48 hours).

Time-Saving Tip: Incubations of bacterial plates are typically done at 37°C overnight. However, cells also grow at lower temperatures, albeit more slowly. Transformed bacterial cells plated on antibiotic-resistant medium and grown at room temperature typically show colonies in 24–48 hours, and can often be left up to 72 hours without cultures being overgrown and forming satellite colonies, which are smaller colonies lacking plasmid that surround the larger, plasmid-containing bacterial colony.

C. Electrophoresis of Ligation Reactions

This should be started during the 60 minutes incubation during transformation.

While the transformation of the control ligations is a much more sensitive method to detect circularization of vector (or vector that was not digested to completion), gel electrophoresis of the ligation products and controls will give us more rapid feedback.

Unlike linear DNA, circularized plasmid DNA has secondary structure, and this secondary structure affects the movement of the DNA through the gel (review Lab Session 3b). Therefore, circular plasmid DNA will often run at an unpredictable molecular weight. Additionally, if ligation has occurred, multiple products may be formed, giving rise to multiple bands. In this way, we can see right away whether ligation was successful. If only a single vector band and a single insert band are apparent in the experimental ligation, it is a strong indicator that the ligation reaction was unsuccessful.

Load your samples as indicated, making sure to add the appropriate amount of loading buffer to samples 2–5.

1. 100 ng 1 kb ladder (premixed with loading buffer)
2. 10 µL reaction 1 (uncut pET-41a)
3. 10 µL reaction 2 (linearized pET-41a)
4. 10 µL reaction 3 (experimental ligation)
5. ~100 ng insert ("purified, digested egfp"). You previously prepared this sample and ran it on a gel in the previous Lab Session 4 d and 4e, but it is helpful to have on this gel as well as a reference.

Run the gel as previously described.

When analyzing your gel, if the experimental ligation (Lane 4) works, you will typically see the band corresponding to cut vector disappear, and the bands lining up with uncut vector appearing, as well as some additional bands. If the experimental ligation appears to contain only linear vector, it is likely that there is a problem with either your ligase or your ligase buffer. If the vector has cut to completion, Lane 3 should show a single band, corresponding to linear DNA.

REFERENCE

1. A. Dugaiczyk, H.W. Boyer, H.M. Goodman, Ligation of *Eco*RI endonuclease-generated DNA fragments into linear and circular structures, J Mol Biol 96 (1975) 171–184.

DISCUSSION QUESTIONS

1. In our ligation, we used a 1:5 molar ratio of vector-to-insert. How could different molar ratios (1:1 or 1:10) affect the ligase reaction? What are pros and cons of excess vector versus excess insert?
2. You perform a ligation with NcoI/NotI-digested, pET-41a vector and the *egfp* insert from pEGFP-N1 (as in lab). You then transform the ligation mix into *E. coli* and plate on LB medium containing kanamycin. Consider the following unexpected outcomes, and suggest controls:
   a. Nothing grows. What controls might you design to determine whether the *E. coli* cells are viable (alive) versus whether the *E. coli* is competent? What would the results of those controls be?
   b. You see lots of growth, but when you isolate plasmid from numerous *E. coli* colonies, all you find is pET-41a with no insert. Suggest the most likely way these colonies arose (aside from contamination). What would the results of your controls look like if your suspicion was correct?

Part II

Screening Transformants

The screening of transformants can be done in a number of ways. In practice, the techniques used by a scientist will often depend on the reagents, time, and budget available. In the next few experiments, you will be performing some of the more common techniques used to determine whether a bacterial clone carries a gene of interest. While the forced cloning experimental design in the previous experiments has ensured the proper results, these screening methods will also confirm that the *egfp* gene was inserted in the correct orientation and reading frame to produce the GST::EGFP fusion protein.

It is important to confirm that a recombinant molecule to be used for large-scale protein expression comprises exactly what is intended: one copy of the vector and one copy of the insert. The restriction mapping experiment will confirm that an insert of the correct size is present. The visualization of green fluorescence will reveal whether the recombinant protein is folded correctly. And finally, sequencing the recombinant plasmid DNA of a putative positive clone will confirm that no single nucleotide errors were incorporated during the PCR amplification of the *egfp* gene during cloning. It is important to know and understand a wide variety of screening techniques to troubleshoot if any problems occur during the cloning process.

Advanced Alternatives Within Part II

ADVANCED ALTERNATIVES, PART II—SCREENING TRANSFORMANTS

Depending on the Advanced Alternatives used in Part I—Manipulation of DNA, the appropriate changes need to be made during this part of the lab project. The overall goal of the project is to confirm the cloning of the gene of interest into the pET-41a vector. In other words, do you have the correct coding sequence present and in frame to enable protein expression to occur in bacteria?

In this part, we describe two ways to do this: restriction analysis of miniprep DNA and DNA sequencing. Depending on whether *egfp* or a different gene was cloned in Part I, different primers and restriction enzymes might be needed for these experiments. In practice, DNA sequencing is the gold standard for this confirmation. To perform sequencing, plasmid DNA needs to be purified from individual clones following transformation. In the time it takes to purify the DNA, send samples for sequencing, and obtain the sequence data, other screening methods can be performed as well. While these other methods are not absolutely necessary, it is always useful to have multiple tools at your disposal. Unlike some of the Advanced Alternatives in Part I, downstream experiments in Part III are not affected by the options suggested here.

AA2.1 Colony PCR Screen

One method that is sometimes used to screen many colonies at once is a whole colony PCR screen. In this technique, one creates a PCR master mix with primers that will amplify the gene of interest to confirm its presence. The template, instead of plasmid DNA, is whole bacterial cells. By creating a PCR master mix, aliquoting it into tubes, and then transferring a very, very small amount of the culture as the template to individual PCR tubes with the master mix, one can assay for whether or not insert is present in the bacteria. In some cases, the bacterial colony is transferred to a small amount water and vortexed first (to lyse the bacteria), then 1–2 microliters of that solution can be used as template. In other protocols, the whole cell is added directly to the master mix, and a longer initial denaturation step (10 minutes) is used to help ensure lysis of the bacteria.

Additional materials

None are needed, although separate primers for amplification could be used. For example, one primer could bind to the GST region of the plasmid while another binds to the *egfp* region. Also, as a cost-saving measure, Taq polymerase could be used instead of Vent, as proofreading activity is not required in a screening procedure.

AA2.2 Choose Different Restriction Enzymes for Analysis

While the most common practice is to simply cut the insert from the plasmid using the same enzymes that were used for cloning, other enzymes could also be used to confirm the presence of insert. In this alternative, allow students to select from a library of restriction enzymes and have them predict gel patterns for both plasmids with insert and plasmids without insert. The most effective method for this is choosing a single enzyme that gives a significantly different pattern in the presence of insert. Furthermore, there are many freely available online tools that analyze given sequences for restriction enzyme sites and generate predicted fragment numbers and sizes. Have students perform the agarose gel electrophoresis to determine if their results support the presence of the desired insert.

Additional materials

Alternative restriction enzymes for students to choose from.

Lab Session 6

Screening of Transformants, Part I

Over the next three lab periods, you are going to use up to three different methods to verify the insert DNA you have cloned into pET-41a. These methods are: miniprep DNA purification and digestion, EGFP fluorescence, and DNA sequencing. To avoid pitfalls, it is important to be prepared and organized as you perform these experiments. You should also have an understanding of what each individual experiment can tell you in terms of your clones.

In real world scenarios, scientists may elect to do all or only some of these screening procedures, based on their individual preferences. Arguably, the most important of these is DNA sequencing, which allows one to directly confirm the sequence of the plasmid from the clones. The other techniques can identify positive clones before sequencing, to avoid having to sequence DNA from a number of clones to find a positive clone. In these next two lab sessions, we provide the framework for three different screening methods: restriction digest analysis of DNA from clones, direct visualization of EGFP fluorescence, and DNA sequencing. Before you can begin these experiments, an interim laboratory session at least one day before your regular laboratory period is required to count and replica plate your colonies and inoculate cultures for minipreps. The two experiments you perform during the regular laboratory period this week will be completed over the next two weeks. The third (DNA sequencing), you will begin next week and analyze the results in a subsequent lab session.

Molecular Biology Techniques. DOI: https://doi.org/10.1016/B978-0-443-26714-7.00005-3

Lab Session 6a

Interim Laboratory Session

COUNTING TRANSFORMANTS, REPLICA PLATING, AND INOCULATING OVERNIGHT CULTURES

Goal: Today you will count and record the number of transformants on each plate. You will also make replica plates of transformants from your experiment. These replica plates will be used to screen for positive clones (those that express the enhanced green fluorescent protein). Finally, you will inoculate cultures from which you will purify plasmid DNA using a miniprep during your regular laboratory period.

I. INTRODUCTION

Replica plating is the technique by which each colony/clone is inoculated onto another plate according to a numbered scheme. This method allows each clone to be tested by a variety of methods, while retaining a master plate from which clones can be picked. Because positive clones should occur at a high rate, we only need to screen tens of colonies, so we will pick each colony individually and inoculate replicas using a sterile toothpick or pipette tip. In cases where hundreds or thousands of colonies must be replicated, a sterile velvet stamp may be touched to the original plate and stamped onto multiple blank plates to grow replicas. The individual fibers of the velvet act as tiny inoculating needles.

II. LABORATORY EXERCISES

A. Counting Transformants and Transformation Efficiency

Count and record the number of transformants on each of your ligation/transformation plates. You should have many colonies on the plate from the uncut vector (positive control) transformation, and several colonies from your experimental ligation. You should have very few, if any, colonies on the plate where the bacteria were transformed with the linear pET-41a (negative control).

Determine whether your controls yielded the expected results (refer to Fig. 5.1 from Chapter 5: DNA Ligation and Transformation of *Escherichia coli*. If they did not, what technical problem could have occurred? If you see numerous colonies with the negative control, it may be difficult to find transformants that have incorporated insert in the experimental ligation.

Based on the number of colonies on your positive control plate (uncut vector), you can calculate the transformation efficiency of your cells. This can be a helpful calculation, particularly in more difficult cloning experiments where you don't expect as many positive clones as we typically observe. The transformation efficiency of cells is usually given to you by a manufacturer, but knowing whether or not you're meeting the efficiency can tell you if your transformation protocol is working as expected.

Transformation efficiency is defined as the number of colony forming units (cfu) per microgram of DNA used to transform the bacteria. The key to the calculation is determining the mass of DNA used when plating cells on selective mediums. For instance, assume you transformed 50 ng of DNA into 50 μL of competent cells. After the heat shock, you added 950 μL of SOC medium. At this point, you still have 50 ng of DNA in your tube with the cells in a total volume of 1000 μL. However, if you only plate 200 μL of the final mixture (1/5 of the total amount) on your LB/kan plate, you are only plating 10 ng of DNA. Thus, if you were to count 100 colonies, your transformation efficiency would be 100 colonies/(0.01 μg DNA) = 1×10^4 cfu.

B. Replica Plating and Inoculation of Overnight Cultures

You will create a replica plate to analyze potential positive clones from your colonies arising from the vector-insert ligations.

1. Obtain an LB/kan plate and adhere a grid sticker to the back. If grid stickers are not available, create a numbered grid of at least 15 squares on the bottom of your plates with a Sharpie marker. (See Fig. 7.1 for an example of a plate with a grid.) *IMPORTANT: Be sure to label the plate with the date and your name/initials. Always label the bottom of the Petri plate, not the lid.*

2. Obtain seven culture tubes, each containing approximately 5 mL of LB/kan media.
3. Obtain a plate containing pBIT (positive control, which is pET-41a containing the *egfp* insert). Using a sterile toothpick or pipette tip, pick up cells from a culture of the positive control (pBIT) by lightly touching a single, well-isolated (not touching another) colony. Lightly drag the toothpick or pipette tip in the shape of a + in one of the squares toward the middle of the plate. Do not puncture or gouge the agar. Place the toothpick or pipette tip into one of the culture tubes (label appropriately as "+ control" or "pBIT"). *Do not go back to the colony with the toothpick or pipette tip in between inoculating the plate and the culture tube*. There are plenty of bacteria present to inoculate both.

   This inoculation method does not utilize the principle of aseptic technique. It works well for inoculating cultures for minipreps because the antibiotics in the medium keep most contaminants from growing, and because the inoculum is so large that it would likely out-compete any contaminant. In cases where a pure culture is critical, such as creating a freezer-stock, do not utilize this method.
4. Repeat the previous step with a colony from the plate containing pET-41a (negative control), except make the shape of a – (negative sign) in a square adjacent to the positive control (+). The + and – signs make it easy to distinguish the controls on your replica plate. Make sure to inoculate the culture tube after the replicate plate.
5. Inoculate the replica plate and liquid cultures with five well-isolated colonies from your experimental ligation plate (*egfp* insert + pET-41a; plate 3). Use a different toothpick or pipette tip for each colony! Be sure to label well so that you know which culture tube corresponds to the appropriate grid on the plate. When plating these cultures on your LB/kan plate, drag the toothpick or pipette tip in a diagonal line across the square.
6. While you will only analyze your positive control, negative control, and five unknown samples using all of the screening assays, fill up the rest of your grids on your LB/kan plate with remaining colonies from your experimental ligation plate. These are additional colonies that can be screened by fluorescence visualization.
7. If you have fewer than five colonies on your experimental ligation plate, pick at least five colonies from a plate from another lab group or see your instructor for a plate of "mixed unknowns." Be sure to record in your notes which plate each clone came from. Do not pick colonies from transformation plates 1 or 2, as these will all be negative.
8. Place your LB/kan replica plate (lid down), in a 37°C incubator overnight (~16 hours). After colonies grow, wrap the replica plate in Parafilm and transfer to a refrigerator (store with lid down).

   Note: Incubation of bacterial plates is typically done at 37°C overnight. However, cells also grow at lower temperatures, albeit more slowly. Transformed bacterial cells plated on antibiotic-resistant medium and grown at room temperature typically show colonies in 24–48 hours. Furthermore, they can often be left up to 72 hours without cultures being overgrown and forming satellite colonies, which are smaller colonies lacking plasmid that surround the larger, plasmid-containing bacterial colony.
9. Culture tubes should be placed in a refrigerator (4°C) until the day before the next lab session. The day before the lab session, they should be placed in a shaking incubator at 37°C overnight.

Lab Session 6b

Isolation of Miniprep DNA From Potential Transformants

Goal: You will isolate the plasmid DNA from several transformants (miniprep). Next week, you will digest the plasmids with the restriction endonucleases NcoI-HF and NotI-HF to confirm the presence of the *egfp* gene.

Clones containing the vector with *egfp* successfully inserted will contain two visible fragments on the gel following digestion: 724 bp (corresponding to *egfp*) and 5.9 kbp (corresponding to the pET-41a vector). Clones containing the vector only will show one visible fragment on the gel of 5.9 kbp (corresponding to the large vector fragment), although another fragment too small to visualize on the gel is also present (73 bp, corresponding to the small fragment of DNA between the NcoI and NotI sites on the vector).

This experiment will be carried out over the course of two weeks. This week, you will purify the plasmid DNA and next week, you will perform the restriction digestion and analysis. The plasmid DNA obtained from this miniprep will also be used for DNA sequencing.

I. INTRODUCTION

Regardless of the initial methods used to identify positive transformants, you will ultimately need to isolate DNA for the final confirmation analysis, DNA sequencing. We will use the same protocol we used at the beginning of the semester to purify plasmid DNA.

Remember, many different products can be formed during your ligation (review Fig. 5.2 from Chapter 5: DNA Ligation and Transformation of *Escherichia coli*). The bottom line in analyzing transformants is if it matches the pattern you expect, keep it. If it doesn't, throw it out.

When analyzing restriction digestion patterns, keep these principles in mind:

- DNA migrates at a rate inversely proportional to the size of the DNA fragment. From previous experience, you know that including a lane of standard molecular weight markers is critical for analysis of your DNA. Because the molecular weight of each of the bands in the DNA standard is known, you can use these molecular weight markers to estimate the size of unknown sample fragments.
- Circular DNA travels at an unpredictable rate through the gel due to secondary structure. Therefore, the size of uncut DNA cannot be estimated using a typical DNA ladder. Whenever you electrophorese restriction digested plasmid DNA, when possible one lane should contain uncut DNA as a reference. Depending on the integrity of the sample, you will see between two and four bands when analyzing uncut plasmid DNA on a gel. Refer to Lab Session 3 for an explanation of the different forms and relative electrophoretic mobilities of uncut DNA.
- The sum of the molecular weight of the fragments generated by digestion of a circular molecule equals the molecular weight of the uncut molecule. (To visualize this, think of cutting a rubber band into pieces; the total length of the pieces is the length of the rubber band.)
- The intensity of band fluorescence is directly proportional to the mass of the DNA loaded. Deviations from this principle are indicative of partial digests or a doublet (two bands of the same size).

II. LABORATORY EXERCISE

Purify plasmid DNA from your seven cultures that you inoculated in your previous lab session (a positive control, a negative control, and five unknowns). *The following protocol is identical to the one you used in Lab Session 2 and is modified from the QIAGEN QIAprep Miniprep Kit handbook (June 2015). You will start at Step 3 of the handbook, which is why the list starts at #3.*

Note to instructor: Numerous other miniprep kits are available for DNA purification. If you choose to use a different kit, follow the protocol for a low-copy plasmid and reagent preparation prior to use.

3. Obtain 1.5–5 mL of the culture in a microcentrifuge tube. We recommend using approximately 5 mL for this experiment to produce a reasonable amount of purified pET-41a plasmid, particularly since you will need to use ~800 ng for DNA sequencing. You can either: (1) split the 5 mL between multiple centrifuge tubes; or (2) fill up one tube, centrifuge as described in Step 4, then add more of the culture on top of the pellet left behind after removing the supernatant. Repeat this until all 5 mL has been pelleted.
4. Harvest the bacterial cells by centrifugation at >8000 rpm (6800 × *g*) for 3 minutes.
   Remove all traces of supernatant by decanting or micropipetting the media into a liquid waste container with 10% bleach.
5. Resuspend the bacterial pellets in 250 µL Buffer P1 (Resuspension Buffer). The bacteria should be resuspended completely until no cell clumps remain. If you have harvested the cells for each sample in multiple tubes, add 250 µL to one tube, resuspend, then transfer the resuspended solution to the next tube. The goal is to have all the pET-41a cells together for each sample in one tube and completely resuspended in Buffer P1.
6. Add 250 µL Buffer P2 to each sample and mix gently but thoroughly by inverting four to six times. Do not vortex, as this will result in shearing of genomic DNA. If necessary, continue inverting the tube until the solution becomes viscous and slightly clear. Do not allow lysis reaction to proceed for more than 5 minutes. *Buffer P2 contains a detergent (SDS) for cell lysis and sodium hydroxide for denaturation of DNA.*
7. Add 350 µL Buffer N3 to the lysate to each sample and mix immediately and thoroughly (but gently) by inverting four to six times. After addition of Buffer N3, a fluffy white precipitate containing genomic DNA, proteins, cell debris, and SDS becomes visible. The buffers must be mixed completely. If the mixture still appears viscous and brownish, more mixing is required to completely neutralize the solution. *Buffer N3 neutralizes the solution, causing plasmid DNA to reanneal, and acts to precipitate the chromosomal DNA and insoluble proteins.*
8. Centrifuge for 10 minutes at 13,000 rpm (17,900 × *g*) in a microcentrifuge.
9. Apply the supernatant (liquid above pellet) from Step 8 on the QIAprep spin column (containing the silica membrane) by pipetting, taking care not to transfer the fluffy white precipitate. Place column in accompanying tube.
10. Centrifuge for 30 seconds. Discard flow-through.
11. Wash the QIAprep spin column by adding 0.5 mL Buffer PB and centrifuge for 30 seconds. Discard the flow-through.
12. Wash the QIAprep spin column by adding 0.75 mL Buffer PE and centrifuge for 30 seconds.
13. Discard flow-through, and centrifuge for an additional 1 minute to remove residual wash buffer.
14. Place the columns in clean 1.5 mL microcentrifuge tubes. To elute DNA, add 50 µL Buffer EB to the center of each column, let stand for 1 minute, and centrifuge for 1 minute. (If necessary, cut the lid off the 1.5 mL tube with scissors so that it will fit in the microcentrifuge.)
15. Remove and discard the column. Be sure the tubes containing the DNA are well-labeled and store at −20°C. The DNA samples will be used in a restriction enzyme digestion next week.

Lab Session 6c

Visualization of Green Fluorescent Protein, Part I

Goal: You will replica plate your transformants (and positive and negative controls) onto LB medium that contains kanamycin and IPTG. This will allow for visualization of green fluorescence in positive clones.

I. INTRODUCTION

As discussed in Lab Session 2, Purification and Digestion of Plasmid (Vector) DNA, IPTG acts to derepress the T7/lac promoter on expression vector pET-41a (Fig. 2.3). By incorporating IPTG into the growth medium, bacteria can express the enhanced green fluorescent protein. Colonies containing the *egfp* gene in the correct reading frame will appear slightly green even under ambient light. Positive clones will fluoresce bright green when UV light is shone on them.

II. LABORATORY EXERCISE

You will use your replica plate to transfer colonies onto the LB/kan/IPTG plate. Essentially, you are creating a copy of your replica plate, except this time on a plate that can induce protein expression.

1. Retrieve your replica plate from the refrigerator.
2. Obtain an LB/kan/IPTG plate and adhere a grid sticker to the bottom. Label with your initials and the date. *Always label the bottoms (agar sides) of Petri dishes rather than the lids to avoid mix-ups.*
3. As you did in Lab Session 6a, use a sterile toothpick or pipette tip to pick a small amount of bacterial growth from each grid square on the master plate and inoculate the same squares on the fresh LB/kan/IPTG plate. Be sure to change toothpicks or tips between each clone you pick.
4. Wrap your original replica plate in Parafilm and place in the refrigerator. Leave your new LB/kan/IPTG plate at room temperature for 24–72 hours. Incubation can be done at 37°C overnight, but we have consistently observed better protein expression at room temperature. After colonies grow on your LB/kan/IPTG plate, it can be wrapped in Parafilm and transferred to the refrigerator until .

DISCUSSION QUESTIONS

1. Will the restriction mapping screen be able to detect transformants with multiple inserts? Why?
2. To 50 μL of competent cells, 10 ng of DNA was added and incubated on ice. After heat shock, 950 μL of SOC media was added and the solution was placed in a 37°C shaking incubator for an hour. Next, 100 μL of this solution was plated on LB/kan plates and incubated overnight in a 37°C incubator. The following day, the plate was observed to have 60 colonies. What is the transformation efficiency of the competent cells based on the observed results?

Lab Session 7

Screening of Transformants, Part II

This week, you will perform restriction digests of the miniprep plasmid purifications from the previous lab session. Then, you will analyze the digested samples using agarose gel electrophoresis. The fluorescent bacterial plate (IPTG-induced) from the previous lab session will also be visualized to confirm your screening results. In addition, you will set up a sequencing reaction on one of your positive samples to verify the nucleotide sequence using Sanger sequencing.

Molecular Biology Techniques. DOI: https://doi.org/10.1016/B978-0-443-26714-7.00023-5

Lab Session 7a

Restriction Digestion of DNA

Goal: You will digest purified DNA (from the last lab session) from your controls and the putative pET-41a/*egfp* transformants in order to confirm the presence of the *egfp* insert.

I. INTRODUCTION

We used the NcoI and NotI restriction enzyme sites to clone the *egfp* gene into pET-41a, and we will use these same two enzymes to release the insert. It will be useful to draw a mock gel resulting from the restriction enzyme digestion to predict the number and size of DNA fragments for both plasmid DNA containing and lacking the *egfp* insert. (See Lab Session 31 for a description of an *in silico* tool that can be used to predict sizes and bands on an agarose gel following digestion.)

You will follow the procedure described here to set up seven double digests with NcoI and NotI. When performing multiple digests with the same restriction enzyme(s), it is helpful to make a master mix containing the common reagents found in each reaction. In this case, the master mix consists of water, restriction enzyme buffer, and the two restriction enzymes (NcoI and NotI). Once this master mix is made, it is aliquoted to separate tubes to which the unique DNA samples will be added for digestion. Creating a master mix provides a number of distinct advantages compared to pipetting all the reagents into separate tubes. First, since you will be pipetting larger volumes, you will be minimizing the error that is inherent in pipetting smaller volumes. Second, using a master mix reduces variability between samples. You can imagine that if you're attempting to pipette 0.5 μL of NcoI in each of the six tubes, some tubes might have 0.4 μL while others might have 0.6 μL. A master mix reduces this variation. Finally, using a master mix saves time. You will be pipetting fewer times than if you were to pipette each component into each separate tube for the digestion. To set up your master mix for your digestion reactions, follow the instructions below:

II. LABORATORY EXERCISE

1. Label eight tubes corresponding to the seven DNA samples you are going to digest, and one labeled "RE Master Mix."
2. Consider the reagent volumes (enzyme, restriction buffers, water) for one enzyme digest (see Table 7.1) and prepare enough master mix for eight reactions (your seven reactions, plus an extra reaction for pipetting variability). Make a table in your lab notebook similar to the one provided, listing the reagents, stock concentrations, final concentrations,

TABLE 7.1 Experimental Setup for Restriction Enzyme Digestions

Master Mix for Double Digests				
Reagent	**Stock Concentration**	**Final Concentration**	**Volume for 1 rxn**	**Master Mix for 8 rxn**
Sterile water				
CutSmart buffer				
NcoI-HF		2 U/rxn		
Not I-HF		2 U/rxn		
DNA	See note[a]		10 μL[a]	NONE
Total volume			20 μL	

[a]DNA is to be added to numbered tubes separately (not in the master mix). The DNA concentration for each sample can be determined using a Nanodrop prior to setting up your restriction digestions. If time does not allow for this, you may assume that your miniprep DNA has a concentration between 50 and 100 ng/μL. Since the goal of this experiment is to simply see if the insert is present, you only need enough DNA to visualize on a gel. Using 10 μL (500–1000 ng) in each digest should be sufficient for this experiment.
No Permission Required.

amount for one reaction, and amount for eight reactions. *Note: the DNA is going to vary in each sample and cannot be added to the master mix.* Each reaction will have a final volume of 20 μL.

3. Combine in order: water, CutSmart Buffer, and enzyme(s) in the RE Master Mix tube. Make sure to mix the tube well and keep on ice.
4. Aliquot 10 μL of master mix into each of the seven tubes for digests. Make sure the tubes are labeled and then add 10 μL of miniprep DNA to each appropriately labeled tube.
5. Incubate the seven reactions for a minimum of 15 min at 37°C. During the incubation, proceed to Lab Session 7b, where you will prepare and pour your agarose gel.

Lab Session 7b

Analysis of Miniprep Digestions Using Agarose Gel Electrophoresis

Goal: Identify positive clones through agarose gel analysis of previous samples.

I. INTRODUCTION

Review Lab Session 6b.

II. LABORATORY EXERCISE

1. Pour a 1% agarose gel as previously described (Lab Session 3).
2. DNA loading buffer needs to be added to each of the miniprep digest reactions (seven tubes in total). Since these digests will not be used for anything other than this experiment, you can add the loading buffer directly to the tubes containing the digested DNA. Each digest should have been performed in a volume of 20 μL. When you are adding loading buffer to the tubes, keep in mind that the final volume of the solution will be changing. For example, if you have a 6× DNA loading buffer solution, you would add 4 μL of loading buffer (6× needs to be diluted sixfold in the final solution, which in this case would be a total of 24 μL). **Do NOT add loading buffer to the tubes containing undigested DNA!**
3. Load your gel as follows:

	Miniprep Digests
Lane 1	150 ng NEB 1 kb Ladder
Lane 2	15 μL positive control
Lane 3	15 μL negative control
Lane 4	15 μL sample 1
Lane 5	15 μL sample 2
Lane 6	15 μL sample 3
Lane 7	15 μL sample 4
Lane 8	15 μL sample 5

4. Run your gel at 85 V (or using the appropriate settings for your setup) for approximately 45 minutes (or until the dye reaches about ¾ of the way down the gel). Visualize the gel on a ultraviolet (UV) transilluminator.
5. Use these results, combined with the results from Lab Session 7c, to determine which DNA sample to send out for Sanger sequencing (Lab Session 7 d).

Lab Session 7c

Visualization of Enhanced Green Fluorescent Protein, Part II

Goal: You will observe the transformants that you plated on LB/kan/IPTG plates using UV light. Positive clones containing the *egfp* gene (and therefore are expressing the GST::EGFP fusion protein) will fluoresce bright green!

I. INTRODUCTION

The enhanced green fluorescent protein has an excitation peak at 488 nm (blue light) and emits light maximally at 507 nm. When you shine UV light on your positive transformants, although you are not exciting EGFP at the excitation peak, there is still enough excitation that occurs, allowing the colonies to appear bright green. This assay, since it is so easy to perform, will be used to confirm the results of the restriction digests of the minipreps from Lab Session 7b. However, most proteins are not fluorescent. In a research lab, it is rare to clone a gene that has such a simple assay for expression. But for our purposes, it serves as an unambiguous positive confirmation assay that can be used to confirm the other screening techniques.

II. LABORATORY EXERCISE

1. Obtain the IPTG replica plate you inoculated last week.
2. Remove the lid and invert the plate (open side down) on a UV transilluminator. *While viewing your transformants on the ultraviolet light box, you must wear a UV protective face-shield, as well as protect any skin from direct exposure to the UV light by wearing gloves and/or a lab coat.*
3. Turn on the ultraviolet light and view the plate. The positive control and positive clones will fluoresce bright green due to the expression of the enhanced green fluorescent protein, as shown in Fig. 7.1. Record which clones are positive and which are negative in your laboratory notebook.

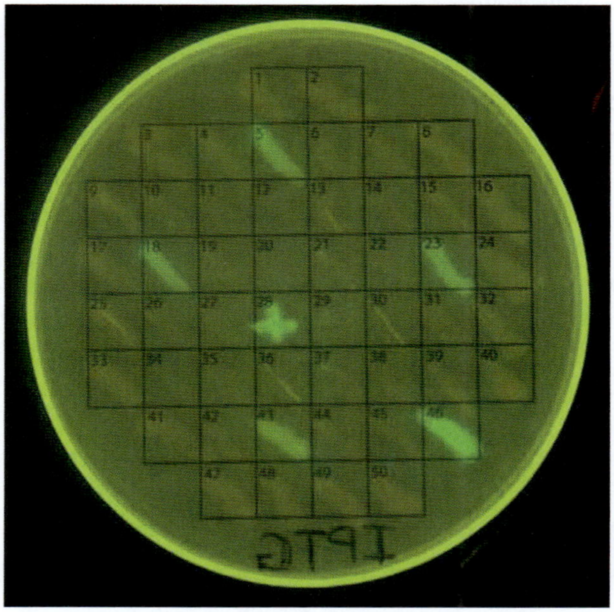

FIGURE 7.1 Fluorescence of *E. coli* expressing the enhanced green fluorescent protein. The positive control, pBIT, is inoculated in grid number 28. Clones numbered 5, 18, 23, 43, and 46 are also positive. The rest of the clones are negative. *No Permission Required.*

Lab Session 7d

Analysis of DNA Sequence from a Positive Clone, Part I

Goal: You will verify the DNA sequence of the *egfp* insert you cloned into pET-41a. This experiment will be done over the course of two weeks. First, you will prepare samples that will then be sent to a DNA sequencing facility. Second, the data obtained from the sequencing run will be analyzed to determine the exact sequence of the *egfp* insert and compared to the known and expected sequence.

I. INTRODUCTION

Cloning by PCR is an effective and straightforward method of cloning a gene into a vector. As discussed in Lab Session 3, Completion of Vector Preparation and Polymerase Chain Reaction Amplification of *egfp*, cloning by PCR has many advantages over traditional cloning. Despite this, there is one main concern in PCR cloning: errors incorporated during PCR. A single base mutation in your PCR product can lead to amino acid changes or early termination of your fusion protein during translation. While high-fidelity DNA polymerases, such as Vent, can be used to minimize these errors, mutations can still occur. To ensure that subsequent experiments are performed with your expected protein and not a mutated version, verifying the DNA sequence of your plasmid is imperative.

DNA sequencing has become a routine practice in molecular biology and is used in numerous applications. DNA sequencing was initially performed in the 1970s, with labor-intensive, expensive techniques. This technology evolved, largely due to the work of Frederick Sanger, who received the second of his two Nobel Prizes in Chemistry for his work in developing the technology of DNA sequencing.[1] This chain-termination method of DNA sequencing, also known as Sanger sequencing, is what you will use to analyze the DNA sequence of a positive transformant. Other DNA sequencing platforms are used for high-throughput applications such as whole-genome sequencing. These methodologies can obtain up to hundreds of millions of base pairs of sequence in a day. Most forms of high-throughput screening are still more expensive per run than the traditional chain-termination method. For sequencing your *egfp* insert, you only need to sequence about 750, not millions of nucleotides.

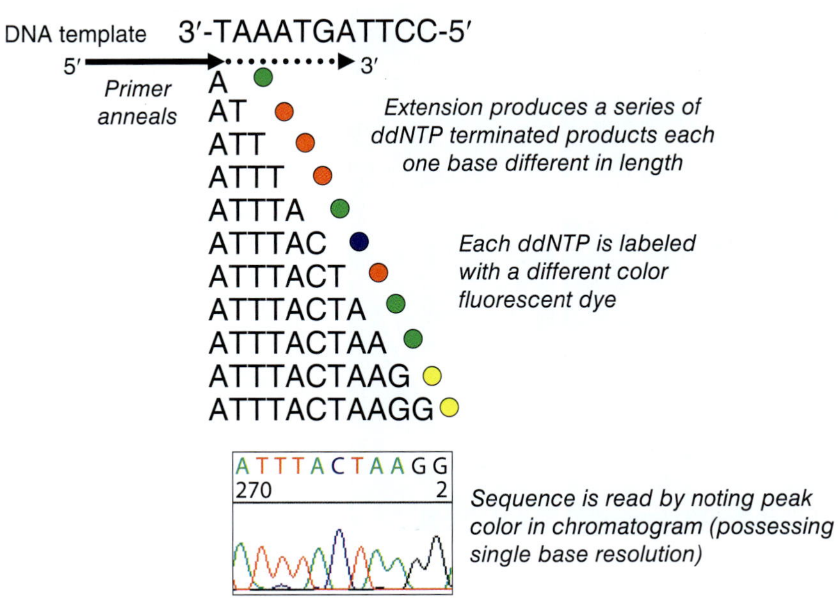

FIGURE 7.2 DNA sequencing using fluorescent ddNTPs, also known as Sanger sequencing. Fluorescently labeled ddNTPs act as chain terminators during the synthesis of complementary DNA fragments. These fragments are separated using capillary electrophoresis and results are obtained in the form of a chromatogram. *Reprinted from Butler JM.* Fundamentals of Forensic DNA Typing. *San Diego, CA: Elsevier Academic Press; 2009. p. 65.*

A typical Sanger sequencing reaction contains DNA template, DNA polymerase, one sequencing primer, deoxynucleotide triphosphates (dNTPs), and dideoxynucleotide triphosphates (ddNTPs). The key players in DNA sequencing by chain termination are ddNTPs. As shown in Fig. 7.2,[2] the DNA to be sequenced serves as a template for synthesis of complementary strands of DNA. The sequencing primer anneals to the template, and DNA polymerase binds to the template/primer hybrid DNA and synthesizes a new DNA strand using the nucleotides present in the reaction mixture. The dNTPs are the "normal" bases (A, C, G, and T) and can be added to each other, one after another, via phosphodiester bonds. Conversely, the ddNTPs lack the 3′ hydroxyl group that is necessary to form a phosphodiester bond. Because of this, if a ddNTP is incorporated into the DNA strand during synthesis, no additional synthesis can occur. In other words, it acts as a chain terminator. The ddNTPs are present in the sequencing reaction in limiting quantities. Thus, the majority of the time a dNTP is added, and synthesis continues and only occasionally does a ddNTP get added. Because the addition of a ddNTP is random, the result of a sequencing reaction is partial copies of the DNA template, varying in size. Each fragment will start at the 5′ end with sequence corresponding to the primer sequence and will terminate wherever a ddNTP is added.

The determination of the DNA sequence is dependent on reading the lengths of these fragments and knowing which ddNTP occupies the 3′ position. This is most typically done by using fluorescently labeled ddNTPs (each ddNTP is labeled with a different color) in conjunction with capillary electrophoresis in automated DNA sequencers. Each sequencing reaction contains a pool of synthesized DNA fragments with different lengths, and each fragment will fluoresce according to which ddNTP was incorporated at its 3′ end. With automated DNA sequencing, these fragments are separated by size using capillary gel electrophoresis, with the smallest fragments traveling through the matrix fastest. As the fragments exit the capillary column, a laser excites the sample and a computer records the fluorescence emitted from the sample. The resulting data, called a chromatogram or electropherogram, shows a series of peaks which reflect the DNA sequence of the synthesized strand of DNA.

An excellent visual representation of DNA sequencing by chain termination can be found here: www.dnalc.org/resources/animations/cycseq.html. [3]

In this lab session, you will assemble sequencing reactions to send to a DNA sequencing facility. When performing DNA sequencing, the selection of your sequencing primers is one of the most critical aspects of the experiment. Typically, a successful sequencing run yields 800–1000 nucleotides of sequence. Since the *egfp* sequence is 720 nucleotides in length, we should be able to sequence the entire gene using a primer that anneals just upstream of the beginning of the *egfp* sequence (Fig. 7.3). It is worth noting that interpretable sequence will start 15–30 nucleotides downstream of the primer, so never design sequencing primers to the very beginning of the region you are trying to sequence.

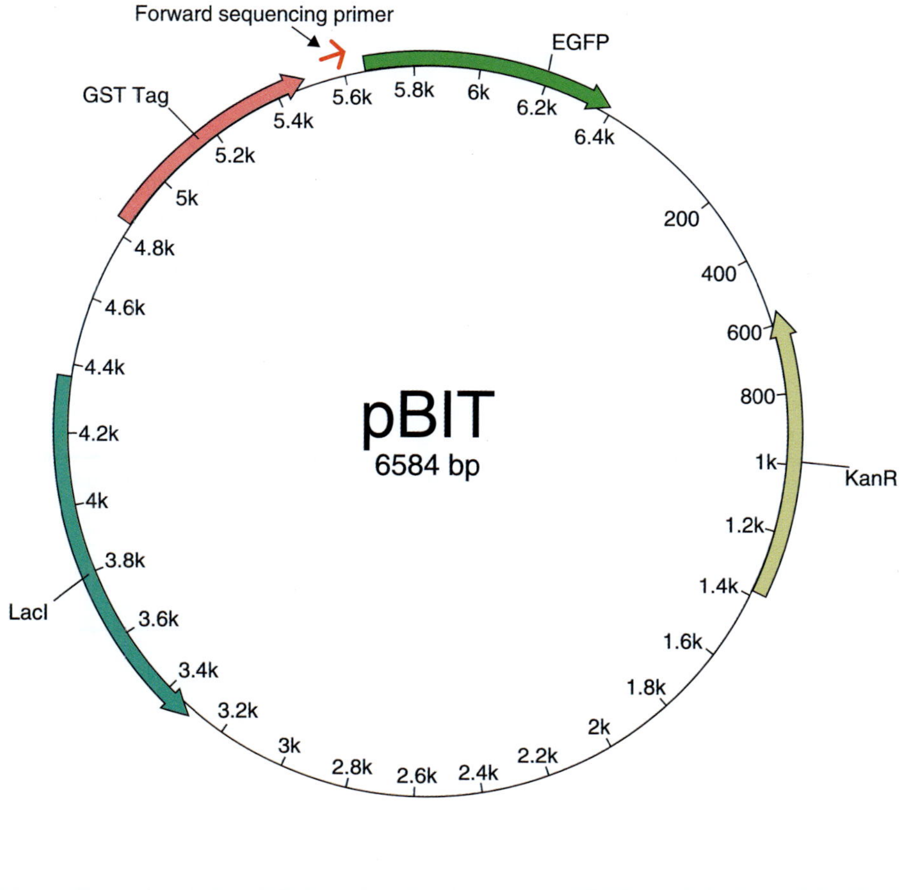

FIGURE 7.3 Schematic of pBIT showing the location of the sequencing primer. *No Permission Required.*

While we are able to sequence the entire *egfp* sequence using one primer, there are many times when using additional primers is necessary. If an insert sequence is over 1000 nucleotides in length, you would need multiple primers to cover the entire sequence. These primers can be designed to either strand of the template; occasionally, researchers will use antiparallel primers flanking a gene and sequence the DNA in both directions (in two separate sequencing reactions). Also, if there are uncertainties or discrepancies in your sequencing results, performing an additional sequencing reaction using a different primer (i.e., one that binds to a different region of the DNA) can clarify the result. Finally, if you are sequencing DNA with an unknown sequence (not confirming a known sequence as we are doing in this exercise), always use multiple primers resulting in overlapping sequence results to most accurately determine the sequence of your DNA.

Next week, you will analyze the results from your sequencing reaction to determine the DNA sequence of *egfp* in your clone.

II. LABORATORY EXERCISE

Note to instructor: The following instructions are for preparing sequencing samples for submission to the GENEWIZ sequencing facility. There are many different sequencing facilities available throughout the world, with many academic institutions having their own facilities. If you use a different facility, follow their specific sample preparation guidelines and instructions.

1. Obtain a thin-walled PCR tube in which you will set up your sequencing reaction. Your instructor will give you details on how to label the tube.
2. Choose which clone you are going to sequence based on your previous results. Choose DNA from a colony that was positive using all of the different screening methods, if possible.
3. Using your undigested miniprepped DNA (purified in Lab Session 6b), determine the concentration of your DNA sample as previously described.
4. The sequencing reaction should contain 800 ng of your plasmid DNA and 25 pmol of sequencing primer. (*Note: If your DNA has too low of a concentration to obtain 800 ng, try choosing a different clone. If all of your clones have a low concentration, choose the clone with the highest concentration and use as much as possible.) Create a table in your notebook as illustrated* (see Table 7.2) *to guide you in setting up your sequencing reaction. If you are instructed to sequence multiple clones, a different tube needs to be set up for each sequencing reaction.*

   Remember, for a sequencing reaction, you only add ONE primer. (You are not performing PCR and amplifying a double-stranded template; you are making partial copies of one strand.)
5. Tap the tube against the bench to be sure reagents mix and are at the bottom of the tube.
6. Give your tube to your instructor for submission to GENEWIZ or another sequencing facility.

III. COMPILED SCREENING DATA

Now that you have performed several different techniques for screening transformants, it is interesting to compare the reliability of each method. Create a chart in your lab notebook similar to the one provided. Leave blank the spaces for transformants that were not tested in specific assays. Keep in mind that the fluorescence test was the most reliable assay method, so use it as your reference for determining which assays gave the most consistent results.

TABLE 7.2 Experimental Setup of Sequencing Reactions

	Stock Concentration	Final Concentration/Amount	Volume to Add
DNA template (Clone__)		800 ng	
pBIT sequencing primer		25 pmol	
dH_2O (to rxn vol of 15 μL)			

No Permission Required.

Transformant Number	Restriction Digestion	Fluorescence Assay

No Permission Required.

REFERENCES

1. F. Sanger, S. Nicklen, A.R. Coulson, DNA sequencing with chain-terminating inhibitors, PNAS 74 (12) (1977) 5463–5467.
2. J.M. Butler, Fundamentals of forensic DNA typing, Elsevier Academic Press, San Diego, CA, 2009.
3. DNA Learning Center. Biology Animation Library: Cycle Sequencing, Cold Spring Harbor Laboratory.

DISCUSSION QUESTIONS

1. Is it possible for a clone to have multiple copies of *egfp* inserted? Could two copies be inserted? Could three copies be inserted? If so, would it have more, less, or the same amount of fluorescence compared to a clone with a single copy of the *egfp* gene present?
2. If your DNA was not cut to completion during the restriction enzyme analysis of your transformants, what would you expect your gel to look like? Would you still be able to distinguish positive from negative clones?
3. When performing DNA sequencing, you only use one primer per reaction. What would happen if you mistakenly added both a forward and reverse primer to a sequencing reaction? How would your data be affected?

Lab Session 8

Analysis of DNA Sequence From a Positive Clone, Part II

Goal: You will analyze the data obtained from the sequencing reactions to ensure your clone has the proper *egfp* sequence.

I. INTRODUCTION

Automated DNA sequencing is a frequently used tool to determine the sequence of DNA in a given sample. While DNA sequencing has evolved from its inception to be an automated process, the data obtained from the computer still requires human validation. Indeed, often the computer analysis of the readout is ambiguous, leaving you to interpret the data and decide the actual sequence of the DNA.

There are a number of types of DNA sequencing platforms currently in use. The type of sequencing used depends on the experiment. Sanger sequencing, which is used in this experiment, is the standard method for evaluating short sequences, such as those from cloning with DNA plasmids. Newer sequencing methods, often referred to as high-throughput sequencing, are used to sequence larger sequences, such as entire genomes and transcriptomes (see Lab Session 32).

In sequencing by chain termination (Sanger sequencing) using fluorescent dideoxynucleotide triphosphates (ddNTPs), the data are obtained in the form of a fluorescent chromatogram. Each peak represents a synthesized DNA fragment with the color representing which ddNTP was incorporated as the chain terminator. As such, the peaks on the chromatogram are read sequentially (from smallest to largest starting closest to your sequencing primer) to determine the sequence of your DNA molecule.

Chromatograms are typically saved as. ab1 files. There are many different programs which can view these files. The instructions in this lab session are specific for Chromas, which is available as freeware via Technelysium Pty Ltd (http://technelysium.com.au/wp/chromas/), but the general use of most programs is similar and the same functions can be performed. For Mac users, Finch TV is popular freeware that works similarly. Upon opening your chromatogram, you'll see a trace similar to what is shown in Fig. 8.1, which is a chromatogram for the pBIT (pET-41a with *egfp* insert) sequence. The box shows the slider used to change the horizontal resolution of the trace, while the circle is used to change the vertical resolution.

In looking at your chromatogram, there are a few different things to consider. First, remember from Lab Session 7d that interpretable sequence begins 15–30 nucleotides downstream from the end of the sequencing primer. As shown in Fig. 8.2, the first ~25 peaks of the trace are rounded and not well resolved compared to the peaks that follow. Despite the fact that DNA fragments will be synthesized starting from the nucleotide immediately after the end of the sequencing primer, you can see from this example why the sequencing primer needs to be designed upstream of the sequence you wish to sequence. Our sequencing primer is designed to anneal 60 nucleotides upstream of the junction between the *egfp* insert and pET-41a. You can see by nucleotide 30, the peaks have become well-defined and equally spaced, as well as having relatively little noise at the bottom of the trace. This is where the sequence can begin to be read from.

Fig. 8.3 shows a screen shot of what the chromatogram should look like toward the middle of the sequencing read. You can see during this part of the trace the peaks are well-defined with little noise present. The peak height in this region is often much higher than the peaks at the beginning of the trace. Most importantly, this sequence is very "clean"; in other words, it is unambiguous with regard to the sequence it represents. Looking above the peaks in the chromatogram, you will see that the software reads the chromatogram and assigns a nucleotide to the corresponding peaks. In the region shown, the clarity between the peaks allows the computer to do so accurately. Looking back at Fig. 8.2, you will notice the beginning of the read contains a number of "N" bases, which means the software was unable to confidently assign a base to a given peak. In the middle part of the sequence (Fig. 8.3), Ns should be infrequent, if present at all.

Each sequencing reaction will yield a different length depending on the quality of the DNA and other factors. Generally speaking, a typical sequencing read will span 800–1000 nucleotides of unambiguous sequence (shown in

Molecular Biology Techniques. DOI: https://doi.org/10.1016/B978-0-443-26714-7.00018-1

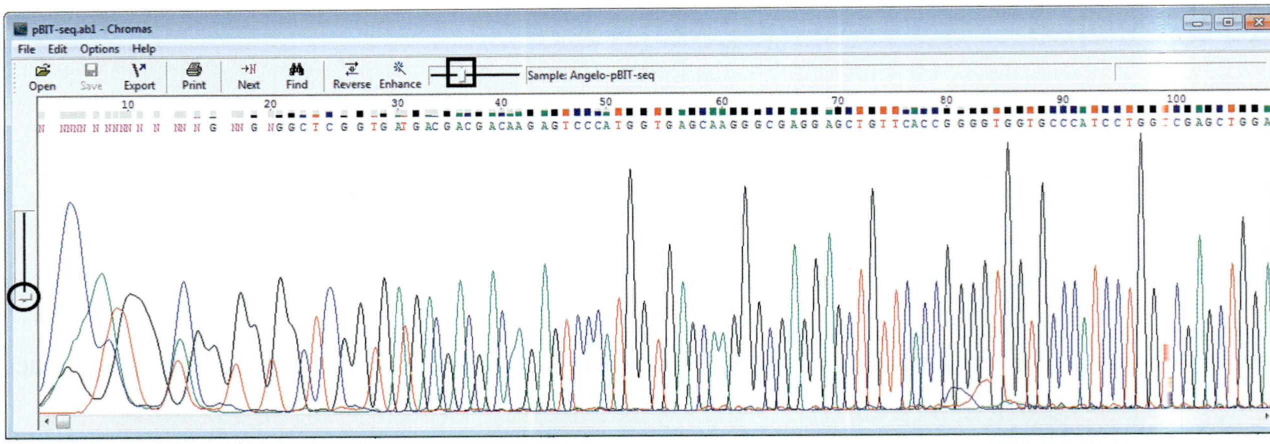

FIGURE 8.1 Sample image of the first ~110 nucleotides of a chromatogram. The box indicates the bar to adjust the horizontal resolution and the circle represents the bar to adjust the vertical resolution. *No Permission Required.*

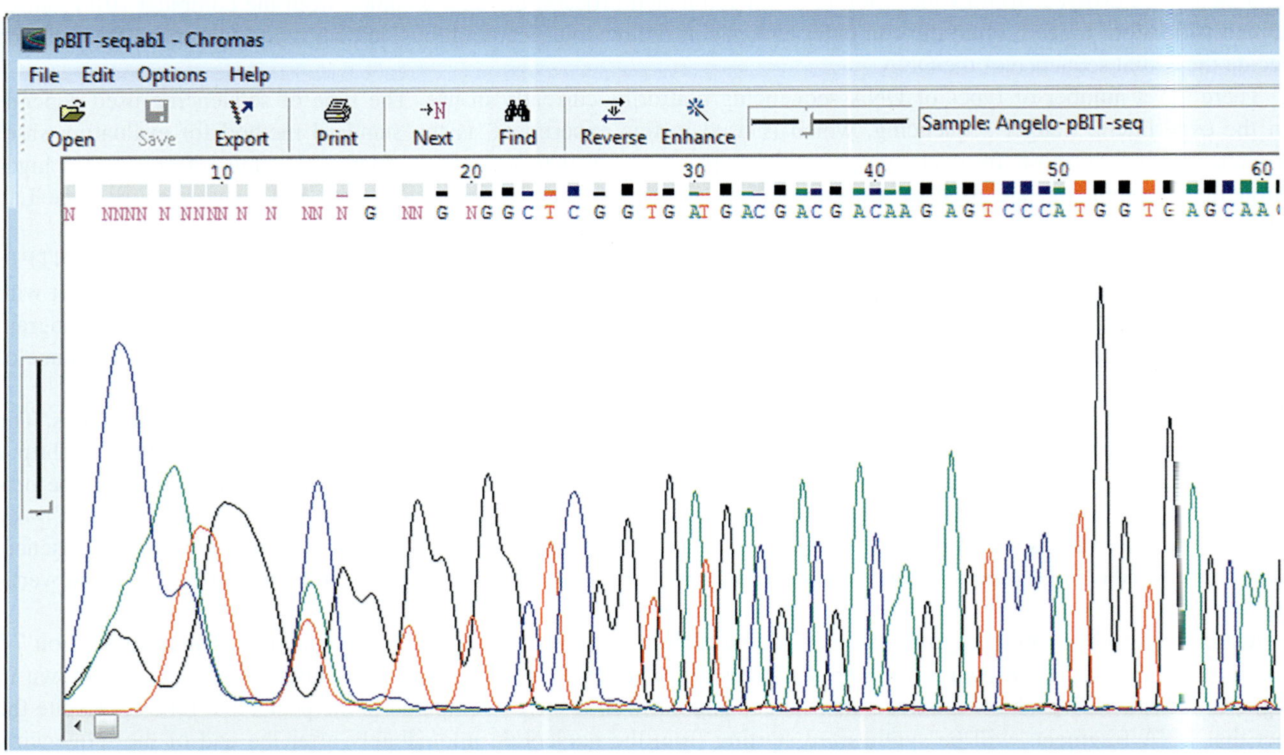

FIGURE 8.2 pBIT sequencing data at the beginning of the chromatogram. Note the broad, rounded peaks in the beginning transitioning to sharp, evenly spaced peaks. *No Permission Required.*

Fig. 8.3) before the signal starts to deteriorate. Eventually, the peaks become broad, more rounded, and less evenly spaced, making assigning the proper base difficult. As shown in Fig. 8.4, the peak height has diminished and there are a few Ns that have crept into the sequencing read. This sequence can still be usable if you're able to confidently assign the bases by visually inspecting the chromatogram, but in general, if there are ambiguous bases such as these, sequencing with a different primer placing these nucleotides in the middle of the sequencing run would be preferred.

In your case, you are sequencing the *egfp* gene cloned into pET-41a to be certain no mutations were made to *egfp* during PCR cloning. Thus, while some of the sequence obtained is ambiguous, you will be able to answer your question as long as the *egfp* sequence can be confirmed. Since you know *egfp* was cloned into the NcoI site (5′-CCATGG-3′) at the 5′ end, you can scan the sequence and look for this site. Looking back at Fig. 8.2, we can see that at nucleotides 48–53 the sequence reads CCATGG, depicting the NcoI site and the beginning of the *egfp* insert sequence. Knowing the *egfp* insert is 720 nucleotides long, this means the sequence should span from nucleotide 50–769. In Fig. 8.4, we see the

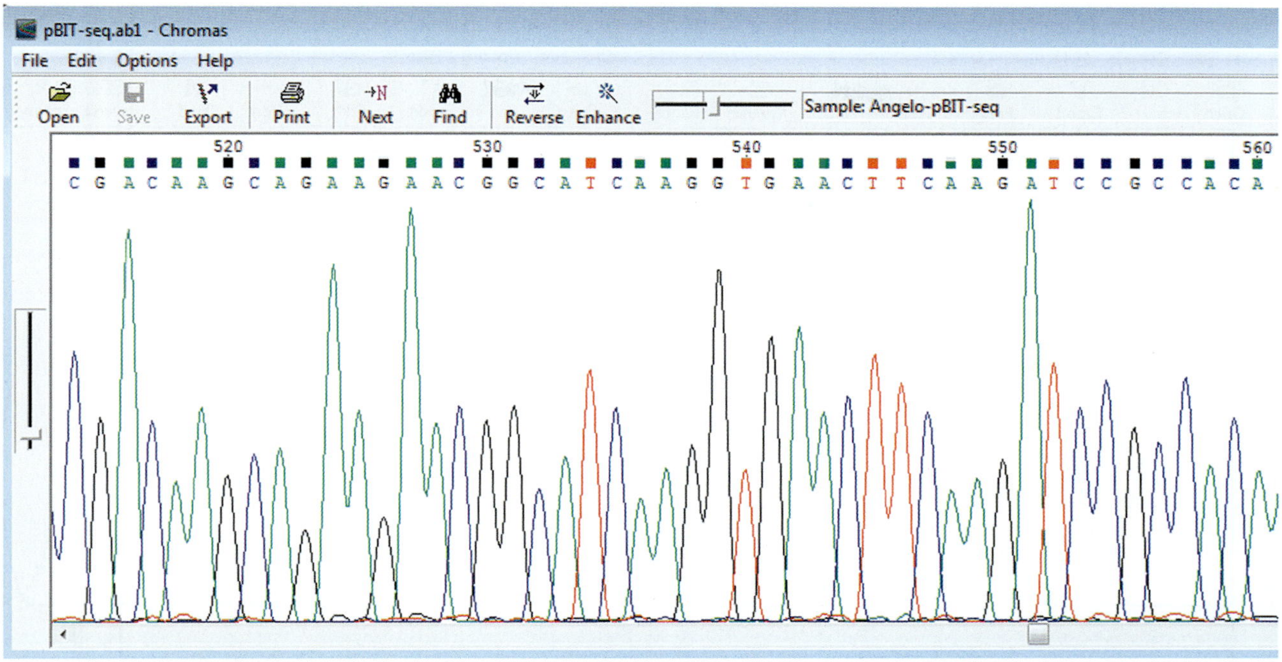

FIGURE 8.3 pBIT sequencing data at the middle of the chromatogram. Note the sharp, evenly spaced peaks throughout and the lack of Ns in the software-generated sequence above the chromatogram data. *No Permission Required.*

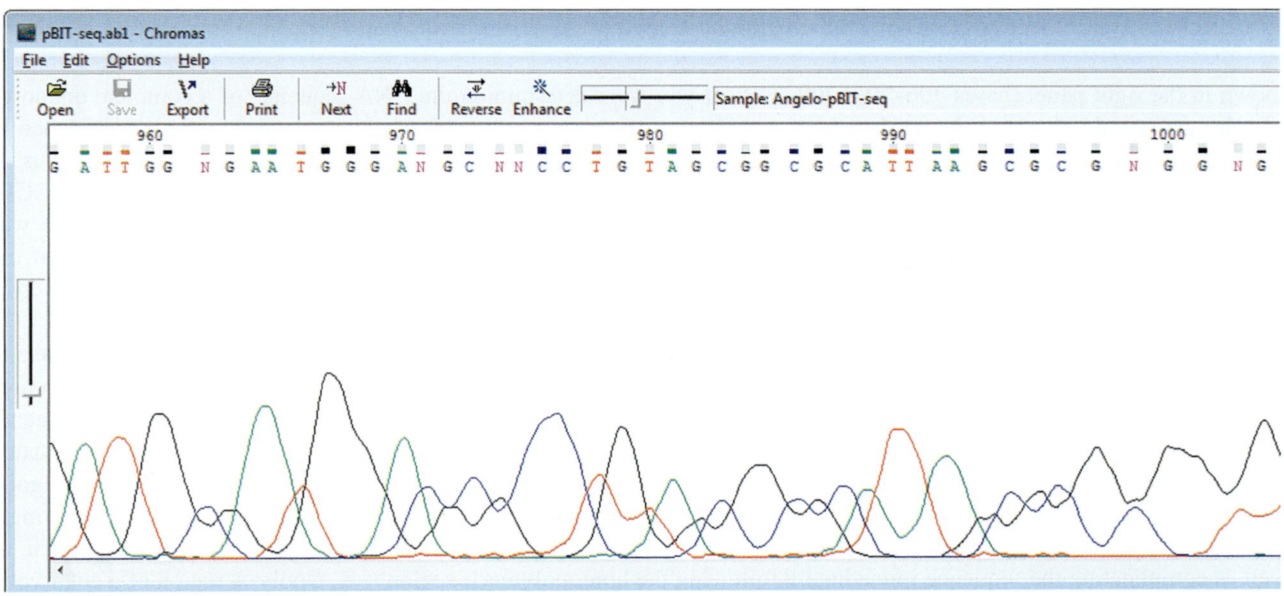

FIGURE 8.4 pBIT sequencing data at the end of the chromatogram. Similar to the beginning of the chromatogram, broad, rounded peaks begin to show up leading to ambiguities in the sequence, depicted by the Ns in the base called sequence. *No Permission Required.*

signal doesn't deteriorate until approximately nucleotide (nt) 970. Thus, we should have an adequate read to look at our sequence.

Scanning through your sequence of interest (in our case, nucleotides 50–769), you want to see if there are any Ns assigned by the software. You can scan through the sequence manually, as well as use the "Next N" button that is circled in Fig. 8.5, left panel. If there are Ns present, you want to use your best judgment to assign these bases. While there were no Ns found within the region of interest in the sample performed, Fig. 8.5 (left panel) shows an example of an N found outside this range, with the horizontal and vertical resolution changed to better view the chromatogram at this exact point. We see at base 827 the presence of a shoulder between the A at nt 826 and the G at nt 828. It appears as though this might be a doublet of consecutive Gs; however, you can compare this doublet to others in the chromatogram, such as those

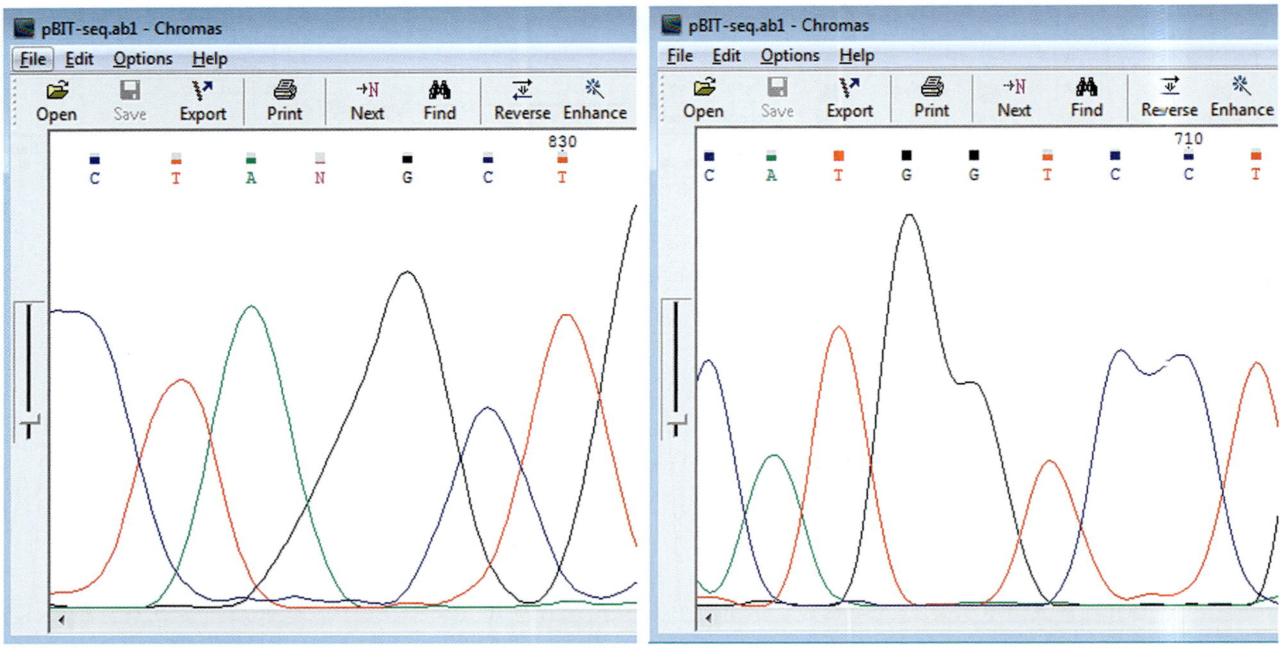

FIGURE 8.5 Zoomed-in view of the pBIT chromatogram. In the left panel, an N at position 827 due to a shoulder in the trace. In the right panel, examples of traditional doublets (not shoulders) are shown at positions 706–707 and 709–710. *No Permission Required.*

shown in the right panel (bases 706–707; 709–710). If you were determining the DNA sequence of a complete unknown sample, you would want to rely on data from a different primer or overlapping sequencing read to have confidence in assigning position 829. In our case, we rely on making an educated guess using the known *egfp* sequence to help us. If the known sequence of *egfp* in the region corresponding to nucleotides 824–829 in the chromatogram was CTAGGC, we could be relatively confident in assigning position 829 as a G. Similarly, if the sequence of *egfp* in that region was CTAGC, we could reasonably assume that despite the visual shoulder, the sequence only contained one G. One of the main differences between verifying a known sequence and determining an unknown DNA sequence is that during verification, one can make these kinds of educated guesses, as random insertions and deletions are infrequent.

Ambiguous nucleotides in a chromatogram can appear for a variety of reasons. In addition to the typical decrease in sequence quality at the beginning and end of a sequencing reaction, other factors can lead to difficult-to-interpret chromatograms. If there are Ns throughout your entire sequence, this is likely due to suboptimal conditions for your sequencing reaction. Low quality or insufficient quantity of DNA or poor-quality sequencing primer (low Tm or a mismatch with the template) are the most common explanations for an overall poor-quality sequencing result. Even within good sequencing reads, ambiguities sometimes arise. One of the most common reasons for ambiguities is a stretch of a single consecutive nucleotide. These can often be problematic, both in deducing the correct number of bases in that stretch, as well as complicating the sequence immediately following the homopolymeric region (i.e., a poly-A region of a eukaryotic cDNA). There are some sequencing "tricks" a sequencing facility can perform to obtain higher-quality sequencing reads if this type of sequence is expected in a given sample.

Comparing your sequencing data to the known *egfp* sequence can be performed in a number of different ways. There are sequence alignment tools available on the internet, such as ClustalW and many others. Another simple method to compare your sequence data to known sequences is by using the Basic Local Alignment Search Tool (BLAST) available from the National Center for Biotechnology Information (NCBI). BLAST (https://blast.ncbi.nlm.nih.gov/Blast.cgi) is a frequently used tool for calculating similarities between sequences.[1] Applications of BLAST include studying evolutionary differences between genes or proteins in different organisms, predicting functions of new genes based on the homology to genes of known function, and identifying particular motifs or domains within a sequence. Instructions for comparing your sequencing data to the positive control (pBIT) sequence using BLAST are detailed below.

II. LABORATORY EXERCISE

1. Open the Chromas software, click the Open button, and find the file containing your sequence to open.
2. Visually scan your chromatogram to determine the length of your sequence containing high-quality data with no (or only a few) Ns. For example, using the data shown in Fig. 8.2 and Fig. 8.4, we could say the quality sequence begins at nt 21 and ends around nt 960. The exact length of this region does not matter; however, it should contain the entire *egfp* sequence, if possible. In this experiment, that region will typically be between approximately nts 50–800. The exact region containing the *egfp* sequence can be found by searching for the NcoI (5'-CCATGG-3') and NotI (5'-GCGGCCGC-3') restriction sites. Including extra sequence (without Ns) on both the 5' and 3' end of the *egfp* sequence will help populate pBIT as your top hit within the BLAST search. Otherwise, you will obtain a number of other vectors containing *egfp*.
3. Perform a BLAST search on your sequence within Chromas by clicking File and selecting BLAST search. The parameters do not need to be altered, with the exception of Region. For Region, you should use the region of the sequence you determined in the previous step. Click OK and a BLAST window should open in your web browser. Click the View Report button. You can also export the sequence from the sequence viewer into a text editor and copy and paste the sequence into a BLAST search (http://blast.ncbi.nlm.nih.gov/Blast.cgi).
4. At the top of the BLAST results page, there will be information relevant to your search (query length, database searched, etc.). A bit further down, a graphic summary will be present and beneath that a tab saying "Descriptions" should appear. The top hit under Descriptions should be pBIT (Accession no. JF275063.1). The E value (or Expected Value) is a measure of the significance of the match. The closer this number is to zero, the more significant the match. In order to view the actual sequence alignment between your input sequence and the known pBIT sequence, click on the name of the construct under Description. This should take you to the alignment of your sequence next to pBIT (Fig. 8.6). You want to confirm the *egfp* insert cloned into pET-41a between the NcoI and NotI has the correct sequence. In pBIT, the NcoI site can be found starting at nucleotide 5674, and the NotI site finishes at nucleotide 6404. These regions are boxed in Fig. 8.6.

In Fig. 8.6, matches are shown by a line between the query (your input) and the subject (pBIT). In this example, the match is 100% within the *egfp* region and shows that we have indeed cloned a non-mutated form of our insert. If the sequence isn't a match, there would be no line between the two. An example of this is at query position 26. This position is outside the *egfp* reading frame, so it doesn't affect our ability to confirm the *egfp* sequence. It would, however, alter the reading frame between the *gst* gene in the pET-41a vector and the cloned *egfp*. While it is highly unlikely that the pET-41a vector incorporated a mutation, we can look back at the chromatogram and analyze the data ourselves. Looking at Fig. 8.2, we see the peak for C, at position 25, is broader than the surrounding peaks, indicative of a potential dimer. One way to confirm this would be to sequence the construct again using a different primer, so this position was in the middle of the read. In our experiment, we can confirm the sequence in that region is actually CTCCGGT due to the fact the transformed bacteria fluoresced green upon induction with IPTG. A frameshift mutation (if one of the Cs were deleted in pET-41a) would have altered the reading frame and prevented the EGFP protein from being properly made in the bacteria.

If there are Ns in your *egfp* sequence, go back to the chromatogram and further examine the data to see if the sequence appears to match. If multiple Ns were present in the sequence, in practice you would probably sequence the region again (perhaps with a different sequencing primer) to obtain more data to more accurately assign the ambiguous bases.

In most cases, your sequencing data will be a perfect match with pBIT. However, on the rare occasion PCR introduces a mutation into the *egfp* insert sequence, this would enable us to detect the mutation and act accordingly. If the mutation was a silent mutation (a nucleotide change that does not change the protein coding sequence), your experiment should not be affected. On the other hand, if the mutation incorporates a missense or a nonsense mutation, you would want to go back and choose a different positive clone to continue with your experiment. You might also notice lowercase letters following the egfp sequence, as in Fig. 8.6. This represents what is referred to as low-complexity sequence, which is often reported with sequences with repeats (in this case, a potential 6X His-tag found within pET-41a).

If a sequencing read is of poor quality and/or short length, you will not obtain a 100% match with pBIT. You may observe partial alignment with that positive control sequence but will not be able to draw a conclusion as to whether the insert is completely correct.

In your lab notebook, record your data for *each* clone sequenced including nucleotide locations of any misalignments, gaps, and/or mutations, and NcoI and NotI sites, if applicable. If a clone does not produce the expected alignment, note that as well.

FIGURE 8.6 Sequence alignment between the input sequence and pBIT. The *egfp* sequence is the region between the boxes depicting the *Nco*I (5'-CCATGG-3') and *Not*I (5'-GCGGCCGC-3') sites. *No Permission Required.*

REFERENCE

1. T. Madden, The BLAST sequence analysis tool, in: J. McEntyre, J. Ostell, (Eds.), The NCBI handbook, National Center for Biotechnology Information, Bethesda, MD, 2003.

DISCUSSION QUESTIONS

1. Why is the DNA corresponding to *egfp* the primary concern of our sequencing in this lab exercise? Why do we not anticipate any mutations occurring outside the *egfp* sequence?
2. When doing your BLAST search, what do you notice about the other hits that show up as significantly similar to your sequence?

Part III

Expression, Detection, and Purification of Recombinant Proteins From Bacteria

In this portion of the course, you will learn how to purify a genetically engineered protein from bacteria. As we have previously discussed, many pharmaceuticals and industrially relevant enzymes are produced using recombinant methods. Furthermore, as biochemists and molecular biologists, we are interested in the events that govern the life of a cell. How is the cell cycle controlled? What happens when a developing cell begins to differentiate? What are the factors that control cell growth? Why and how do cells die? If the organism is a pathogen, how is it able to infect the host? Before we can answer these questions in detail, we must know more about the actual molecules that participate in these processes. One classic biochemical solution to this problem is to take the cell apart and isolate each of its components for study. For proteins highly expressed in the cell, this approach is generally feasible. However, nothing short of herculean efforts have been necessary to purify proteins or peptides that are present in extremely small quantities. This represents an important limitation, since highly active gene products such as enzymes or hormones are typically present at extremely low levels in biological materials.

With the advent of recombinant DNA techniques and the abundance of sequence information, it became much easier to clone the gene and express it in bacteria than it was to purify the protein from the native organism. You have now finished cloning the *egfp* gene into the pET-41a expression vector. As you will see, the GST::EGFP fusion protein may constitute as much as 20% of the total bacterial protein. The abundance of the fusion protein reduces the proportion of contaminating host proteins, but it is still necessary to have a method for its fractionation and subsequent purification. Many of the pET vectors have been engineered to allow for a one-step fractionation method called affinity chromatography, which can be used to isolate pure protein in relatively large quantities.

In this part of the course, you will induce the expression of the fusion protein from your plasmid clone. Following the induction of your bacterial cultures, you will visualize the protein expression pattern by sodium dodecyl sulfate polyacrylamide gel electrophoresis (SDS-PAGE) and western blot. After confirming high levels of recombinant protein expression, you will grow a larger-scale culture for purification by affinity chromatography. You will conduct assays to monitor the quantity of the fusion protein in different stages of the purification procedure and to ensure the fusion protein is correctly folded and retains functional integrity. The purity of the final preparation will be analyzed by SDS-PAGE.

Advanced Alternatives Within Part III

ADVANCED ALTERNATIVES, PART III—EXPRESSION, DETECTION, AND PURIFICATION OF RECOMBINANT PROTEINS FROM BACTERIA

This part is designed to purify the recombinant fusion protein from bacteria and analyze it. With protein purification, there are numerous alterations that can be performed in an attempt to optimize the procedure. Similarly, depending on the downstream applications with the protein, certain methods of purification are more desirable than others, while some might not work at all in the context of future experiments. For example, the GST fusion tag can diminish the activity of certain enzymes or block important protein interactions.

AA3.1 Omit Performing a Western Blot on Whole Cell Lysates and Perform a Western Blot on Purification Fractions

This is useful if you want to include a western blot as an experimental technique and still have time to perform experiments from Part IV, Analysis of mRNA Levels. Instead of performing western blots using whole cell lysates in this part (Lab Session 9), begin by performing the GST purification (Lab Session 11). The purification fractions can then be analyzed by both non-specific protein visualization (using a stain-free gel or GelCode Blue) and western blot.

AA3.2 Semi-quantitative Western Blot

Similar to "Semi-quantitative Agarose Gel Electrophoresis," western blots can be performed in a semi-quantitative way. While precise, accurate experimental design (with replicates) can allow one to acquire truly quantitative western blot data, what is proposed here is a semi-quantitative method. This alternative is useful if a western blot is performed following the purification of the GST::EGFP protein as described in AA3.1 "Omit Performing a Western Blot on Whole Cell Lysates and Perform a Western Blot on Purification Fractions."

As mentioned in the text, there are numerous ways to visualize western blots. Depending on the method you are using, the details of the experiment will differ. The main idea, however, remains the same. You will purchase recombinant EGFP (available from a variety of suppliers) and prepare a dilution that will enable you to load a range of masses of EGFP protein within the linear detection range of your assay. This can be determined empirically. For chloronaphthol development (as described in Lab Session 10), a starting point for range of standards is 50 ng–5 μg. For ECL, 1–500 ng is a reasonable starting range (depending on the sensitivity of the ECL reagents used). Load as many different masses of standards as possible on your gel, and you can compare intensities of bands in your samples to your standards. Many imaging systems include software tools to perform the analysis. NIH Image J (described in Lab Session 22) can also be used. Remember that you are looking at the mass of protein loaded within that lane of the gel. To determine the concentration of your sample, you need to divide the mass by the volume of the sample that was loaded on the gel (don't forget to subtract the amount of sample buffer). One caveat to this experiment is that the antigen in the standards (recombinant EGFP) is slightly different than the experimental samples (GST::EGFP fusion protein). It is possible that the antibody could bind worse (or possibly even better) to the fusion protein due to the presence of the GST tag.

Additional materials

Recombinant EGFP. Utilization of 12- or 15-well gels might be preferred to enable the loading of more samples on the gel.

AA3.3 Cleaving the GST Fusion Tag

In many cases, it is desirable to remove the fusion tag from the protein of interest following purification. Most vectors, including pET-41a, make this possible through the incorporation of protease cleavage sites within their sequence. Protease recognition sites for both thrombin and enterokinase are encoded within the vector DNA, so that when the fusion protein is made, the tag can be removed by adding protease.

There are a few common ways to remove the fusion tag from the recombinant fusion protein, both of which require extra steps compared to regular purification. The first method involves purifying the protein using a glutathione affinity column as described in the text. Once the purified protein is eluted, it can be treated with either thrombin or enterokinase. After treatment with the protease, further purification must be performed to remove both the protease and GST from the EGFP protein. The GST can be readily removed by passing the solution over another glutathione column. This time, the GST will bind and your EGFP protein (since it is no longer attached to the GST) will flow through the column. The protease will also pass through the column. If you want to further separate the protease from the EGFP, you will need to perform additional chromatography methods, such as cation or anion exchange chromatography. Note that size-exclusion chromatography would not be useful to separate thrombin or enterokinase from EGFP due to the similarity in the molecular weight.

Another common alternative involves adding the protease directly to the column while the protein is still bound. In this case, instead of adding elution buffer, protease is added. As the protease cleaves the peptide bond and breaks the fusion protein, EGFP is released and the GST remains bound to the column. After elution with protease, separation of the EGFP and the protease would still require an additional step.

Many companies offer thrombin Sepharose beads, in which thrombin is covalently linked to beads in the form of a column. Using these beads, one could take purified GST::EGFP fusion protein and pass it over the column and the elution would contain GST and EGFP. A second step of passing the elution over a glutathione column would result in purified EGFP. Enterokinase-binding resin, which can specifically bind enterokinase, is a useful product for removing enterokinase from a solution. These beads are conjugated with soybean trypsin inhibitor, which binds to enterokinase with high affinity. Enterokinase can be removed from EGFP by passing the solution over this resin.

Additional materials

Thrombin (or thrombin Sepharose beads). Enterokinase and enterokinase-binding resin.

Lab Session 9

Expression of Fusion Protein From Positive Clones, SDS–PAGE and Western Blot: Part I

Goal: This week, you (or your instructor) will induce small-scale cultures of your clones with isopropyl-β-d-thiogalactopyranoside (IPTG) and prepare protein samples for gel electrophoresis.

You will harvest actively growing, logarithmic-phase bacterial cultures to prevent proteolysis of the fusion protein, which can occur during the stationary phase of growth. IPTG is added just a few hours before the bacteria are harvested since the induction of recombinant protein is often detrimental to bacterial growth and viability.

Today, you will make lysates of the cultures, separate them by sodium dodecyl sulfate–polyacrylamide gel electrophoresis (SDS–PAGE), and transfer them to a nitrocellulose membrane. You will save the nitrocellulose membrane for next week to finish the western blot.

Molecular Biology Techniques. DOI: https://doi.org/10.1016/B978-0-443-26714-7.00012-0

Lab Session 9a

Interim Laboratory Session

INOCULATE CULTURES FOR SDS-PAGE

I. LABORATORY EXERCISE

1. Obtain four culture tubes containing 1 mL LB/kan.
2. Heavily inoculate these tubes with transformants from your master plate corresponding to your positive control (pBIT), negative control (pET-41a), and two positive clones as determined from your screening assays. Use a separate toothpick/pipette tip to inoculate each of the LB/kan tubes.
3. Label the LB/kan cultures with sample information, your initials, and the date.
4. Place the inoculated cultures in designated racks.
5. The LB/kan tubes containing your inoculated cultures will be refrigerated at 4°C to retard growth. The afternoon before your lab period, they will be placed in a shaker at 37°C and incubated overnight.

Lab Session 9b

Expression of Fusion Protein from Positive Clones, SDS–PAGE and Western Blot, Part I

I. INTRODUCTION

SDS–PAGE is used to separate proteins by molecular weight. SDS–PAGE of proteins has numerous applications, including molecular weight determination, determining sample purity, quantifying expression, western blotting (aka immunoblotting), and isolating proteins for peptide sequencing or for generating antibodies.

You are already familiar with DNA agarose gel electrophoresis, and SDS–PAGE shares some similarities with this method. Both methods require a matrix to separate molecules by size and use electrical charge differences to cause migration. However, the structural and biochemical differences between DNA and proteins lead to a number of variations in their separation by electrophoresis.

1. Proteins are generally smaller than DNA. Thus, while DNA (larger than 100 bp) is routinely separated on agarose gels, proteins are generally run on polyacrylamide gels, as polyacrylamide matrices have a smaller pore (sieve) size than agarose. DNA fragments smaller than 100 bp are often separated using polyacrylamide.
2. All DNA is negatively charged, but proteins have varying charges depending on the amino acid content of the specific polypeptide and the pH of the buffer. Some proteins are positively charged, while some carry a net negative charge.
3. DNA, especially linear DNA, has little secondary structure, while proteins can be globular or linear and have quaternary structure, such as dimers and other multimers.

Because of numbers 2 and 3, if proteins were run on a native or non-denaturing polyacrylamide gel (i.e., run without SDS), protein migration would depend on at least three factors: size, charge, and shape.

SDS–PAGE allows proteins to migrate by size alone, through the use of SDS and a reducing agent. SDS is an ionic detergent that denatures (unfolds) proteins by wrapping around the polypeptide backbone forming a micelle, and thus conferring a net negative charge in proportion to polypeptide length. SDS also disrupts most non-covalent interactions, such as electrostatic interactions and hydrogen bonds, thereby decreasing protein folding. A reducing agent, such as β-mercaptoethanol or dithiothreitol, is added to reduce disulfide bonds (cystine bonds) and further unfold the proteins.

After boiling a protein sample in SDS and β-mercaptoethanol, proteins act as negatively charged linear molecules and can be electrophoretically separated by size alone (Fig. 9.1).

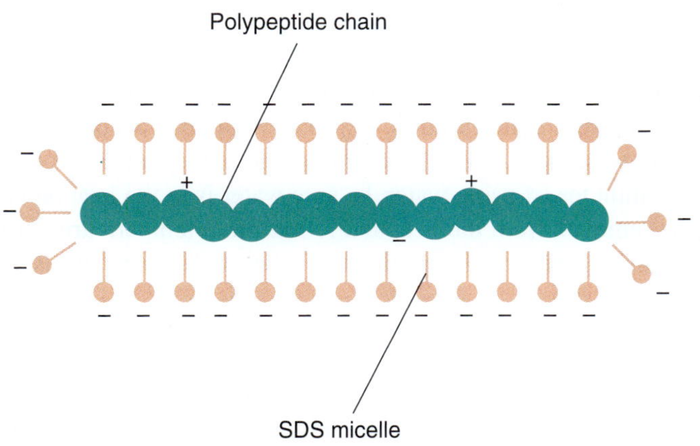

FIGURE 9.1 Polypeptide treated with β-mercaptoethanol and SDS. β-Mercaptoethanol breaks the covalent disulfide (cystine) bonds. SDS disrupts non-covalent interactions, further denaturing the protein, and forms a micelle around it, conferring a negative charge. *SDS*, Sodium dodecyl sulfate. *No Permission Required.*

After running the gel, it can either be stained non-specifically to visualize the protein bands using reagents such as Coomassie Blue, GelCode Blue, or silver stain; or the proteins can be transferred to a nitrocellulose membrane for western blotting (immunoblotting) to visualize a specific protein of interest. In Lab Session 12, Analysis of Purification Fractions, we will run an SDS–PAGE gel and stain it using GelCode Blue to visualize protein bands.

In today's lab session, we will begin a western blot (to be completed in the following laboratory session). The first step of this process is to prepare the protein samples and separate them using SDS–PAGE. Then, the proteins from the polyacrylamide gel are transferred to the nitrocellulose membrane. Total protein on the nitrocellulose membrane may be visualized at this point using the water-soluble Ponceau stain. After the proteins are transferred, a monoclonal antibody against GFP is used to specifically visualize your GST::EGFP fusion protein (more information on this in Lab Session 10: Expression of Fusion Protein from Positive Clones, SDS–PAGE, and Western Blot: Part II). This portion of the western blot will be completed in the next laboratory session.

The molecular weight of the GST::EGFP fusion protein can be estimated, assuming the average weight per amino acid is equal to 114 Da. The *gst* gene is 660 bp, encoding 220 amino acids: 220 × 114 = 25,080 Da. The *egfp* gene is 720 bp, encoding 240 amino acids: 240 × 114 = 27,360 Da. There are 174 additional nucleotides between *gst* and *egfp*, encoding 58 amino acids: 58 × 114 = 6612 Da. The weight of the fusion protein can therefore be approximated as: 25,080 + 27,360 + 6612 = 59,052 Da or ~59 kDa. You can determine the actual molecular weight (using the molecular weight for each amino acid) using free online software; the exact molecular weight of the GST::EGFP fusion protein is 58,500 Da. You might be able to non-specifically visualize a protein band of this approximate size in your positive clones using the Ponceau stain. The completion of the western blot exercise next week will use an antibody specific for EGFP to confirm that the band is indeed GST::EGFP. In the negative clones, after Ponceau staining, you may see a band of approximately 25 kDa, corresponding to the GST protein alone.

II. LABORATORY EXERCISES

A. Sample Preparation

Steps 1–3 will be done for you.

1. Add 2 mL of 2 × YT/kan/IPTG medium to each overnight culture.
2. Incubate at 37°C for 3–5 hours to induce protein expression. (Overnight induction is possible, but not preferred.)
3. Set heat block to 95°C or higher (be sure heat block has water/sand/lab armor beads in the holes for heat transfer).
4. Harvest cultures. Centrifuge 200 μL of each culture (positive control, negative control, and your individual clones) for 2 minutes in a microcentrifuge at max speed. Remove and discard supernatants, saving pellets in the tube. Invert tubes over paper towels to remove residual liquids.
5. To each tube, add 200 μL B-PER complete reagent (this reagent contains both DNase and lysozyme) and pipette up and down (or vortex) until pellet is dispersed. Incubate at room temperature for 10–15 minutes. During this incubation, assemble your gel apparatus. *Note: The manufacturer's B-PER protocol calls for using 5 mL of B-PER reagent per gram of cell pellet. This protocol uses an excess of B-PER reagent, creating a more diluted protein sample appropriate for direct loading on the gel for your western blot.*

   Money-Saving Tip: Instead of lysing the cells with B-PER, you can add 100 μL of dH_2O and 100 μL of 2× SDS–PAGE sample buffer directly to the pellet, resuspend, and centrifuge for 5 minutes. The resulting supernatant solution will be highly viscous; therefore, it is recommended to sonicate the samples first and then place them on ice for 2 minutes prior to loading on the gel.
6. Centrifuge the samples for 5 minutes at max speed in the microcentrifuge.
7. Add 25 μL of the supernatant of each sample to a separate microfuge tube containing 5 μL of 6× SDS–PAGE sample buffer. (If your SDS–PAGE sample buffer is supplied at a different concentration, adjust the volumes accordingly.) Make sure that you are using SDS–PAGE sample buffer, not previously used DNA loading buffer, as they are both blue!
8. Incubate the tubes at a minimum temperature of 95°C for 10 minutes (if using water, be sure water is in the heat block holes—incubating in boiling water is fine). Samples can be removed from the heat and left at room temperature until they are loaded on the gel. If the lids to your tubes pop open during heating, microcentrifuge tube cap locks can be used.

Gel Apparatus Preparation

Note: The instructions for assembling the gel apparatus and transblot assembly are specific for the Bio-Rad Mini-Protean 3 or Mini-Protein Tetra Electrophoresis unit and the Bio-Rad Mini Trans-Blot Cell. If you are using equipment made by another manufacturer, your instructor will provide you with assembly and usage instructions.

We are using precast polyacrylamide gels. However, if you must make and pour your own polyacrylamide gels in the future, keep in mind that non-polymerized acrylamide in both the powder and liquid forms are potent neurotoxins. Great care must be taken to avoid inhalation, ingestion, and skin exposure to this substance. Solidified polyacrylamide does not pose the same safety hazard.

Time-Saving Tip: Bio-Rad currently offers stain-free TGX polyacrylamide gels (catalog number 4568033) that allow proteins to be visualized by exposure to UV light without a staining step. These are Tris-glycine gels with an extended shelf life. If you are planning on using stain-free gels in Lab Session 12, Analysis of Purification Fractions, the same gels can be substituted in this lab session. Additionally, if using these stain-free gels, you can verify the efficiency of your transfer at the end of the Lab Session 9 this week by visualizing the membrane under UV light to confirm the proteins transferred from the gel to the membrane (the proprietary stain travels with the transferred proteins). You can then omit the Ponceau S membrane staining steps in Lab Session 10.

1. Two gels must be run per unit. If you are not paired with another group/student and only have one gel to run in the apparatus, you will need to use a buffer dam in place of a second gel.
2. Assemble the Bio-Rad apparatus as demonstrated by your instructor, with the short plates toward the center.
3. Add 1× Tris-glycine running buffer to the top of the center buffer chamber (be sure the buffer covers the wells) and also to the outer chamber (only about 5 cm. deep). Check for leaks.
4. Wash wells with Tris-glycine running buffer or dH_2O.

B. Loading Samples on the Gel

1. When pipetting your samples, do not try to expel all of the liquid from the pipette tip. An air bubble will result, which could distort the migration of your sample and could also force your sample out of the well. We recommend gel-loading tips that are specially designed for SDS–PAGE.
2. Add 10 μL of Bio-Rad Kaleidoscope molecular weight marker (Fig. 9.2), 10 μL of the positive control (pBIT), 10 μL of the negative control (pET-41a), and 10 μL of each transformant fusion protein sample to individual lanes according to the following scheme, which contains samples for two lab groups on a single gel (group A, Lanes 1–5; group B, Lanes 6–10).

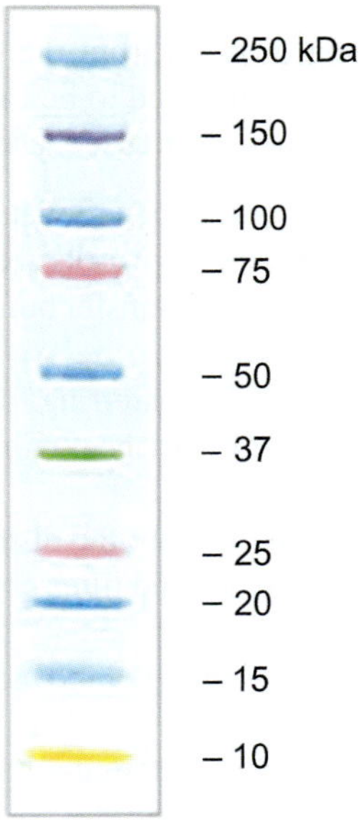

FIGURE 9.2 Bio-Rad Precision Plus Protein Kaleidoscope Prestained Protein Standards, catalog no. 1610375. *Used with permission.*

Lane 1	Lane 2	Lane 3	Lane 4	Lane 5	Lane 6	Lane 7	Lane 8	Lane 9	Lane 10
MWM	Positive control	Negative control	Sample 1	Sample 2	MWM	Positive control	Negative control	Sample 1	Sample 2

MWM, Molecular weight marker.

C. Electrophoresis

1. Two gels will likely be run per apparatus: coordinate with another group before you start.
2. Attach the electrodes to the power supply. *CAUTION*: Be sure to match the anode (red) with the positive lead and the cathode (black) with the negative lead.
3. Turn on your power supply. Run your gel at constant voltage (100–110V). If you have multiple gels attached to your power supply, you do not need to change the voltage. Every 10 minutes during electrophoresis, check the current (amperage) on your power supply. The current should not exceed 60 mA per gel at any time during the experiment.
4. Run the gel until the bromophenol blue tracking dye reaches the bottom of the gel. This typically takes 30–45 minutes, but you check your gel every 10 minutes to ensure it is running properly.

D. Stopping Electrophoresis

1. Turn off the power supply.
2. Detach the leads from the power supply. Remove your gel from the apparatus. If the other gels connected to the power supply have not finished, turn the power supply back on.
3. Pour the buffer into the sink.
4. Disassemble the gel from plates as demonstrated.
5. Cut the wells off the gel after disassembling.

It is possible to stain the gel to visualize the proteins at this point, but for a western blot the gel cannot be irreversibly stained. Instead, you will first transfer the protein to a nitrocellulose membrane, and then stain the membrane with a water-soluble dye prior to completing the western blot.

E. Transferring to Nitrocellulose

We will use the Bio-Rad Mini Trans-Blot Cell for the transfer of proteins from the polyacrylamide gel to the nitrocellulose membrane (Fig. 9.3). If you are using a polyvinylidene fluoride (PVDF) membrane instead of nitrocellulose, refer to the manufacturer's instructions for transferring.

Time-Saving Tip: Other methods of transferring proteins from the gel to the nitrocellulose, such as using dry and semi-dry systems, are typically much faster. These systems tend to be more expensive, but they can be as effective as the wet transfer method. Refer to the manufacturer's instructions to use these instruments for transferring.

1. Prepare 1.5 L of transfer buffer by mixing 3.7 g of sodium bicarbonate into 1.5 L of deionized water.
   Cost-Saving Tip: The bicarbonate transfer buffer frequently works well with low molecular weight (less than 100 kDa) and basic proteins. Traditional Tris/glycine/SDS/methanol transfer buffers can be used, but for transfer of the GST::EGFP fusion protein, the bicarbonate buffer works equivalently.[1]
2. Be sure to wear gloves from this point forward to avoid transferring fingerprints to the nitrocellulose membrane. Obtain a glass dish, two sheets of Whatman filter paper and one sheet of nitrocellulose. Label nitrocellulose on top left in PENCIL.
3. Pour a small amount (about 1 cm deep) of transfer buffer into the dish.
4. Thoroughly wet the nitrocellulose membrane, fiber pads, and filter paper in the dish of transfer buffer. *Do not allow the nitrocellulose to touch the gel.*
5. Prepare a gel sandwich as described here and depicted in Fig. 9.4. Under no circumstances should you shift the position of the nitrocellulose on the gel once they are in contact.
   - Place the cassette, gray side down, in the glass dish containing the transfer buffer.
   - Place one pre-wetted fiber pad on the gray side of the cassette.

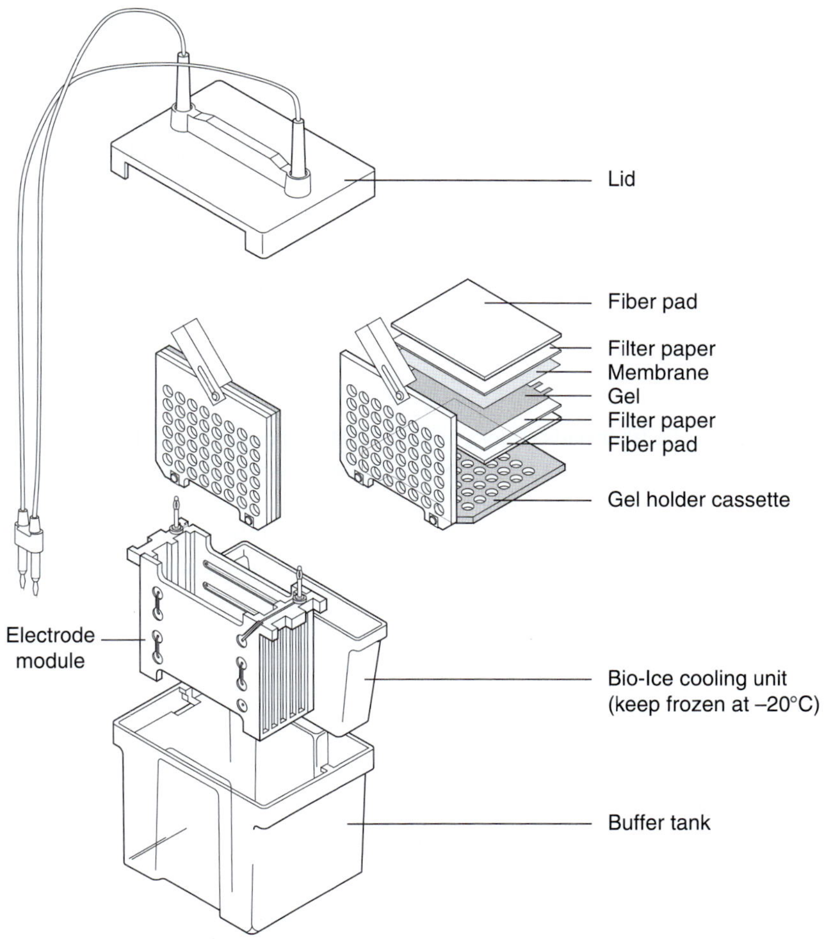

FIGURE 9.3 Mini trans-blot cell description and assembly of parts. *©2004 Bio-Rad Corporation. Used with permission.*

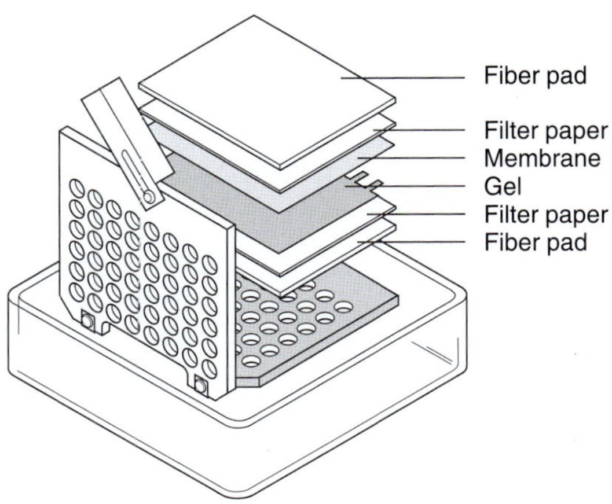

FIGURE 9.4 Close-up of "gel sandwich." *©2004 Bio-Rad Corporation. Used with permission.*

- Place a sheet of filter paper on the pad.
- Place the gel on the filter paper. Remove bubbles by gently rolling a glass tube over it.
- Place the pre-wetted nitrocellulose membrane on the gel, labeled side down at the top of the gel. Remove bubbles by gently rolling a glass tube over it.
- Place a filter paper on the membrane. Remove bubbles by gently rolling a glass tube over it.
- Add the last fiber pad and close the sandwich.

6. Place the cassette in the module. The gray side of the cassette should face the black side of the module, and the clear side of the cassette should face toward the red side of the module.

7. Fill with transfer buffer (approximately 1 L) to the top (if you run out, use the transfer buffer you used for soaking your gel or make more).
8. Put on the lid of the assembly so that red matches red and black matches black.
9. Turn on the current to 300 mA for 40 minutes. Ensure the system is running correctly by noting the voltage on your power supply. If the voltage reads 0 V, check your setup, particularly making sure you have enough transfer buffer in the system.

   Time-Saving Tip: If there is not enough time to finish the transfer, many power supplies have a timer that you can set to supply current for a fixed amount of time. Alternatively, gels can be transferred for 30 mA overnight at 4°C. The gels sandwiches should be disassembled, and the nitrocellulose stored the next day.

F. Disassemble and Store

1. Disassemble the apparatus.
2. Discard the gel and two sheets of filter paper and rinse the fiber pads (fiber pads are NOT disposable) and the rest of the apparatus unit with plenty of dI water and put away. As before, be sure to not touch the nitrocellulose membrane with your fingers!
3. Place the blot between two sheets of fresh Whatman filter paper and hand into the instructor to be wrapped in foil and refrigerated at 4°C until next week. *Note: The nitrocellulose membrane can also be stored in a container with a lid in Tris-buffered saline until next week.*

REFERENCE

1. S.D. Dunn, Effects of the modification of transfer buffer composition and the renaturation of proteins in gels on the recognition of proteins on Western blots by monoclonal antibodies, Anal Biochem 157 (1986) 144–153.

DISCUSSION QUESTIONS

1. Following SDS–PAGE, what is an advantage of protein detection by immunoblotting (western blotting) over a non-specific gel-staining procedure? Under what circumstances would it be desirable to non-specifically stain the gel, rather than immunoblotting?
2. An alternative procedure for this lab procedure involves lysing the bacteria directly in 1× SDS–PAGE sample buffer instead of the B-PER reagent. While this works, the resulting solution is highly viscous (mucous-like) and difficult to pipette. What components in the bacteria and the SDS–PAGE sample buffer cause the viscous solution? Why is this problem solved by using the B-PER reagent?

Lab Session 10

Expression of Fusion Protein From Positive Clones, SDS–PAGE, and Western Blot: Part II

Goal: Today you will stain your membrane with Ponceau Red to confirm the transfer of your proteins to the nitrocellulose and complete the western blot that was started in the last laboratory session.

I. INTRODUCTION

Antibodies are a multifaceted tool used in the study of proteins in a variety of assays. Western blot, immunofluorescence, enzyme-linked immunosorbent assay (ELISA), immunoprecipitation, cell and tissue labeling, and fluorescence-activated cell sorting are just a few of the antibody-based techniques commonly employed by scientists. Antibodies are proteins made by the immune system in mammals that specifically bind other proteins. For research purposes, most antibodies are raised in rabbits (polyclonal antibodies) or mice (monoclonal antibodies) by injecting a peptide, protein, or protein fragment of interest into the animal, waiting some time for the immune system to naturally produce antibodies, then purifying the antibody from the serum of the animal. Some antibodies are more specific than others; for instance, the antibody used in today's experiment was designed to bind GFP, but also binds EGFP, yellow fluorescent protein, cyan fluorescent protein, and nearly every other colored variant of GFP. This is because the antibody recognizes a region of the protein that is constant between these variants. Importantly, antibodies allow scientists to visualize and distinguish a specific protein of interest from the thousands of other proteins present in a given sample. In today's experiment, the antibody will allow you to specifically visualize the GST::EGFP fusion protein among all the other proteins present in the bacterial cell.

The cellular proteins of each clone were separated by SDS–PAGE in the last lab session, and today you will visualize the specific protein band corresponding to the GST::EGFP fusion protein by performing a western blot. Following protein transfer to the nitrocellulose membrane, the membrane is blocked with a protein that will not interact with the primary antibody (such as casein found in nonfat milk or bovine serum albumin). Primary antibody is added and allowed to bind to the specific epitope on the protein and the unbound antibody is washed away.

Nomenclature for antibodies is often represented with the Greek letter "alpha" (α) or the prefix "anti." For example, an antibody that specifically binds to the EGFP protein may be referred to as "anti-EGFP" or "α-EGFP." The secondary antibody must be raised in a species different than that of the primary antibody and is designed to specifically react with the antibody itself (not your protein of interest). In your experiment today, secondary antibody conjugated to horseradish peroxidase (HRP) (goat anti-mouse peroxidase; GAMP) is added and allowed to bind to the primary antibody, and then the excess is washed away. The HRP is a key component to the secondary antibody that allows for visualization on the blot upon the addition of a substrate in the next step. The most commonly used visualization method for western blots is enhanced chemiluminescence (ECL). In ECL, luminol is oxidized by the HRP enzyme into a chemiluminescent product (which is often stabilized by an additional chemical for longer light duration). Bands on ECL-developed blots are visualized by exposure of the blot to film or by specific imagers with cameras equipped to detect the luminescence. A less sensitive alternative to ECL is colorimetric detection. The principle is similar in that the substrate, chloronaphthol, is converted to a purple product by HRP. The result is purple bands on your membrane where the antibody is bound; for the GST::EGFP fusion, this band should appear at a molecular weight of approximately 59 kDa. Clones that did not express EGFP (negative clones) will show no band. Fig. 10.1 shows a flowchart of the steps involved in performing a western blot.

Molecular Biology Techniques. DOI: https://doi.org/10.1016/B978-0-443-26714-7.00001-6

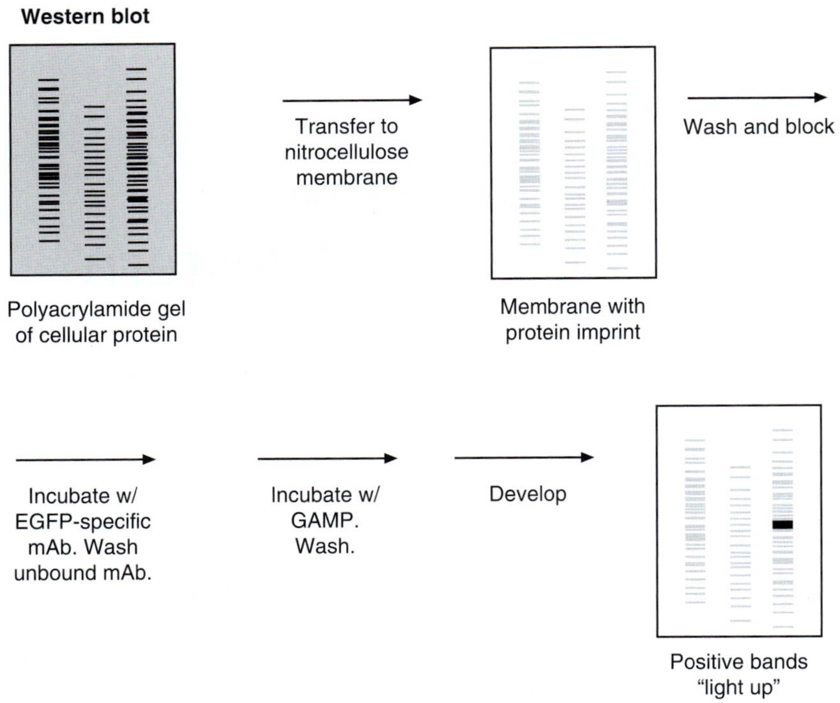

FIGURE 10.1 Western blot experimental flowchart. Samples are separated by SDS–PAGE, then transferred to a nitrocellulose membrane. The membrane is washed and blocked, then incubated with the primary monoclonal antibody (mAb), α-EGFP. Excess primary antibody is washed off, and the secondary antibody bound to horseradish peroxidase (goat anti-mouse conjugated to horseradish peroxidase or GAMP) is added. Excess secondary antibody is washed off, then the blot is developed using the peroxidase substrate, chloronaphthol, or enhanced chemiluminescence. *No Permission Required.*

II. LABORATORY EXERCISES

A. SDS–PAGE and Western Blot, Part II

Ponceau Stain

In addition to the western blot you will be performing, you will visualize the total protein banding pattern for each sample. You are staining with a soluble dye called Ponceau Red (the same dye used in Lab Session 1: Getting Oriented; Practicing With Micropipettes) that will be removed during the blocking step due to the detergent (Tween 20) present in the blocking solution. You can complete this laboratory exercise with the group who shares your membrane. Alternatively, your instructor may have you carefully cut the membranes after the Ponceau stain. As a reminder, always wear gloves. Never handle the blot directly with your bare fingers, as proteins from your fingers can adhere to the blot and be visualized during the blotting process.

1. Remove the blot from the refrigerator.
2. Place the blot (labeled side up) in a square Petri dish and just cover (do not fill the dish) with Ponceau Red stain. Incubate for 2 minutes with gentle agitation.
3. After the 2-minute incubation, rinse the blot three times with dH_2O with agitation to remove dye that has bound the membrane. At this point, you will see protein bands. You might see a band of approximately 59 kDa molecular weight in the positive clones that is not present in the negative control or negative clones.
4. Record observations of the Ponceau stain in your lab notebook. In particular, note the following *Were lanes evenly loaded? Did any air bubbles appear on the blot? Did you see the expected band at 59 kDa?*

As an alternative to Ponceau staining, if stain-free gels were used in Lab Session 9, you can directly visualize the blot under UV light after transfer to confirm that proteins are present.

Blocking

The blocking step of the western blot is critical to prevent antibody from sticking non-specifically to the nitrocellulose membrane. In research settings, protocols for the blocking step range from as short as 5 minutes to as long as overnight. In our case, for the sake of time, we will be doing a short blocking step. The shortened blocking step has been shown to work well specifically with the primary and secondary antibodies we are using in subsequent steps.

1. Pour the water out of the Petri dish and add 10 mL of blocking solution (5% powdered milk in TBS-T).
2. Incubate for 5 minutes on a platform rocker or orbital shaker (50 rpm).
3. Wash twice with fresh TBS-T (5 minutes each).

Incubation With α-EGFP (Primary Antibody)

1. Obtain an aliquot of the α-EGFP (anti-EGFP). Store the antibody on ice.
2. Make a 1:2000 dilution of the α-EGFP by mixing 4 μL antibody with 8 mL TBS-T and add the solution to the washed membrane. Incubate on a platform rocker or orbital shaker for 1 hour. Prepare your replica plate (Part B of this Lab Session) during this incubation.
3. Wash twice with fresh TBS-T (5 minutes each).

Incubation With Goat Anti-Mouse Peroxidase (Secondary Antibody)

1. Discard the final wash.
2. Make a 1:1000 dilution of the goat anti-mouse antibody conjugated to HRP (GAMP) by mixing 8 μL with 8 mL of TBS-T. Add to the blot and incubate for 45 minutes on a platform rocker or orbital shaker. *Note: if you are using ECL detection, a more diluted secondary antibody concentration is likely necessary. Please see the recommendation of the ECL reagent you are using for guidelines. For the Bio-Rad Clarity ECL Reagent, a 1:3000 dilution of the secondary antibody was sufficient for detection using a Bio-Rad ChemiDoc™ MP imaging system.*
3. Wash twice with fresh TBS-T (5 minutes each).

Detection of Peroxidase Activity

Choose the appropriate method for detection depending on the instrumentation you have available.

Option A—Colorimetric Detection

1. Discard the TBS-T solution from the blot.
2. Add about 10 mL of peroxidase substrate to the tray containing the blot. Color development (dark purple) should be evident within seconds or minutes.
3. Stop the reaction by rinsing with dH_2O.
4. To preserve color, dry the blot between two pieces of filter paper.
5. Record the results in your lab notebook.

Option B—Enhanced Chemiluminescence

The protocol provided is for the Bio-Rad Clarity ECL reagent. Most other reagents work similarly, but be sure to check the manufacturer's instructions for the product used.

1. Prepare the working solution by mixing 3.5 mL of Western Peroxide Reagent with 3.5 mL of Western Luminol/Enhancer Reagent in a 15 mL conical tube. Be sure to change pipette tips between pipetting the two solutions! Mix well.
2. Discard the TBS-T and add the working solution to cover the blot. The incubation can be done in the same tray as your antibody staining or the blot can be placed on a piece of clear plastic wrap and the working solution can be gently added over the blot. Incubate for 5 minutes (in the dark). Placing an empty ice bucket over the membrane will keep the light out.
3. Using forceps, remove the membrane from the solution and allow excess solution to drip off.
4. If exposing by film, wrap the membrane carefully in plastic wrap before exposing to film in a darkroom. If using a digital imager, place the membrane on the surface of the imager and follow the instructions for the system to acquire the image.
5. Record the results in your lab notebook.

Positive clones should show a band with a molecular weight of approximately 59 kDa, while negative clones should show no bands. In the positive clones, you may see some smearing or distinct bands below the 59 kDa band. This is most likely due to partial degradation of the fusion protein.

B. Replica Plate Positive Clone

Replica plate one of your positive clones onto LB/kan so you will have a fresh inoculum for next week. Label and place the inverted plate in the 37°C incubator. Your instructor will save it in the refrigerator after an overnight incubation.

DISCUSSION QUESTIONS

1. After performing the Ponceau stain, you may see a distinct band of approximately 25 kDa in your negative control that is not present in your positive control or your other positive clone lanes. What does this band likely represent?
2. If the proteins used in SDS–PAGE are denatured, how would the primary antibody bind to them? Is secondary structure important for the antibody to bind, or are antibodies created to bind to the denatured version of proteins? *Outside resources will be necessary to think about this question.*

Lab Session 11

Extraction of Recombinant Protein From *Escherichia coli* Using a Glutathione Affinity Column

Goal: This week you will purify the GST::EGFP fusion protein by affinity chromatography. You will need to inoculate one of your clones into liquid culture at least one day before your lab. You or your instructor will induce protein expression with isopropyl-β-d-thiogalactopyranoside (IPTG) 3–4 hours before your regular lab period.

Molecular Biology Techniques. DOI: https://doi.org/10.1016/B978-0-443-26714-7.00024-7

Lab Session 11a

Interim Laboratory Session

INOCULATE CULTURES FOR PROTEIN PURIFICATION

I. LABORATORY EXERCISE

Inoculate (heavily) your positive clone from your most recent plate into 2 mL LB/kan broth in a snap-cap tube. Do this by picking bacteria from your most recent master plate with a sterile toothpick or pipette tip and dropping it into the tube containing LB/kan. Be sure to label the tube with your initials and date. If you are not completely confident that you have a positive clone, you should use your positive control (pBIT) as your clone for the protein purification. Place the tube in a designated rack. Inoculated cultures will remain refrigerated until the evening before your lab period, when your instructor will place them in the 37°C shaking incubator overnight.

Lab Session 11b

Extraction of Recombinant Protein From Escherichia coli and Purification Using a Glutathione Affinity Column

Goal: Today you will lyse bacterial cells induced with IPTG to express the GST::EGFP fusion protein and prepare a crude cellular homogenate. You will then use affinity chromatography to purify the expressed GST::EGFP fusion protein. This is achieved by taking advantage of the affinity between the glutathione-*S*-transferase (GST) moiety of the fusion protein and the small tripeptide, glutathione.

I. INTRODUCTION

The first step in purifying a cellular protein is lysing (breaking open) the cells to have access to the protein. There are a number of different ways to achieve cell lysis, and these methods vary between cell types depending on the composition of the cell wall (in some organisms) and cell membrane. Physical methods such as sonication, freeze-thawing, and French press commonly disrupt and homogenize cell membranes though the physical shearing of the membrane. Chemical methods involve treating cells with a reagent that reacts with, breaks open, and solubilizes the membrane to release the intracellular proteins. In most cases, peripheral proteins associated with membranes will be released as well. Depending on the reagent being used, integral membrane proteins are often solubilized and released as well when using chemical methods. In your experiment today, you will use either a physical (sonication and freeze-thawing) or chemical method (Bacterial Protein Extraction Reagent with lysozyme and DNase) to lyse the bacterial cells and prepare your crude protein homogenate. Following this homogenization of cellular components, our target protein, the GST::EGFP fusion protein, can be purified by the technique of affinity chromatography. The overall procedure for preparing cell lysate from your bacterial culture is outlined in Fig. 11.1.

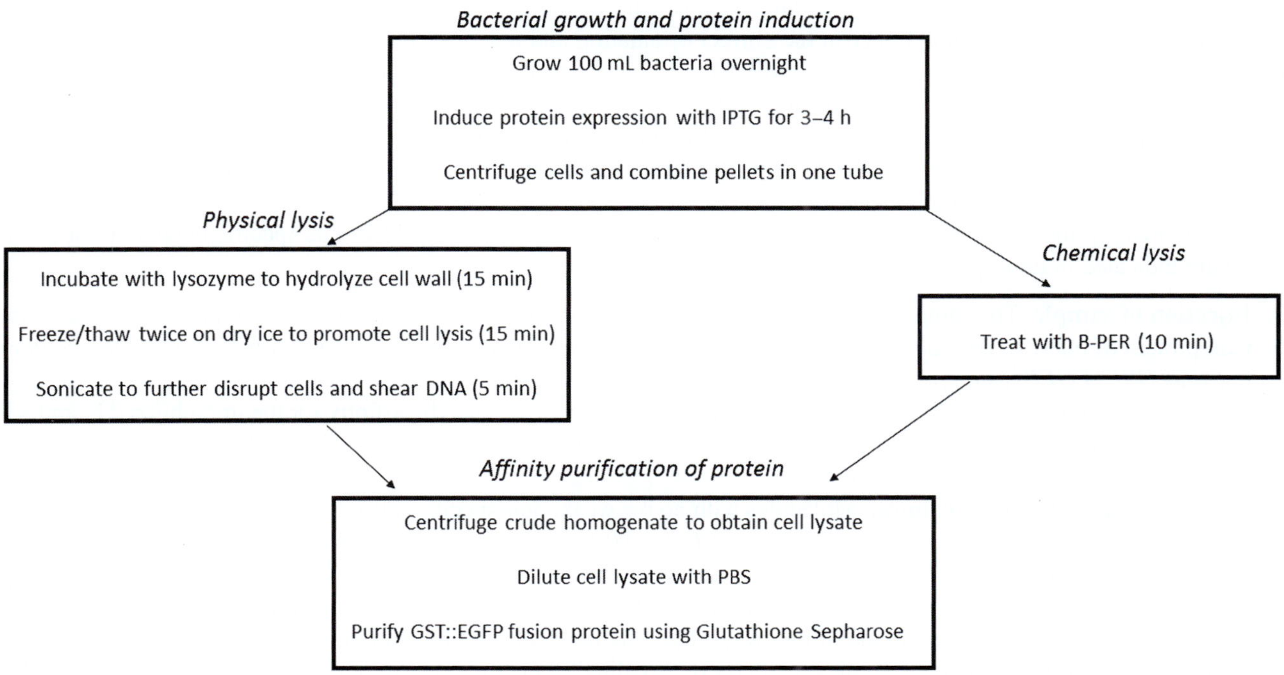

FIGURE 11.1 Overview of preparing cell lysate from bacterial cells. The times shown indicate the approximate total times for each of the steps involved in lysis. *No Permission Required.*

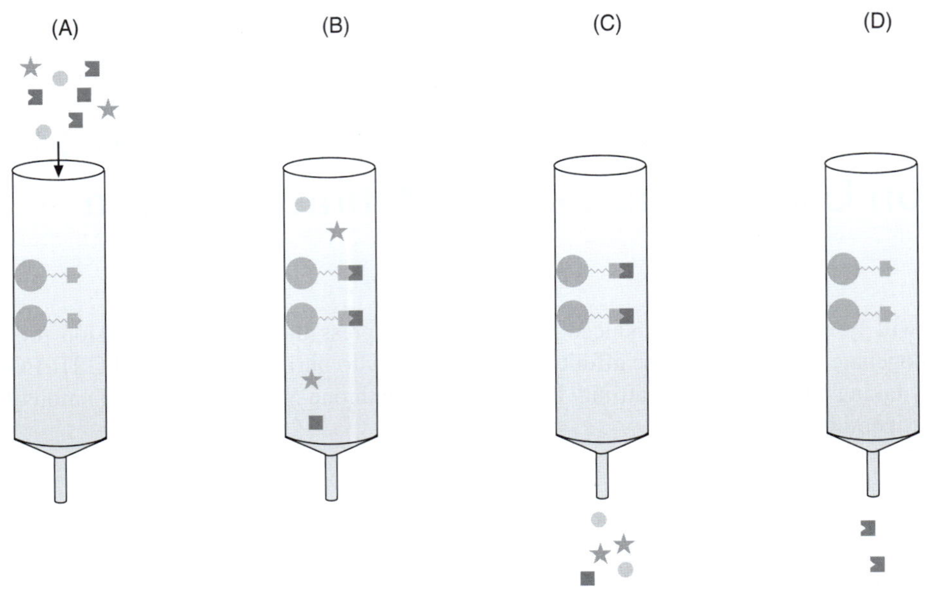

FIGURE 11.2 The principal steps of affinity chromatography. (A) Sample injection, (B) adsorption of target molecule(s), (C) washing of impurities, and (D) elution of target molecule(s).

Affinity chromatography is a commonly used method for purifying recombinant proteins. The use of affinity chromatography has become widespread due to the development of numerous expression vectors containing sequences that may be used as affinity tags. These expression vectors contain DNA sequence encoding an affinity tag (also called a fusion tag) either directly upstream or downstream of the multiple cloning site. The gene of interest must be inserted in the correct orientation and reading frame with respect to the affinity tag if you wish to create a fusion protein of your gene of interest and the affinity tag. Once protein expression is induced, a fusion protein will be produced. The fusion protein will contain two polypeptide moieties: one corresponding to the affinity tag and one corresponding to the protein of interest. Numerous affinity tags exist, with two of the most common being GST (the tag we will use) and hexahistidine (6X-His; a series of six histidine residues).

pET-41a contains the gene for GST (*gst*) directly upstream of the polylinker region. Once a gene of interest (in our case, *egfp*) is inserted into the polylinker in the correct orientation and reading frame, a fusion protein can be expressed. The fusion protein we will express is GST::EGFP. We will take advantage of the GST moiety of the GST::EGFP fusion protein in order to purify it via affinity chromatography.

The principle of affinity chromatography relies on the binding of a biospecific ligand to the molecule of interest. The ligand for GST is glutathione, which is covalently attached to a support matrix to allow separation of glutathione binding molecules from other cellular components.

The following principal steps of affinity chromatography are illustrated in Fig. 11.2. The specific details of our purification are indicated in italics.

1. **Injection of sample.** The sample contains a mixture of proteins and other cellular molecules along with the recombinant protein of interest. *The sample you will use is cleared crude homogenate of the induced bacterial culture containing your GST::EGFP fusion protein.*
2. **Adsorption of molecules with affinity for the ligand.** Molecules with specific affinity for the immobilized ligand will bind to the affinity resin while the rest of the molecules will flow through. *Only the GST::EGFP fusion protein will bind to the glutathione-linked affinity resin.*
3. **Wash impurities from the column.** Molecules with no (or weak, non-specific) affinity for the ligand are washed from the column. *Cellular molecules other than the GST::EGFP fusion protein will be washed from the column, and the GST::EGFP fusion protein will remain bound to glutathione in the column.*
4. **Elution of the target molecule(s) from the column.** A biospecific elution buffer reagent releases the affinity-bound molecules. This usually occurs in one of two ways. Specific molecules in the elution buffer can bind to the resin, outcompeting the affinity-bound molecules, essentially "stealing" their binding site and forcing them to elute through the column. Alternatively, the biomolecules in the elution buffer can bind directly to the affinity-bound molecules,

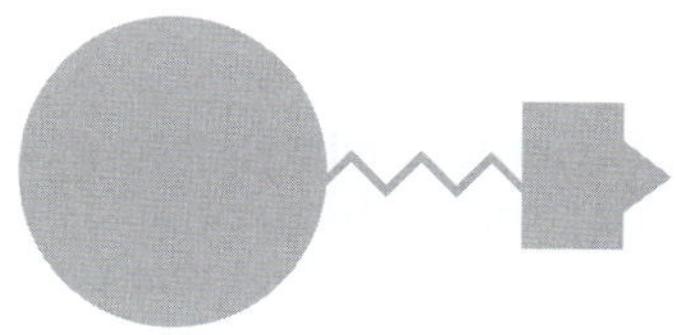

FIGURE 11.3 The affinity resin is composed of a biospecific ligand, a biochemically inert spacer, and a support matrix.

outcompeting the sites on the resin. When this happens, the molecule is released from the resin and elute through the column. *The GST::EGFP fusion protein (aka target molecule) will be eluted from the column using an elution buffer containing a solution of reduced glutathione, which binds directly to the GST::EGFP fusion protein and removes it from the affinity resin.*

The affinity column will be packed with a slurry, what we will refer to as the "affinity resin," illustrated in Fig. 11.3. The substance that makes up the affinity resin has three covalently bound components: the matrix, the spacer, and the ligand.

The support matrix functions to suspend the ligand in the column. The matrix must be rigid, biochemically inert, and have a high surface-to-volume ratio. The most commonly used support matrix is Sepharose, which is composed of small, crosslinked agarose beads.

The spacer is a carbon chain that links the ligand to the matrix. The purpose of the spacer is to present the ligand to the molecule of interest. If no spacer was present, or if it was too short, the ligand would be embedded in the matrix and might not be available for binding to the molecule of interest. Conversely, a spacer that is too long may have an affinity for undesired molecules via hydrophobic interactions. The ideal spacer would be just long enough to present the ligand, but short enough to remain biochemically inert.

The biospecific ligand of the affinity resin specifically binds the molecule of interest. Depending on the type of molecule to be purified, various classes of ligands may be used. For example, if an enzyme is to be purified, the ligand may be a substrate, inhibitor, or cofactor of the enzyme. If an antibody is to be purified, the cognate antigen may be used as the ligand. If a nucleic acid is to be purified, ligands could be complementary nucleic acid sequences. (For example, an oligo-dT column is used to purify eukaryotic messenger RNA (mRNA).) Affinity resins can be made in the laboratory by chemically crosslinking reactive resins with ligand, but in this experiment, you will use a commercially available affinity matrix. Because your protein is tagged with the enzyme GST, the biospecific ligand you will use is the enzyme's co-substrate, glutathione.

A final consideration when designing an affinity purification protocol is the issue of column capacity. Column capacity refers to the number of ligand molecules available for binding. In practice, this is related to the amount of resin used for purification. Different types of ligands will have different binding capabilities. If there is not enough ligand for all of the target molecules in the sample to bind, some of the target molecules will flow through the column and be lost. Conversely, if there are an excessive number of ligand binding sites compared to target molecules in the sample, non-target sample molecules with a lower affinity for the ligand may be able to bind to the affinity resin along with the target molecules. If these non-target molecules remain bound during the wash steps, they will be eluted together with the target molecules during the elution. This, of course, leads to low purity. Optimally, the number of ligand binding sites would be exactly equal to the number of target molecules in the sample. In practice, this is impossible to achieve. Column capacity needs to be optimized empirically in most cases because expression levels of various proteins expressed from various promoters under varying growth conditions differ. The diagnostic SDS–PAGE you will be performing next week will help you analyze the purity of your sample. If the eluate does not appear to be pure, or if you lost a great deal of target protein in the washes, you would modify the purification protocol either by adjusting column capacity or the stringency of the washes.

II. LABORATORY EXERCISES

A. Growing Bacterial Suspension Cultures for Fusion Protein Purification

Your instructor will inoculate your overnight culture into 50 or 100 mL of 2× YT/kan/IPTG medium and incubate at 37°C for 3–4 hours. We recommend using 50 mL cultures for purification using GST spin columns and 100 mL for gravity column purification.

B. Harvesting IPTG-Induced Cultures

1. Transfer your culture (40–45 mL) to a large centrifuge tube (or two tubes, if 100 mL of culture is present) capable of withstanding 10,000 × g force. Excess culture can be discarded in the bacterial graveyard. (*Note: if you are going to later use the chemical method for cell lysis (Part C below), tare empty centrifuge tube(s) before adding the resuspended cells.*)
2. Pellet the bacteria by spinning for 5 minutes at 10,000 × g in a high-speed centrifuge at 4°C.
3. Pour off the clear supernatant and discard into the bacterial graveyard.
4. Transfer the pellet to a microcentrifuge tube by resuspending the pellet with about 0.75 mL of 2× YT medium (or LB) to the bacterial pellet. Pipette up and down to resuspend the pellet. Transfer this cell suspension to a 1.5 mL microcentrifuge tube. If centrifuging in two tubes, after adding 0.75 mL of 2× YT medium (or LB) to one bacterial pellet and resuspending, transfer this cell suspension to the second pellet and pipette up and down to resuspend. Now transfer the resuspension to a 1.5 mL microcentrifuge tube. Balance the weight of the bacteria-containing tube with that of another (water-containing) tube and spin for 2 minutes in a microcentrifuge at 13,000 rpm at 4°C to pellet the bacteria. Remove the clear supernatant by decanting or pipetting.
5. Invert the bacteria-containing tube(s) over a paper towel and leave for about 2 minutes to remove residual liquid.

C. Breaking Open Bacterial Cells

Two different methods of breaking open the cells and preparing the crude homogenate are described, a physical method (sonication and freeze-thawing) and a chemical method (treatment with B-PER reagent). Choose only one of these methods.

Option A—Sonication and Freeze-Thawing (Physical Method)

1. Add 0.75 mL of ice-cold 1× phosphate buffered saline, PBS (with 2 mM Pefabloc, a protease inhibitor), to the pellet. Vortex to suspend the pellet.
2. Add 15 μL of 10 mg/mL lysozyme (final concentration 200 μg/mL). Vortex and incubate on "wet" ice (H_2O) for 15 minutes.
3. Place the microcentrifuge tube on a bed of "dry" ice (CO_2) to freeze.
4. Thaw the tube by warming it in the palm of your hand. As soon as the sample melts, freeze it again on dry ice.
5. Thaw the tube as described in Step 4 and store it in liquid form on wet ice. Freezing and thawing helps release the fusion protein by disrupting the bacterial cell wall.

   CAUTION: Put on protective ear guards before activating the sonicator in Step 6.
6. You will use a sonicator with an immersible tip (probe). Sonication helps release the fusion protein from the bacterial cells and reduces the viscosity of the solution by shearing chromosomal DNA. *Note: There may be a maximum power setting for this tip; make sure you do not exceed this setting. The conditions are not specified here because they vary depending on the particular sonicator used—most likely you will use a low setting because of the small volume being sonicated. Rinse the tip with distilled water and use Kimwipes to dry. Immerse the tip into the bacterial solution, but do not let the tip come into contact with the sides of the tube. Sonicate the solution for 15 seconds on ice. Sonication will heat up the solution; thus it is important to sonicate for short periods of time. Perform a total of three cycles of sonication, waiting 30 seconds between each pulse. If you cannot sonicate the solution on ice, keep the tube on ice as much as possible, and lengthen the time in between pulses to 1 minute to help keep the sample cold. This mixture represents the crude homogenate.*

Option B—Bacterial Protein Extraction Reagent, B-PER (Chemical Method)

1. Prepare the B-PER complete reagent by adding Pefabloc to a final concentration of 2 mM. Add 5 mL of B-PER reagent for each gram of cell pellet and resuspend the pellet by pipetting up and down until the solution is homogenous.
2. Incubate the solution at room temperature for 15 minutes with gentle rocking. This mixture represents the crude homogenate.

D. Removing Insoluble Debris From the Crude Homogenate

1. Spin the crude homogenate at 13,000 rpm in a refrigerated microcentrifuge or in a microcentrifuge kept in a cold box (4°C) for 15 minutes to remove insoluble debris. The insoluble debris represents cells that were not thoroughly sonicated and bacterial membranes.

2. Transfer all of the supernatant to a fresh microcentrifuge tube and centrifuge at 13,000 rpm in a refrigerated centrifuge or in a microcentrifuge kept in a cold box (4°C) for 2 minutes. It is critical to ensure the sample you load onto the affinity column is free of particulate debris. Insoluble debris from your cell pellet loaded onto your affinity column can make the column flow slowly. *This cleared crude homogenate will be referred to as the cell lysate.*
3. Transfer the cell lysate into a 15 mL conical tube and add approximately 2 volumes of PBS. If you have approximately 0.5 mL of lysate, you would add 1.0 mL of PBS. Diluting your sample before adding it to the column will help to increase the binding of your sample to the column. *Note: If you are performing purification with a GST Spin Column, you can also dilute your sample with the provided Equilibration/Wash Buffer.*
4. Record the approximate volume of the cell lysate (after dilution) in your lab notebook.
5. Remove a 50 µL aliquot of the diluted cell lysate for analysis by SDS–PAGE. Add an appropriate amount of SDS–PAGE sample buffer to the sample, label the tube, and store at −20°C.
6. Transfer an additional 200 µL of the cell lysate (for fluorescence or protein analysis) to a labeled microcentrifuge tube and store at −20°C.

E. Purifying Protein by Affinity Chromatography

There are two options depending on the amount of bacterial culture grown. For 50 mL of bacterial culture, we recommend using 0.2 mL glutathione spin columns. For 100 mL (or larger), the capacity of the spin columns is not large enough, so we recommend using gravity chromatography.

Option A (50 mL bacterial culture): Thermo Scientific Pierce GST Spin Purification Kit, 0.2 mL (Thermo Scientific Catalog Number 16106).

The affinity columns and buffers should be allowed to equilibrate to room temperature before use.

1. Twist off the bottom tab from the spin column and place the column into a collection tube.
2. Centrifuge for 2 minutes at 700 × *g* to remove the residual buffer. Discard the flow-through. Do not centrifuge at full speed, as this can damage the resin!
3. Add 400 µL of Equilibration/Wash buffer to the column, then centrifuge again for 2 minutes at 700 × *g*. Discard the flow-through.
4. Add your protein solution to the column, let sit for 3 minutes, then centrifuge for 2 minutes at 700 × *g*. If your entire volume of protein solution does not fit on the column at once, repeat this step until all the protein solution has been centrifuged through the column. Save the flow-through from this step in a tube labeled "flow-through."
5. After all the flow-through samples are combined in one tube, remove 50 µL of the sample and add it to a new tube along with the appropriate amount of SDS–PAGE sample buffer. Label the tube and store at −20°C. Remove an additional 200 µL of the flow-through sample and add it to a second tube. The latter tube will be saved for use in fluorescence or protein analyses at a later stage. Store at −20°C.
6. Add 400 µL of Equilibration/Wash buffer to the column, then centrifuge for 2 minutes at 700 × *g*. This serves to wash away any loosely attached proteins that might be non-specifically associated with the column. Discard the flow-through. Repeat this step two times (for a total of three washes). After the last wash, aliquot the final flow-through into two tubes: 50 µL for SDS–PAGE (along with the appropriate amount of SDS–PAGE sample buffer) and 200 µL for fluorescence/protein analysis). Label accordingly and store at −20°C.
7. Add 200 µL of 1X Elution buffer (10 mM reduced glutathione) to the column, then centrifuge for 2 minutes at 700 × *g*. *Reduced glutathione is the competitive agent in the elution buffer that displaces your GST::EGFP fusion protein from the affinity column.* Save the flow-through in a separate tube as Elution 1. Repeat this two more times to have a total of three elution samples, all saved in separate tubes for further analysis. Store samples at −20°C.

Option B (100 mL bacterial culture): Gravity chromatography with bulk resin (GE Healthcare Catalog Number 17513201).

The affinity resin and buffers should be allowed to equilibrate to room temperature before use.

1. Your instructor will provide you with a slurry of Glutathione Sepharose resin (affinity resin). You should have ~0.75 mL of settled bead volume. *Note: Be gentle! The beads are fragile.*
2. Assemble the chromatography column. Attach the stopcock to the bottom of the column. Mount the assembled column on a ring stand. (If you do not have a stopcock, many disposable columns are supplied with plugs that can be attached to the bottom of the column to stop the flow of liquid through the column.)
3. Make sure the stopcock is closed. Invert the Glutathione Sepharose resin to gently mix. Add the uniform slurry of affinity resin material to the column. Open the stopcock to allow the buffer to drip through into a waste beaker; do not allow

the buffer to drop below the top of the settling beads. Close the stopcock once the buffer level is just above the affinity resin. Also, make sure your column does not leak. If it does, you may have to replace the stopcock.

4. Wash the column with 5 mL of PBS. Keeping the stopcock wide open, allow the liquid level to drop until just above the bed of resin. Close the stopcock.
5. Add your cell lysate to the column and open the stopcock. Collect the flow-through fraction in a 15 mL conical tube labeled "FT." Close the stopcock once the buffer level is just above the settled resin. The majority of *Escherichia coli* proteins will not bind to the glutathione affinity resin and thus will be found in this flow-through fraction. *In affinity chromatography, the binding of the fusion protein to the ligand requires a slow flow rate. Allowing the sample to flow through the column by gravity flow achieves this. Forcing the sample through the column at a more rapid rate will decrease the binding efficiency of the target molecule to the ligand. Special formulations of affinity resins are required for high-pressure purifications, such as fast liquid protein chromatography or high-performance liquid chromatography (HPLC).*
6. Reapply the flow-through to the column, collecting the flow-through in the same 15 mL tube. Store the flow-through on ice when done. Passing your sample over the column a second time will increase the amount of GST::EGFP fusion protein that binds to the affinity resin, and therefore will increase your yield.

   Time-Saving Tip: The re-application of the flow-through to the column step can be omitted with only a small loss in overall protein recovery.
7. Remove a 50 µL aliquot of flow-through sample for analysis by SDS–PAGE. Add an appropriate amount of SDS–PAGE sample buffer to the sample to a labeled tube and store at −20°C.
8. Transfer an additional 200 µL of the flow-through (for fluorescence or protein analysis) to a labeled microcentrifuge tube and store at −20°C.
9. Add 5 mL PBS to the column, open the stopcock and collect the wash fractions, 1 mL per tube labeled W1–W5, and freeze at −20°C. Close the stopcock once the buffer level is just above the settled beads. *In the first wash fraction, you will actually be collecting the remainder of the flow-through fraction due to the void volume of the column. The void volume refers to the volume of liquid trapped within the beads that gets displaced as more liquid is added to the column. Hence, your first wash fraction will have a substantial amount of total protein. In the rest of the washes, you will likely have little total protein present. You hope to have little GST::EGFP fusion protein in either the flow-through or the washes, as it should be bound to the column at this stage.*
10. Wash the column with an additional 5 mL of PBS to remove any non-GST proteins left behind. Collect an aliquot of the end of this wash (~300 µL) and split into two microcentrifuge tubes labeled "last wash" (50 µL for SDS–PAGE and 200 µL for fluorescence/protein analysis). This will be used to make sure there is no GST::EGFP displaced from the column before elution. Close the stopcock once the buffer level is just above the settled beads. Discard the flow-through solution from this 5 mL wash step (except the saved aliquot).
11. Label six microcentrifuge tubes for the eluted samples (eluates) as E1–E6.
12. Add 3 mL of GST elution buffer (PBS + 10 mM reduced glutathione) to the column. Collect ~0.5 mL fractions in each of your labeled tubes. *Reduced glutathione is the competitive agent in the elution buffer that displaces your GST::EGFP fusion protein from the affinity column.*
13. Place the tubes with your washes and elution on the UV transilluminator in a dimly lit room. Be sure to wear a UV-protective face shield. Record the fractions that have green fluorescence in your notebook.
14. Store your fractions (crude homogenate, flow-through, washes, elutions) at −20°C for further analysis.

Time-Saving Tip: If you have time today, look ahead to Lab Session 12, Analysis of Purification Fractions, to see which samples you will be loading on your SDS–PAGE gel. Prepare your samples for next week's gel analysis by adding SDS–PAGE sample buffer to the necessary samples in separate tubes.

DISCUSSION QUESTIONS

1. What is likely to happen if the column capacity of the column were too high? Too low?
2. You accidentally spill the Glutathione Sepharose resin as you are attempting to add the beads to the column. Unfortunately, there is no extra Glutathione Sepharose resin anywhere to be found! What other methods of affinity chromatography could you use to purify your fusion protein?

Lab Session 12

Analysis of Purification Fractions

This week you will analyze your purification fractions by sodium dodecyl sulfate–polyacrylamide gel electrophoresis (SDS–PAGE) and quantify the amount of purified GST::EGFP fusion protein. You will also make a fresh replica plate of one of your positive clones for use next week.

Molecular Biology Techniques. DOI: https://doi.org/10.1016/B978-0-443-26714-7.00019-3

Lab Session 12a

Analysis of Purification Fractions

Goal: Today you will analyze your purification fractions by SDS–PAGE and quantify the amount of purified GST::EGFP fusion protein using either fluorescence or a Bradford protein assay. SDS–PAGE will allow you to estimate the purity of the eluted fractions and determine whether any of the recombinant fusion protein was lost during the wash steps. The fluorescence or Bradford protein assay will allow you to quantify how much GST::EGFP fusion protein was purified.

I. INTRODUCTION

The purpose of this laboratory session is to quantify how much GST::EGFP fusion protein you purified, as well as to determine the purity of your elution samples. You will also examine your fractions from affinity purification to see if your fusion protein is present in your unbound or wash fractions. These experiments will use two complementary methods: SDS–PAGE followed by total protein staining and fluorescence or Bradford protein detection.

In Lab Sessions 9 and 10, you performed SDS–PAGE on lysates from your transformants, then transferred the proteins to a nitrocellulose membrane and probed with an anti-GFP antibody in order to confirm the presence of the GST::EGFP fusion protein. In that particular case, we were not concerned with the presence or absence of any other protein(s) in the sample; we simply wanted to confirm that GST::EGFP was expressed. The purpose of the SDS–PAGE in this laboratory session is to determine which fractions contain fusion protein and to determine whether the fusion protein is free of contaminating proteins in the eluted fractions. We will not use antibodies because we want to visualize all of the proteins in the wash and eluted fractions. To achieve this, we will use GelCode Blue reagent, a G-250 colloidal Coomassie-based dye that binds to all proteins in the gel. The cell lysate (CL) includes all cellular proteins, including the fusion protein. Ideally, the flow-through (FT) and washes will contain all cellular proteins EXCEPT the fusion protein, and the eluate fractions will contain ONLY the fusion protein. The presence of fusion protein in FT and/or wash fractions indicates that the column did not have enough affinity resin or that the flow rate was too fast for the fusion protein to bind to the resin. The presence of additional protein bands in the eluate fractions can result from a number of possibilities: incomplete washing of the column; proteolysis of the fusion protein; nonspecific binding of proteins to the affinity resin; or co-purification of endogenous proteins from *Escherichia coli* associated with the fusion protein.

To quantify the amount of GST::EGFP protein you purified, you will use one of two methods: fluorescence measurements or Bradford protein analysis. These two methods are complementary, in that both allow you to determine the amount of the fusion protein purified. The analysis of fluorescence specifically measures the presence of the GST::EGFP fusion protein, whereas Bradford protein analysis quantifies all proteins present in a given sample.

A. Protein Quantification by Fluorescence

To determine the amount of purified GST::EGFP fusion protein, you can directly measure EGFP fluorescence in your samples obtained during purification. Additionally, by analyzing the amount of fluorescence in your other fractions, that is, your FT and washes, you will be able to quantitatively determine how much of your initial starting material you lost during the purification protocol. As discussed previously, the EGFP protein will fluoresce upon excitation by light at specific wavelengths. While EGFP can be minimally excited by UV light, as you did in Lab Sessions 7c and 11b, it is optimally excited by light at 488 nm (blue light). Upon excitation, EGFP emits light at 507 nm, which corresponds to the green color you observed when visualizing your colonies and purification fractions. To quantify the fluorescence of EGFP in your samples, you can use a fluorescence microplate reader. Although your fusion protein is actually GST::EGFP, the fluorescent properties should be identical to that of EGFP alone.

The microplate reader is equipped with filters that allow only light of a certain wavelength to pass through. In your experiment, the excitation filter is a 485/20 bandpass filter. This means that light ranging between 465 and 505 nm will go through the filter, while the rest will be blocked. This light will serve to excite the EGFP present in your sample. Upon excitation, the EGFP will fluoresce and the light emitted will pass through the emission filter before being detected by the instrument. The emission filter is a 528/20 bandpass filter, thus allowing light from 508 to 548 nm to pass through. As you can see from comparing the filters, light from the excitation will be filtered out by the emission filter, thereby ensuring that the light detected by the plate reader corresponds specifically to fluorescence from your sample.

Similar to other standard protein assays (such as the Bradford assay), to quantify fluorescence, you must first make a standard curve using known amounts of recombinant fluorescent protein, EGFP. The amount of fluorescence in your unknown samples is then determined by comparison to the standard curve. Keep in mind, when quantifying fluorescence, instead of measuring total protein, you will be measuring only the amount of the GST::EGFP fusion protein in your samples. By combining the fluorescence data with your results from SDS–PAGE, you will determine both the relative purity of your sample, as well as your yield of GST::EGFP fusion protein.

B. Protein Quantification by Bradford Assay

As an alternative to using a fluorescence plate reader, which is a relatively specialized piece of equipment, overall protein quantification can be determined using a general protein quantification assay, such as the Bradford assay. Protein quantification assays measure the total amount of protein in a sample. A number of assays exist for protein concentration. For each of these assays, protein concentration is estimated by comparing the value of an unknown sample to a standard curve created by assaying a series of known concentrations of a reference protein such as bovine gamma globulin (BGG) or bovine serum albumin (BSA). The most commonly used assays are the Bradford assay and the bicinchoninic acid (BCA) assay. Each assay has its own strengths and weaknesses.

The Bradford assay is the one we will use to determine the protein concentrations. This assay uses Coomassie G-250, which binds primarily to basic (especially arginine) and aromatic amino acid residues. Coomassie-based dyes are brownish red in color (absorbance max of 465 nm) on their own, but upon binding to proteins in acidic environments they turn blue (absorbance max of 610 nm). The largest difference between the absorbance of the free and protein-bound forms of the dye is at 595 nm, and this is where dye–protein complexes are usually monitored. This assay is able to detect and quantify protein concentrations in the range of 1–1500 μg/mL. The disadvantage of this assay is that it is incompatible with most detergents and some common protein buffers. If you lysed your cells using the chemical method (with B-PER reagent), the detergent present in that reagent could interfere with your Bradford assay, particularly in your CL sample. However, the sample was diluted with PBS before affinity chromatography and will be diluted again prior to analysis in the assay. These dilution steps will alleviate any issues resulting from the detergent interfering with the assay. Reagent compatibility charts for protein assay reagents are commonly supplied by the manufacturer.

The BCA assay has the advantage of being compatible with various detergents and buffers. However, this assay is based on interactions with tyrosine, cysteine, and tyrosine residues in the protein sample, with tyrosine having the strongest interaction. This is useful for a mixed sample of protein, but can be problematic for individual proteins. A protein sample that has a below-average amount of those three amino acid residues (particularly tyrosine) will give an artificially low reading, and a protein sample that has an above-average amount of these residues will give an artificially high reading. While the same can be true for the Bradford assay, the Bradford assay depends on more amino acids than the BCA assay. For this reason, the BCA assay is not advised for measuring the concentration of a single protein, especially if the amino acid composition is not known. This assay has a sensitivity range of 20–2000 μg/mL. In our particular experiment, the BCA assay cannot be used because it reacts strongly with the reduced glutathione that is used to elute the protein from the column. Glutathione is a tripeptide consisting of glutamate, cysteine, and glycine, with the cysteine being strongly reactive in the BCA assay. Because of this, the Bradford assay is the best choice for protein concentration determination in our experiment.

II. LABORATORY EXERCISES

A. Sodium Dodecyl Sulfate–Polyacrylamide Gel Electrophoresis (SDS-PAGE) of Purified Fusion Protein

Time-Saving Tip: *Note: If you use stain-free TGX polyacrylaminde gels, you can forgo the need to stain with GelCode Blue. Simply visualize protein bands by exposing the gel to UV light according to the manufacturer's instructions. Save a picture of the gel image and proceed to analyze the results.*

1. Retrieve last week's samples from the −20°C freezer. Prepare 30 μL of each of your samples by adding the appropriate amount of SDS–PAGE sample buffer to each of the samples listed in the table. *The molecular weight marker does not need SDS–PAGE sample buffer added to it.* If you noticed fluorescence in any additional wash or elution samples, please

discuss with your instructor, as you may want to include those samples on your gel. The goal is to load your gel with samples that will allow you to most effectively analyze your purification. The table below shows a sample scheme for loading samples that were generated using the gravity column purification scheme. If you performed the purification with a GST spin column, your scheme can be adjusted for the samples you have.

Lane 1	Lane 2	Lane 3	Lane 4	Lane 5	Lane 6	Lane 7	Lane 8	Lane 9	Lane 10
Molecular weight marker	Cell lysate	Flow-through	Wash 3	Wash 4	Last wash	Elution 2	Elution 3	Elution 4	Elution 5

No Permission Required.

2. Once the samples have been mixed with SDS–PAGE sample buffer, heat the samples (except the molecular weight marker) for 10 minutes at 95°C or higher before loading. If the lids to your tubes pop open during heating, microcentrifuge tube cap locks can be used.
3. Perform a quick spin in the microcentrifuge to bring liquid contents to the bottom of each tube.
4. Load the 12% gel (or whatever percentage is supplied by your instructor) by adding 5 μL of the molecular weight marker (Bio-Rad Kaleidoscope marker used in Lab Session 9b unless otherwise instructed) and 20 μL of each sample mixture into the appropriate lanes on the gel.
5. Run the gel at constant voltage (100–110 V) until the dye front reaches the bottom of the gel (as described in Lab Session 9b).
6. Proceed to Part B (Protein Quantification Assays) while the gel is running.

Staining the Gel

CAUTION: Wear gloves while handling gel.

1. Turn off the power supply and remove your gel from the apparatus. Be careful not to squeeze the gel.
2. Gently pull apart the two pieces of plastic. The gel should stick to one plate.
3. Cradle the plate in your hand with the gel side facing up. While holding it over your staining tray, direct a stream of distilled water (from a squirt bottle) under the gel to loosen it. Tilt the plate to allow the gel to drop into a square Petri dish.
4. Place the gel in the dish and rinse three times (5 minutes each) with dH_2O with gentle shaking. Label the tray with a piece of tape.
5. Mix GelCode Blue Reagent by inverting gently, immediately before using.
6. Add ~20 mL of stain reagent (enough to cover the gel, but do NOT fill the tray with it!) and rock the dish slowly on a shaker or orbital mixer for an hour.

   Time-Saving Tip: GelCode Blue can be left on the gel overnight, if it is more convenient. The next day, the stain can be poured off and the gel can be photographed or destained with dH_2O.
7. Replace the stain reagent with dH_2O and continue shaking. This step enhances stain sensitivity and weak protein bands will continue to develop. The water may need to be changed several times for optimal results. The gel can be left in water for as long as necessary: 15 minutes to overnight.

   Time-Saving Tip: Destaining is not necessary with GelCode Blue but does provide for a better final image. Alternatively, destaining can be sped up by adding a few rolled-up Kimwipes to the edge of the Petri dish in the water during the destaining procedure.
8. Photograph your gel and analyze the results.

B. Protein Quantification Assays

Choose one of these two protocols to quantify the amount of protein present in your purification samples: fluorescence or Bradford assay.

Protein Quantification Using Fluorescence Analysis

Start this assay while your SDS–PAGE is running.

1. Prepare dilutions of recombinant EGFP for a standard curve as further described.

   You will be given a tube with 3.2 μL of a 1 mg/mL recombinant EGFP solution in PBS. Prepare dilutions of the EGFP protein in three tubes as shown in Table 12.1. Attach a new pipette tip to the micropipette each time you make the dilutions. Be sure to mix each tube well after performing the dilution.

2. Before adding your samples to the microtiter plate, you will dilute them with PBS. Failure to dilute the samples, particularly with your cell lysate (CL) and some of your elution fractions, will result in fluorescence too high for the microplate reader to detect. For your CL, you will make two different dilutions to help ensure that your reading falls within the standard curve. For the other samples, you will prepare 10-fold dilutions of your samples. Table 12.2 provides the details for the dilutions. Depending on the details of your purification, your exact samples may differ, but 10-fold dilutions are recommended for everything except the cell lysate.
3. Add 100 µL of PBS (for background), EGFP standards (for your standard curve), and your unknown samples from your purification to the 96-well microtiter plate, in triplicate. Table 12.3 provides a suggestion of how samples could be added to the 96-well plate.
4. Measure the fluorescence in the wells using the microplate reader. Your instructor will set up the instrument to properly read the fluorescence. This setup will vary between different microplate readers. In general, you should be using filters comparable to: excitation 485/20; emission 528/20. Furthermore, fluorescence sensitivity should be scaled, if possible, such that the ST5 (highest concentration of standard) wells are at the upper range of the scale.

To Calculate the Concentration and Amount of Protein in Each Sample

1. Background subtraction: Open up your data from the microplate reader in a computer program such as Microsoft Excel and average your three PBS wells. This is your background, and this value should be subtracted from all your wells.
2. Making a standard curve: Graph a standard curve derived from your EGFP standards. Your graph should have the concentration of EGFP on the *X*-axis and fluorescence reading on the *Y*-axis. Don't forget to include 0.0 as a point on the graph from your blank!

TABLE 12.1 Recombinant EGFP Dilutions for Standard Curve

Tube	Dilution	Protein Concentration (µg/mL)	Microplate Abbreviation
1	3.2 µL of stock (1 mg/mL)+636.8 µL PBS	5.0	ST5
2	175.0 µL from tube 1+175.0 µL PBS	2.5	ST2.5
3	125.0 µL from tube 1+500.0 µL PBS	1.0	ST1
4	175.0 µL from tube 3+175.0 µL PBS	0. 5	ST0.5

PBS, Phosphate buffered saline.
No Permission Required.

TABLE 12.2 Dilution Scheme for Samples

Sample	Microplate Abbreviation	Sample Volume (µL)	Volume of PBS for Dilution (µL)
Cell lysate (10-fold dilution)	CL10	35	315
Cell lysate (30-fold dilution)	CL30	12	348
Flow-through	FT	35	315
Last wash	LW	35	315
Eluate 1	E1	35	315
Eluate 2	E2	35	315
Eluate 3	E3	35	315
Eluate 4	E4	35	315
Eluate 5	E5	35	315
Eluate 6	E6	35	315

PBS, Phosphate buffered saline.

TABLE 12.3 Schematic of Microtiter Plate

	1	2	3	4	5	6	7	8	9	10	11	12
A	PBS	PBS	PBS									
B	ST0.5	ST0.5	ST0.5	ST1	ST1	ST1	ST2.5	ST2.5	ST2.5	ST5	ST5	ST5
C	CL10	CL10	CL10	CL30	CL30	CL30	FT	FT	FT	LW	LW	LW
D	E1	E1	E1	E2	E2	E2	E3	E3	E3	E4	E4	E4
E	E5	E5	E5	E6	E6	E6						
F												
G												
H												

PBS, Phosphate buffered saline; *FT*, flow-through; *LW*, last wash.
No Permission Required.

3. Determine the protein concentration of diluted samples: Calculate the fluorescence average of each triplicate reading for your unknown samples. If you suspect that you made a mistake in pipetting, and that particular well's value supports that, you may discard that reading. Use a computer program such as Microsoft Excel to calculate the slope of the standard curve: $m = \Delta Y/\Delta X$ or $Slope = \Delta\text{fluorescence}/\Delta\text{concentration}$. Then plug in $y = mx + b$ or $fluorescence = slope \times concentration$ ($b = 0$ in this case). The concentration you calculate here is the concentration of the sample in the assay well (C_a). Recall that you diluted your samples for the assay, so to determine the protein concentration in the original sample, further calculations are necessary.
4. Determine the concentration in the original sample:

$$C_0\left(\frac{\mu g}{\text{mL}}\right) = \frac{C_a\left(\frac{\mu g}{\text{mL}}\right)}{d_1}$$

where C_o is the protein concentration in the original sample (μg/mL), C_a is the protein concentration in the assay microtiter well, and d_1 is the dilution of your sample prior to reading fluorescence. For example, for your FT last wash (LW), and eluate samples, $d_1 = 35/(35 + 315) = 0.1$. For your CL sample, you will determine which dilution factor to use based on which reading falls within your standard curve. Record concentrations in your laboratory notebook.

5. Calculate total protein (in μg).

$$\text{Total protein} = \text{protein concentration} \times \text{total volume}.$$

You recorded the total volume of the CL in Lab Session 11, and collected ~500 μL fractions of your eluates. (If the volume of your elutions is different, adjust accordingly.)

After calculating your total amount (mass) of GST::EGFP fusion protein in your various samples, answer these questions:

What percentage of your total fusion protein did you successfully purify? (Sum of the amount of eluates/total amount in CL × 100.)

Proteins are purified for a number of reasons. Some proteins are used as therapeutic agents, that is, insulin and growth hormones. Another reason is to sell the proteins commercially for scientific use. The recombinant EGFP protein you used to make your standard curve is an example of a commercially available fusion protein (the recombinant EGFP contains a 6 × -His tag). Currently, the price of recombinant EGFP is ~$364 for 100 μg of pure protein. How valuable was your work in the lab over the semester? Do the math to figure out how much the recombinant protein you purified is worth!

Protein Quantification Using Bradford Protein Assay

Note: The protocol below is specific for the Bio-Rad Quick Start™ Bradford Protein Assay. If using a different reagent or kit, please follow the manufacturer's protocol.

Note: The following steps are for a 1 mL standard assay in a spectrophotometer. However, you may use the microplate-scale setup instead. See the manufacturer's instructions for protocols and data analysis.

1. Your professor will provide you with standards (known concentration) of bovine gamma globulin (BGG). The standards range from 0 to 1500 μg/mL and will be used to make your standard curve.

   Money (and Time) -Saving Tip: Prepare a single set of BGG standards for the entire class by mixing 20 μL of each standard with Bradford working reagent. The standards can be made by the instructor and each group can read the absorbances independently or the absorbance values of the samples can be measured by the instructor and given to the students. Note that the absorbance values of the samples will increase over time once they are mixed with the Bradford working reagent. The most accurate data will be obtained if samples are read after incubation with the dye for a fixed amount of time (between 5 and 60 minutes).
2. Label a microcentrifuge tube or disposable cuvette for each of the following purification fractions you saved: CL, FT, LW, and all eluates (E1, E2, etc.). Also include tubes for each of the standard curve BGG samples if they are not prepared by your instructor (blank, 125, 250, 500, 750, 1000, 1500 μg/mL).
3. Pipette 5 μL of your CL and FT into the correct tubes/cuvettes.
4. Pipette 15 μL of dH_2O into both the CL and FT tube to dilute it. What fold dilution did you make?______
5. Pipette 20 μL of all of your other fractions into a labeled tube/cuvette. For BGG protein standards, also pipette 20 μL into each tube/cuvette.
6. Add 1 mL of 1× Bradford dye reagent to each of your tubes/cuvettes. Vortex to mix.
7. Incubate for 5 minutes at room temperature (or longer, up to 60 minutes) and record the absorbance from the spectrophotometer. It is important that all samples incubate for as close to the same amount of time as possible, as the absorbance of the samples change over time.
8. Create a spreadsheet with your data and prepare a standard curve with the absorbance values from your BGG standards. Remember to include 0,0 as a point on the graph corresponding to your blank. (See the manufacturer's data analysis instructions concerning blanks if using the microplate-scale setup)
9. Using the slope from the line generated from your standard curve, calculate the total protein concentration in each of your fractions. Remember that your CL and FT fractions were diluted in Step 4 and be sure to account for those dilutions in your calculation. (Only your CL and FT fractions were diluted, so others will not need a dilution factored in.)
10. Calculate the total protein in μg in each of your original fractions:

$$\text{total protein}(\mu\text{g}) = \text{total protein concentration}\left(\mu\frac{\text{g}}{\text{mL}}\right) \times \text{volume of fraction}(\text{mL})$$

You recorded the total volume of the CL in Lab Session 11 and collected ~500 μL fractions of your eluates. (If the volume of your elutions is different, adjust accordingly.)

After calculating your total amount (mass) of GST::EGFP fusion protein in your various samples, answer these questions:

What percentage of your total fusion protein did you successfully purify? (Sum of the amount of eluates/Total amount in CL × 100.)

Proteins are purified for a number of reasons. Some proteins are used as therapeutic agents, for example, insulin and growth hormones. Another reason is to sell the proteins commercially for scientific use. The recombinant EGFP protein you used to make your standard curve is an example of a commercially available fusion protein (the recombinant EGFP contains a 6 × -His tag). Currently, the price of recombinant EGFP is $364 for 100 μg of pure protein. How valuable was your work in the lab over the semester? Do the math to figure out how much the recombinant protein you purified is worth!

Lab Session 12b

Replica Plating (Only if Performing Lab Session 13)

I. LABORATORY EXERCISE

Replica plate one of your positive clones onto LB/kan so you will have a fresh inoculum for next week. Label, tape shut, and place inverted in the 37°C incubator. Your instructor will save it in the refrigerator after an overnight incubation.

DISCUSSION QUESTIONS

1. After running the SDS–PAGE, why do some of the lanes with eluate not show any bands? Why are we collecting the eluate in fractions, rather than all into one tube?
2. Assume that you performed both fluorescence analysis and a Bradford assay on your purified fractions. In which assay would you expect to calculate a higher protein concentration for your cell lysate fraction? For your elution fractions? Why?

Part IV

Analysis of mRNA Levels

At this point, you have successfully engineered a recombinant DNA molecule and expressed and purified the recombinant fusion protein, GST::EGFP. The experiments described in Part IV of this manual aim to measure the levels of *gst::egfp* messenger ribonucleic acid (mRNA). As described in the Central Dogma of Molecular Biology, "DNA makes RNA makes protein." DNA, whether it's genomic DNA or plasmid DNA, is transcribed into RNA by an enzyme called RNA polymerase (Figure IV.1). In prokaryotes like *Escherichia coli*, one RNA polymerase is responsible for transcribing different types of RNA. The most abundant type of RNA is ribosomal RNA. This type of RNA is a major component of ribosomes. Transfer RNAs are also involved in translation, as they help add amino acids to the growing polypeptide chain. Protein-coding genes are transcribed into mRNAs, which eventually get translated into proteins. Hence, mRNAs serve as messengers to take information from DNA and deliver it to protein-synthesizing ribosomes. While these three types of RNAs were once believed to constitute all of a cell's total RNA, we now know that there are many types of RNA, including microRNAs, small nuclear RNAs, PIWI-interacting RNAs, and many others. Their discoveries have expanded our view on just how diverse RNA functions are.

In your project, the PCR product you cloned into pET-41a is the coding sequence for *egfp*. pET-41a already encodes the coding sequence for *gst*. Therefore, when transcription of your positive clone occurs, *gst::egfp* mRNA will be produced. In this part of the course, you will induce gene expression (transcription and translation) from your plasmid clone as you did in Part III. This time, you will not be analyzing protein levels; rather, you will be analyzing complementary DNA (cDNA) levels, and by inference, mRNA levels. To accomplish this, you will learn how to purify and quantify total RNA (which includes mRNA) from *E. coli* and convert it into cDNA by a reverse transcription reaction. In addition, you will be comparing induction with IPTG or lactose to no induction at all. Purified total RNA is a common starting material for many methods that analyze RNA levels. These include RT-PCR, northern blotting, microarrays, and RNA-sequencing.

CHALLENGES OF WORKING WITH RNA

Compared to DNA, RNA is much less stable and should be handled with special care. This difference in stability can be attributed to several features. All RNAs have a 2′ hydroxyl group on the ribose sugar. This hydroxyl group can be involved in a nucleophilic attack of a 5′ phosphorous atom. In alkaline (basic) conditions, the hydroxyl group acts as a stronger nucleophile. This results in cleavage of the RNA's phosphodiester bond. DNA contains deoxyribose instead of ribose, so this reaction does not occur in DNA. DNA has smaller grooves in its double helical structure, whereas RNA has larger grooves (and is usually not double-stranded). These larger grooves provide more sequence that can be susceptible to enzymatic cleavage. It should be noted that degradation of bacterial mRNA occurs quite quickly, often with half-lives of only several minutes.

In addition to these structural characteristics of RNA, the existence of a certain group of enzymes called RNases renders RNA vulnerable to degradation. Microorganisms are sources of RNases which are found in the air, on surfaces, on our skin, and in bodily fluids. These ubiquitous enzymes degrade RNA and can be a nuisance to someone trying to isolate and work with intact RNA. In addition, lab techniques that use RNases, such as nuclease protection assays and plasmid purification protocols, may have been previously performed in the same lab, with the same instruments and reagents that you are using. The best solution is prevention: minimize the amount of RNases you expose your samples to. Several steps should be taken when working with RNA in the following lab sessions to accomplish this. First, always wear gloves and change them often. RNases from hands can easily be transferred to pipettes, tubes, and reagents. Second, use pipette tips, tubes, and water that are certified by the manufacturer as "RNase-free." Autoclaving may be fine to sterilize your tips for

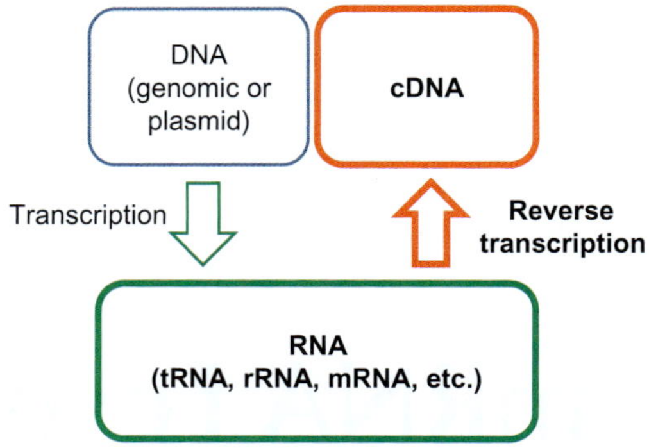

FIGURE IV.1 Central dogma of molecular biology. This figure depicts reverse transcription (first-strand cDNA synthesis only) in a prokaryotic cell using random primers. Therefore, mRNA, as well as rRNA and tRNA, are reverse transcribed into cDNA. *cDNA*, complementary DNA; *mRNA*, messenger RNA; *rRNA*, ribosomal RNA; *tRNA*, transfer RNA.

general use, but it does not guarantee that RNases have been denatured. Pipette tips that are certified "RNase-free" may also contain filters (barriers) that prevent aerosols like RNases from contaminating the pipette shaft. Purchasing sterile, RNase-free tips, tubes, and water is the easiest way to prevent contamination from these sources. Finally, it is best to set aside pipettes and gel electrophoresis systems that are dedicated solely to RNA work. However, if this is not practical in your lab, you may clean your pipettes and gel electrophoresis systems with a commercially available reagent that inactivates RNases, such as RNaseZAP™. You should also use this reagent to clean your bench top before starting a lab session.

Unlike DNA, RNA samples are best stored at −80°C. If an RNA sample is to be stored for more than several weeks, adding a salt (sodium acetate) and ethanol solution to the sample before storage at −80°C is recommended. RNA samples should be kept on ice while at your bench; however, each protocol described here will specify if a reaction including RNA should be set up on ice or at room temperature. In general, it is best to work quickly (but carefully) when handling RNA.

The lab sessions in this part of the manual were designed for you to use one of two methods for analyzing mRNA levels. If you have a thermal cycler equipped for quantitative PCR (a real-time system), it is recommended that you perform Lab Sessions 13, 14, and 15. If you do not have a real-time system, you should perform Lab Sessions 13, 16, and 17. Both methods will generate comparable results, although real-time PCR is much more sensitive. Both methods begin with the same steps (induction, total RNA purification, DNase digestion, quantification, and reverse transcription). After that, the PCR and data analysis steps diverge.

Lab Session 13

Total RNA Purification

Goal: This week, you will inoculate cultures of a positive transformant, and then your instructor will induce these cultures with IPTG, lactose, or no inducer. You will then purify total RNA from these cells, treat each RNA sample with DNase, and quantify the RNA. Samples will be saved at −80°C until the next lab session.

Molecular Biology Techniques. DOI: https://doi.org/10.1016/B978-0-443-26714-7.00002-8

Lab Session 13a

Interim Laboratory Session

I. LABORATORY EXERCISE

1. Inoculate a positive clone into 3 mL LB/kan broth in a snap-cap tube. Do this by picking bacterial growth from your most recent master plate with a sterile pipette tip and dropping it into the tube containing LB/kan.
2. Label the tube with your initials and date.
3. Label three empty snap-cap tubes with your initials, date, and either "no," "IPTG," or "lactose."
4. Place the inoculated culture and the three empty, labeled tubes in the designated racks. The LB/kan tubes will be refrigerated to retard growth. The night before your lab, they will be placed in a shaking incubator at 37°C. After growing overnight, the cultures will be ready for induction. Your instructor will use the empty, labeled tubes for this induction.

Lab Session 13b

Total RNA Purification

I. INTRODUCTION

Recall that in previous lab sessions, IPTG was added to your bacterial cultures (liquid broth or agar plate) to induce gene expression. This induction was measured at the protein level (visualization of GST::EGFP by UV light, SDS–PAGE, and western blotting). Gene expression commonly refers to both transcription and translation, with the end result being protein production. To analyze induction at the *mRNA level*, cultures will similarly be induced with IPTG to derepress the lac operator of pET-41a. IPTG is a non-hydrolyzable analog of lactose; therefore, you will try inducing with IPTG and then you will compare that to induction with the same concentration of lactose. These two conditions will serve as your experimental conditions and no inducer at all will serve as your negative control.

Total RNA will be purified from the *E. coli* cells following these treatments. Total RNA refers to all the RNA content of a cell including, but not limited to, mRNA, rRNA, and tRNA. The most abundant type of RNA is rRNA, while the type of RNA we are interested in analyzing, mRNA, makes up an extremely small fraction of total RNA. We will use the QIAGEN RNeasy Mini Kit to purify total RNA (Fig. 13.1).[1] First, *E. coli* cells are incubated with lysozyme to degrade the bacterial cell wall. Then, a lysis buffer is added that contains guanidine thiocyanate. This chemical is a denaturant that quickly inactivates RNases (enzymes that degrade RNA). Next, the lysate is vortexed and passed through a column called a QIAshredder homogenizer. These homogenization steps will serve to reduce viscosity (large molecular weight genomic DNA is sheared) and remove insoluble material from the bacterial lysate. The QIAshredder is quicker and easier than homogenization methods involving beads or a needle and syringe. Next, ethanol is added to create conditions favorable for RNA binding to the RNeasy column membrane. The final purification steps are similar to the silica adsorption steps used in this manual to purify plasmid DNA. The sample is applied to an RNeasy column, contaminants are washed away, and total RNA is eluted in RNase-free water. The column membrane will selectively bind RNA larger than 200 nucleotides; therefore the eluted RNA does not include tRNA or other low-molecular weight RNA. This is beneficial in our case, since we are interested in analyzing mRNA.

RNeasy columns will remove most of the DNA found in the lysates; however, for downstream steps, efficient DNA removal is required. Even trace amounts of either *E. coli* genomic DNA (found naturally in all *E. coli* cells) or plasmid DNA (transformed into *E. coli* cells) can be detected with a technique as sensitive as RT-PCR. The template that you are interested in amplifying in the PCR step is cDNA directly made from mRNA. Therefore, to achieve accurate quantification of mRNA levels, you must eliminate any DNA in the RNA sample that may also serve as a template for amplification. Although DNase digestion can be performed "on-column," treating the eluted RNA separately is more efficient. Finally, DNase will be inactivated, and the salts will be removed by a final purification step on another RNeasy mini column. This step is analogous to the column purification of your linearized plasmid DNA.

It is often valuable to assess the integrity (lack of degradation) of purified RNA before proceeding. In methods such as northern blotting, microarray, or RNA-sequencing sample preparation, intact RNA is vital for successful experiments and subsequent analysis. One way that RNA integrity can be analyzed is by agarose gel electrophoresis. Due to RNAs single-strandedness, it has the ability to fold back on itself and form secondary structures through complementary base pairing. To prevent this secondary structure from affecting migration through an agarose gel, denaturing conditions are used. Rather than a native agarose gel like those you prepared for analyzing DNA, denaturing agarose gels (usually containing formaldehyde) are often used to accurately estimate molecular weight. RNA samples are also heated in a loading dye that contains the denaturant formamide before loading them in the gel. RNA is not stable in alkaline conditions, so a neutral buffer is used in both the gel and running buffer. MOPS (3-[*N*-Morpholino]-propanesulfonic acid) buffer is most commonly used for electrophoresis of RNA. These steps ensure that RNAs are kept denatured (single-stranded) and migrate based on size alone. RNA molecular weight markers are run next to RNA samples, since fragments of a DNA molecular weight marker would migrate differently than RNA fragments of the same size due to their double-strandedness.

Results of a typical denaturing agarose gel loaded with total RNA purified from *E. coli* are shown in Fig. 13.2. Remember that total RNA was purified, so only the most abundant types of RNA will be detected using ethidium bromide or a similar fluorescent nucleic acid dye like GelRed. That is why mainly rRNA bands are visible on the gel. In *E. coli*, 23S

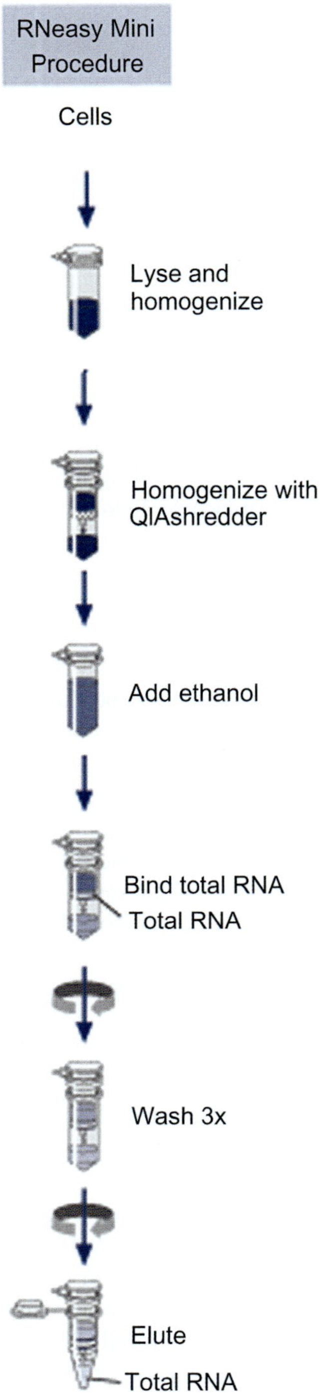

FIGURE 13.1 RNeasy Mini Kit flowchart. *Copyright 2012 QIAGEN Corporation. Used with permission.*

rRNA and 16S rRNA are clearly seen migrating at 2.9 and 1.5 kb, respectively. Notice in lane 2 that there is approximately twice as much 23S rRNA as 16S rRNA. This 2:1 ratio is one indication of intact RNA. There are also sharp bands seen in lane 2, demonstrating little degradation. If RNA is degraded, smearing will be evident. Lane 3 shows RNA that has been treated with DNase. Three differences arising from DNase digestion can be seen. First, there is slight smearing in lane 3 below each rRNA band, indicating some degradation. If the RNA was heavily degraded, no distinct bands would be apparent at all, and the entire lane would be a smear. Second, the 2:1 ratio is not maintained. Lane 3 looks as if 23S and 16S rRNA are equal in abundance. Third, notice that the rRNA bands in lane 3 have migrated further than the rRNA bands in lane 2. This means that the rRNAs are smaller than normal, consistent with degradation. These results are not surprising, since the DNase treatment and subsequent purification are additional physical manipulations to the sample. Recall that RNA,

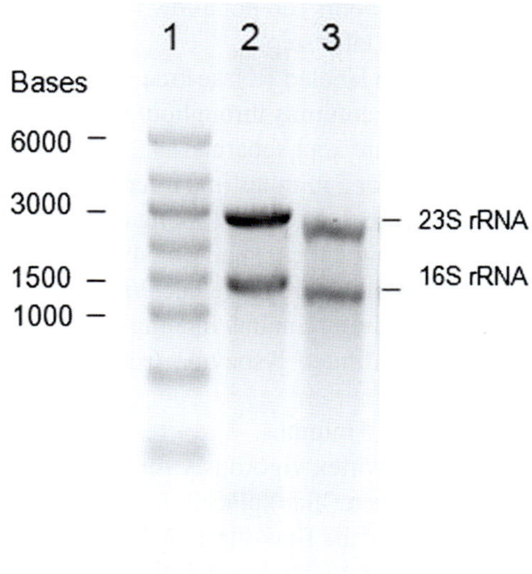

FIGURE 13.2 Denaturing 1% agarose gel electrophoresis of total RNA from ***Escherichia coli***. Lane 1: Fermentas RiboRuler High-Range RNA Ladder. Lane 2: 1 μg total RNA without DNase digestion. Lane 3: 1 μg total RNA with DNase digestion. *No Permission Required.*

especially bacterial RNA, can degrade rapidly. For our RT-PCR experiment, we will include the DNase treatment but skip the denaturing agarose gel electrophoresis. Even though the amounts of 23S and 16S rRNA are apparently equal, this is not the only indicator of RNA integrity. Furthermore, the amount of RNA degradation that seems to have occurred due to the DNase treatment steps is still tolerable and will not adversely affect the reverse transcription and PCR steps. The need for eliminating contaminating DNA in this experiment outweighs any minor degradation the RNA will undergo.

Finally, RNA will be quantified by a spectrophotometer (or Nanodrop microvolume spectrophotometer). Like DNA, RNA also absorbs UV light maximally at 260 nm. However, RNA has a different average extinction coefficient than double-stranded DNA due to its single-strandedness. An absorbance of one unit at 260 nm corresponds to approximately 40 μg of single-stranded RNA/mL. When calculating the concentration of an RNA sample, the correct extinction coefficient must be used. If you use a Nanodrop, you will need to specify that you are analyzing RNA, not DNA samples. That will allow the software to automatically calculate concentration using the appropriate extinction coefficient.

The purity of an RNA sample can also be estimated by a spectrophotometer. Recall that pure DNA should have an A_{260}/A_{280} ratio of approximately 1.8. Pure RNA should have an A_{260}/A_{280} ratio of approximately 2.0. Anything less than 2.0 could indicate possible protein contamination; however, the amount of contamination would have to be extreme to significantly affect this ratio. More likely, buffer components may be present that absorb UV light and decrease the ratio. If the ratio is extremely low, it may simply be that the concentration of the sample itself is low. As both DNA and RNA absorb UV light at 260 nm, an acceptable A_{260}/A_{280} ratio cannot rule out DNA contamination. Remember that we are including a DNase digestion to minimize the amount of contaminating DNA. Future steps will serve to calculate the actual contaminating DNA levels. Ultimately, ratios of 1.9–2.1 are generally acceptable for RNA.

II. LABORATORY EXERCISES

A. Purification of Total RNA

We will purify total RNA from bacteria using the QIAGEN RNeasy Mini Kit. QIAshredder spin columns can be purchased separately or obtained as part of the QIAGEN RNeasy Plant Mini Kit. A lysozyme solution is also required that is not supplied by the Mini Kit or Plant Mini Kit. Another option is to use the RNeasy Protect Bacteria Mini Kit. The following protocols have been modified from the manufacturer's handbook (QIAGEN RNeasy Mini Handbook 06/2012).

B. Sample Preparation

Steps 1–3 will be done for you.

1. Pipette 1 mL of the overnight culture into each of the three empty, labeled, culture tubes.
2. Pipette 2 mL of the appropriate 2 × YT broth into each tube: 2 × YT/kan/IPTG, 2 × YT/kan/lac, or 2 × YT/kan *(no inducer)*.

3. Incubate in the shaking incubator at 37°C for 2 hours (do not allow the incubation to proceed any longer than 2 hours).
4. Set your heat block or water bath to 37°C for a later step.
5. Pipette 1.5 mL of each *E. coli* culture into three labeled RNase-free microcentrifuge tubes.
6. IMPORTANT: Clearly label all tubes and spin columns throughout this protocol to be sure your control (no inducer) and experimental samples (IPTG and lactose) are kept separate.
7. Harvest the cells by centrifugation at 300 × *g* for 5 minutes in a microcentrifuge.
8. While the cells are spinning, dilute the lysozyme stock 1:100 (taking a 40 mg/mL stock and diluting it down to 0.40 mg/mL) by pipetting 5 μL into a new microcentrifuge tube and adding 495 μL of water. Keep on ice.
9. Carefully remove (decant) all supernatant into an appropriate bacterial liquid waste container and invert the tube over a paper towel to be sure no residual liquid medium remains. Incomplete removal of the medium will inhibit lysis.
10. Resuspend each bacterial pellet in 100 μL of diluted lysozyme from Step 8. Pipette up and down until no clumps remain.
11. Allow the cells to lyse at room temperature for 5 minutes.
12. Pipette 350 μL Buffer RLT into each tube and vortex vigorously for 1 minute.
13. Transfer the lysate to a QIAshredder spin column (lilac) placed in a 2 mL collection tube and centrifuge for 2 minutes at full speed. IMPORTANT: Do not throw away the flow-through. You will need it in the next step!
14. Carefully transfer the *supernatant of the flow-through* to a new RNase-free microcentrifuge tube without disturbing the cell debris pellet in the collection tube. Use only this supernatant in subsequent steps.
15. Add 250 μL of ethanol (96%–100%) to the saved supernatant and mix immediately by pipetting. Do not centrifuge. Proceed immediately to the next step.
16. Transfer the sample (usually 700 μL), including any precipitate that may have formed, to an RNeasy spin column (pink) placed in a 2 mL collection tube.
17. Close the lid gently and centrifuge for 15 seconds at ≥8000 × *g* (≥10,000 rpm). Discard the flow-through.[1] Reuse the collection tube in the next step.
18. Add 700 μL Buffer RW1 to the RNeasy spin column. Close the lid gently, and centrifuge for 15 seconds at ≥8000 × *g* (≥10,000 rpm) to wash the spin column membrane. Discard the flow-through.[2] Reuse the collection tube in the next step.
19. Add 500 μL Buffer RPE to the RNeasy spin column. Close the lid gently and centrifuge for 15 seconds at ≥8000 × *g* (≥10,000 rpm) to wash the spin column membrane. Discard the flow-through. Reuse the collection tube in the next step.
20. Add 500 μL Buffer RPE to the RNeasy spin column. Close the lid gently and centrifuge for 2 minutes at ≥8000 × *g* (≥10,000 rpm) to wash the spin column membrane. The long centrifugation dries the spin column membrane, ensuring that no ethanol is carried over during RNA elution. Residual ethanol may interfere with downstream reactions.
21. Place the RNeasy spin column in a new, labeled RNase-free microcentrifuge tube. Add 50 μL RNase-free water directly to the center of the spin column membrane. Close the lid gently and centrifuge for 1 minute at ≥8000 × *g* (≥10,000 rpm) to elute the RNA.

C. DNase Digestion

Do not vortex the reconstituted DNase I. DNase I is especially sensitive to physical denaturation. Mixing should only be carried out by gently inverting the tube.

1. To each tube of eluted RNA (50 μL), add the following for a total volume of 100 μL:
   - 30 μL RNase-free water
   - 10 μL Buffer RDD
   - 10 μL DNase (2.73 Kunitz units/μL stock, 0.273 Kunitz units/μL final concentration)
2. Mix by flicking and place in the 37°C heat block or water bath for 30 minutes.
3. Add 350 μL Buffer RLT and mix well by vortexing.
4. Add 250 μL ethanol (96%–100%) to the diluted RNA and mix well by pipetting. Do not centrifuge. Proceed immediately to Step 5.

---

[1]Flow-through contains Buffer RLT and is therefore not compatible with bleach. Dispose of it according to your institution's waste-disposal guidelines.
[2]Flow-through contains Buffer RW1 and is therefore not compatible with bleach. Dispose of it according to your institution's waste-disposal guidelines.

5. Transfer the sample (approximately 700 µL) to an RNeasy Mini spin column placed in a 2 mL collection tube. Close the lid gently and centrifuge for 15 seconds at ≥8000 × *g* (≥10,000 rpm). Discard the flow-through.
6. Add 500 µL Buffer RPE to the RNeasy spin column. Close the lid gently and centrifuge for 15 seconds at ≥8000 × *g* (≥10,000 rpm) to wash the spin column membrane. Discard the flow-through.[3] Reuse the collection tube in Step 7.
7. Add 500 µL Buffer RPE to the RNeasy spin column. Close the lid gently and centrifuge for 2 minutes at ≥8000 × *g* (≥10,000 rpm) to wash the spin column membrane.
8. Place the RNeasy spin column in a new, labeled RNase-free microcentrifuge tube. Label each tube with your initials, date, and the inducer used for that sample (no, IPTG, or lac). Add 50 µL of RNase-free water directly to the center of the spin column membrane. Close the lid gently and centrifuge for 1 minute at ≥8000 × *g* (≥10,000 rpm) to elute the RNA.

D. Quantification of RNA

Depending on the equipment available in your lab, you will use a spectrophotometer or microvolume spectrophotometer such as a NanoDrop to quantify each of your DNase-digested RNA samples. In addition, you will assess each sample's purity.

1. Refer to Lab Session 2 for details on how to use either instrument.
2. When blanking the instrument, use the same RNase-free water that was used to elute your samples.
3. IMPORTANT: If using a microvolume spectrophotometer, you must set the software to read RNA by choosing "RNA" from the drop-down list. RNA has a different extinction coefficient than DNA, so the concentration of your sample will be inaccurate if the software uses DNA's extinction coefficient in calculations. If you are calculating the concentration manually from a spectrophotometer reading, use RNA's extinction coefficient (40 µg/mL) in your equation.
4. Record the concentration of each sample in your notebook and write it on the side of the tube. Be aware of the units you are using.
5. Calculate and record the 260/280 ratio of each sample. If the ratio is not in the 1.9–2.1 range, let your instructor know.
6. Save samples in a −80°C freezer until your next lab session.

Reminder: If you are going to perform quantitative PCR with a real-time-capable thermal cycler, proceed to Lab Session 14, Analysis of *gst::egfp* mRNA Levels by RT-qPCR: Part I. If you are going to perform semi-quantitative PCR with a standard thermal cycler, proceed to Lab Session 16, Analysis of *gst::egfp* mRNA Levels by Semiquantitative RT-PCR: Part I.

REFERENCE

1. QIAGEN. RNeasy® Mini handbook. 2012.

DISCUSSION QUESTIONS

1. Could you quantify just the amount of *mRNA* in our samples by using a spectrophotometer or NanoDrop microvolume spectrophotometer? Why?
2. Will the 260/280 ratio differ between intact RNA samples and degraded RNA samples? Why?

[3] Flow-through contains Buffer RW1 and is therefore not compatible with bleach. Dispose of it according to your institution's waste-disposal guidelines.

Lab Session 14

Analysis of *gst::egfp* mRNA Levels by RT-qPCR: Part I

Goal: Total RNA that you purified in the last session will be converted to complementary DNA (cDNA) by a reverse transcription (RT) reaction. Then, cDNA will be used as a template for quantitative (real-time) PCR. You will amplify cDNA with primers specific to *egfp* (your test gene) and *23S* rRNA (your reference gene).

I. INTRODUCTION

Several techniques have been developed to detect RNA and analyze its abundance. Northern blots are similar to the western blots you performed, but an RNA sample is loaded on an agarose gel. The samples are then transferred to a nylon membrane and incubated with a labeled DNA or RNA probe specific to a gene of interest. The intensity of bands detected on the membrane is used to quantify RNA levels. This technique, although still used, is labor-intensive and requires a relatively large amount of fully intact RNA. A type of nuclease protection assay called an RNase protection assay (RPA) also relies on a labeled probe to detect an RNA it has hybridized to. This method is also laborious, although it is more sensitive than a northern blot. In contrast, newer methods based on PCR (which you also performed in this course) are less time-consuming, more sensitive, and are able to tolerate partially degraded RNA samples. In this Lab Session and Lab Session 15, RT-PCR will be performed. The first part of this method is reverse transcription (also called cDNA synthesis). This is the process of converting RNA to DNA using an enzyme called reverse transcriptase (see Fig. IV.1). The resulting DNA is complementary to the RNA template, so it is called cDNA. In the second part of this method, the cDNA is used as a template in PCR. In this way, the amount of PCR product is indicative of the starting amount of input cDNA (and by inference, RNA that it was generated from). RT-PCR can be performed in two steps, as described here, with an RT reaction occurring first. Then, that product is used as a template in a PCR reaction. Alternatively, RT-PCR can be performed in one step, with both reactions occurring in one tube.

This Lab Session and Lab Session 15 use a highly sensitive version of RT-PCR called quantitative RT-PCR (RT-qPCR). A few clarifications concerning nomenclature should be made at this point. Although other acronyms can be found for this method, RT-qPCR (with q in front of PCR) will be used in this manual because it is the PCR step that is quantitative. The RT stands for reverse transcription. Quantitative PCR (qPCR) can also be performed on DNA as the starting material, so there is no need for RT in that case. This is simply abbreviated qPCR. Both types of qPCR mentioned here can also be referred to as real-time PCR because the abundance of product is measured in real-time, as the products are being amplified. Lab Sessions 16 and 17 are included as an option for a less sensitive and less quantitative version of RT-PCR.

A. Reverse Transcription

What is required for an RT reaction? The enzyme that catalyzes this reaction is a reverse transcriptase, and there are several commercially available enzymes usually isolated from viruses called retroviruses. These viruses have an RNA rather than a DNA genome, so they must encode a reverse transcriptase to convert their genome to DNA. Human immunodeficiency virus (HIV) is an example of a retrovirus. The reverse transcriptase we will use is isolated from Moloney murine leukemia virus. Along with its supplied buffer, primers are necessary to prime cDNA synthesis. Instead of forward and reverse primers such as those that are used in PCR, RT reactions use only one primer because only one strand of cDNA is made. Primers fall into three categories: randomers (ours are hexamers), oligo-dT, and gene-specific. Oligo-dT primers will specifically anneal to poly-A tails found on most eukaryotic mRNAs. We cannot use this type of primer since we have purified prokaryotic RNA, which lacks poly-A tails. We want to prime cDNA synthesis from all the RNAs in the cell, so gene-specific primers will also not be suitable in this case. We will use random hexamers that have the ability to anneal to all types of RNA,

Molecular Biology Techniques. DOI: https://doi.org/10.1016/B978-0-443-26714-7.00021-1

without knowledge of sequence. Random hexamers are a pool of primers designed to represent all possible combinations of six-nucleotide stretches. This design allows for the primers to bind to all RNA sequences. Finally, dNTPs (identical to those used in PCR) will be included in the RT reactions. These will be incorporated into the newly synthesized cDNA strand.

Whenever RT-PCR is performed, an important aspect to control for is contaminating DNA. As previously mentioned, RT-PCR uses RNA as starting material. The goal is to synthesize cDNA from that RNA, followed by amplification of the cDNA with gene-specific primers. However, if DNA is still present in the RNA sample, it can be amplified during PCR as well, leading to inaccurate quantification of mRNA levels. To control for any contaminating DNA remaining in your samples even after DNase treatment, negative control reactions should be included that do not include reverse transcriptase. Without this enzyme present, the only DNA in the sample will be contaminating DNA (either genomic or plasmid), which will result in a PCR product. However, if you have effectively eliminated DNA in your starting material, amplification using a "−RT" control should show no product. If there is amplification in this negative control, it can be quantified and accounted for. Essentially, it reveals the presence of contaminating DNA. The corresponding "+RT" reaction should result in a product. These reactions have cDNA synthesized from RNA, and it is that amplification you aim to quantify.

B. Quantitative PCR

qPCR does not rely on visualization of products on agarose gels. Instead, quantification of amplified DNA is achieved using fluorescence measurements. Rather than examining fluorescence at the end of the PCR experiment, fluorescence is monitored as it is produced (in "real-time"). There are two main types of detection chemistries that can accomplish this. A fluorescent dye called SYBR Green binds to double-stranded DNA and, upon excitation by a light source, fluoresces approximately 1000 times brighter than when it is in its unbound form. Therefore, the amount of fluorescence in a reaction is proportional to the amount of double-stranded DNA. The second type of detection chemistry commonly used in qPCR is a 5′ nuclease assay called a TaqMan assay. This involves not only gene-specific primers, but also a gene-specific DNA probe. Although the TaqMan assay is more sensitive and eliminates formation of non-specific products, it requires a probe for each gene you want to quantify. SYBR Green has the disadvantage of binding unintended primer dimers (short DNA fragments that can form during PCR) and non-specific products as well as specific products (because it cannot discriminate between different double-stranded DNAs). However, it is more cost-effective, especially if you are assaying several different genes. We will use SYBR Green in this manual as a way to quantify mRNA levels.

RT-qPCR is a powerful, sensitive technique for analyzing gene expression, but designing a successful RT-qPCR experiment requires attention to many details. First, since we wish to quantify the level of *gst::egfp* mRNA (our test gene), we need primers that will specifically amplify this sequence from cDNA. In general, it is suggested that primers used in qPCR follow these guidelines:

- Amplify a sequence (amplicon) between 50 and 150 bp.
- GC content should be 20%–80%.
- Avoid runs of the same nucleotide (runs of four or more guanidines, especially).
- *Tm* should be close to 60°C.
- *Tms* of forward and reverse primers should not differ by more than 1°C.

Primer design software can be useful when default settings are changed to specify the guidelines outlined. In addition to our test gene, a second gene will be assayed. This provides an internal control that will be used to normalize experimental results. It is referred to as a reference gene. The reference gene used here is endogenous *Escherichia coli 23S* rRNA. Reference genes must be expressed in each sample, but, in order to be considered valid, expression levels cannot vary. The only gene we expect to change with treatment (induction with IPTG or lactose) is *gst::egfp*. Due to the limitations of SYBR Green previously mentioned, it is imperative that qPCR primers only produce one specific amplicon. Remember that we will not be visualizing the PCR results on an agarose gel. We will be relying on fluorescence produced during the reaction to indicate the amount of DNA present. If fluorescence can be attributed to some amplification of primer dimers or other non-specific products, we will not have an accurate quantification of *gst::egfp*. The primers you will use have been analyzed to verify that only one amplicon is produced. The most common way to accomplish this is through a melt curve analysis. A melt curve (also called a dissociation curve) is a graph that displays information on how DNA dissociates (melts) upon heating. This analysis can be performed at the end of the qPCR program by gradually increasing the temperature in the thermal cycler. The changes in fluorescence are plotted against temperature. An example of a melt curve is shown in Fig. 14.1. A single, narrow peak indicates one specific amplicon was formed. The *23S* rRNA amplicon shown has a *Tm* of approximately 86°C. If present, unintended primer dimers will produce a peak with a *Tm* lower than that of the desired amplicon due to their shorter length.

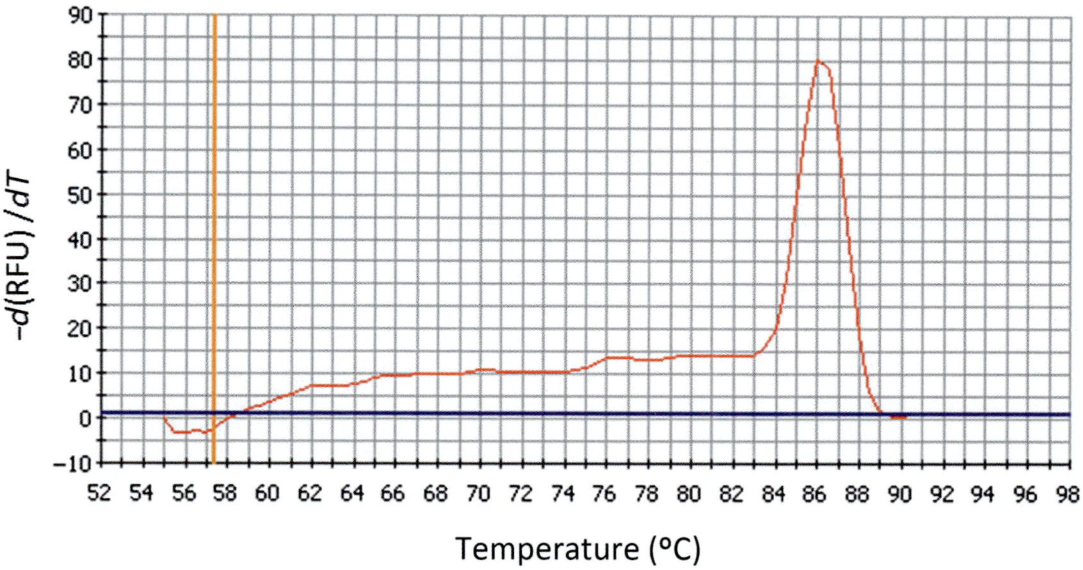

FIGURE 14.1 Melt curve analysis of 23S rRNA amplicon. The x-axis displays the temperature and the y-axis displays the change in fluorescence (relative fluorescence units, RFU) over change in temperature. *No Permission Required.*

To produce truly quantitative results, both biological and technical replicates are required. In our experiment, biological replicates would consist of RNA samples from independently induced bacterial cultures. Technical replicates would consist of multiple PCR reactions, each with the same cDNA template. However, so that the entire class's samples will fit in one thermal cycler, we will not perform replicate experiments.

The main disadvantage of RT-qPCR is expense, because it requires a thermal cycler capable of detecting fluorescence. However, if a capable thermal cycler is available to use, it has many advantages over semiquantitative RT-PCR in gene expression studies. It is a relatively short procedure, since no gel electrophoresis is required. Use of a sensitive fluorescent dye (much more so than ethidium bromide or GelRed) enables detection of genes expressed even at extremely low levels. In addition, RT-qPCR has a broader dynamic range of detection (the range of input RNA that the assay can detect). These favorable qualities make RT-qPCR a powerful method in gene expression studies, and it can be used in validating microarray and transcriptomics results. Beyond the basic research lab, RT-qPCR and qPCR are also used in clinical and industrial settings. Determining viral load, detecting pathogens, genotyping, determining drug therapy efficacy, and quality control can all be achieved with these technique.

For additional resources on qPCR, please visit the following website: https://www.thermofisher.com/us/en/home/life-science/pcr/real-time-pcr/real-time-pcr-learning-center.html[1]

II. LABORATORY EXERCISES

A. Reverse Transcription

The following protocol has been adapted from Bio-Rad's iScript Select cDNA Synthesis Kit. Note that we are setting up the reactions in PCR tubes. However, this step is an RT reaction, not a PCR. We are simply using a thermal cycler as a programmed heat block to conveniently carry out the required timed temperature steps of the RT in appropriate tubes. Make sure to continue to use RNase-free tips and tubes and use the strategies presented in Part IV "Challenges of Working with RNA" to minimize RNA degradation. Wearing gloves at all times is essential. In addition, use your best pipetting techniques. Realize that when a protocol calls for 5 μL, a P10 should actually be set to 5.00 μL. Likewise, a P20 may show 05.0 μL in the window. Always set the micropipette as close as possible to the desired volume, accounting for all digits shown.

1. Remove your three RNA samples from the −80°C freezer and immediately place them on ice to thaw.
2. Label six RNase-free PCR tubes with: −no, −IPTG, −lac, +no, +IPTG, and +lac.
3. Calculate the volume of RNA needed for 1 μg based on your quantification from last week. Use the equation:

TABLE 14.1 Total RNA Volumes Needed for Reverse Transcription

RNA Sample	Volume Needed for 1 μg	Volume of Nuclease-Free Water to Bring Total to 13 μL
No inducer		
IPTG		
Lactose		

No Permission Required.

TABLE 14.2 Reverse Transcription Master Mixes

Reagent	For 1 RT Reaction (μL)	+RT Master Mix (μL)	−RT Master Mix (μL)
5× iScript select reaction mix	4	16	16
Random primer	2	8	8
iScript reverse transcriptase	1	4	0
Nuclease-free water	0	0	4
Total	7	28	28

No Permission Required.

$$\text{Volume needed } (\mu\text{L}) = \frac{\text{Amount desired } (\mu\text{g})}{\text{Concentration } (\mu\text{g}/\mu\text{L})}$$

Do this for each of the three samples and record the volumes in your notebook in a table similar to Table 14.1.

4. Pipette this volume into the appropriate PCR tube on ice. Each RNA sample will get pipetted twice: once into a "–" tube and once into a "+" tube. Change tips between each of your three samples.
5. Calculate the amount of nuclease-free water for each sample to bring the total volume to 13 μL. This volume may be different for each of the three samples. Record in your notebook. Pipette these volumes of nuclease-free water into the correct tubes.
6. To ensure that multiple samples are treated equally with each reagent and for ease of pipetting, you will make two separate master mixes, as described in Table 14.2. One master mix will contain reverse transcriptase (+RT) and the second master mix will just contain nuclease-free water (−RT). Before proceeding, calculate the necessary volumes of each component to complete a similar table (Table 14.2) in your lab notebook. Then add the described components, in the order shown, to a microcentrifuge tube on ice.
7. Vortex to mix, perform a quick (2–3 seconds) centrifugation to ensure all material is in the bottom of the tube, and keep on ice.
8. Pipette 7 μL of the correct master mix into each of the three PCR tubes on ice. Pipette up and down to mix. IMPORTANT: Change pipette tips before taking from the master mix again.
9. Keep reactions on ice until all groups are done.
10. Place PCR tubes in a thermal cycler programmed at:
    a. 25°C 5 minutes
    b. 42°C 30 minutes
    c. 85°C 5 minutes
    d. 4°C ∞.
11. Once 4°C is reached, stop the thermal cycler and place your six PCR tubes on ice.

B. Quantitative PCR

The cDNA that you synthesized will now be used as a template in qPCR reactions. You will amplify *gst::egfp* cDNA (from the mRNA that was produced from your expression vector) in each of your samples. The primers specifically anneal to the *egfp* sequence. You will also amplify *E. coli 23S* rRNA (a gene found on the chromosome of all *E. coli* cells). This amplicon will serve as a reference gene (an internal control for different amounts of input RNA in each reaction). Next week, you will quantify the level of *gst::egfp* cDNA (and by inference, mRNA) in your positive clone after no induction, induction with IPTG, or induction with lactose.

Note to instructor: A thermal cycler with 96 wells will hold PCR reactions for up to eight pairs of students (12 reactions per pair). If you have more than eight pairs of students, you will need to either: use a second thermal cycler, keep some reactions covered in foil and in the refrigerator until the first run is complete and then perform a second run, or eliminate reactions in order for one thermal cycler to accommodate all the reactions. Preferably, eliminate the lactose-induced samples to leave eight reactions instead of 12. The thermal cycler will then accommodate up to 12 groups. Since both gst::egfp and 23S need to be amplified for quantification purposes and −RT reactions ensure contaminating DNA is not contributing to the quantification of mRNA, it is important to include these if at all possible.

The primers used to specifically amplify *gst::egfp* from your positive clone are:

- egfp14 F: 5′-CGACGGCAACTACAAGACC-3′
- egfp14 R: 5′-GTCCTCCTTGAAGTCGATGC -3′

The primers used to specifically amplify *23S* rRNA from your positive clone are:

- 23S14 F: 5′-GACGGAGAAGGCTATGTTGG-3′
- 23S14 R: 5′-GTTGCTTCAGCACCGTAGTG-3′

This protocol has been adapted from Bio-Rad's iQ SYBR Green Supermix instructions:

1. The cDNA that you just synthesized will now be used as a template in a qPCR assay. An extremely small amount of cDNA is needed. However, you do not directly quantify the cDNA produced. Rather, you assume that the same mass amount of RNA produces the same mass amount of cDNA during the RT reaction. Since you do not directly quantify cDNA, you can state the equivalent mass amount of RNA. For these qPCR reactions, you want to use the equivalent of 10 pg of RNA template, so you will need to dilute each cDNA sample as shown in Table 14.3. Using a P10, pipette the indicated RT reaction volume into a new microcentrifuge tube labeled "cDNA dilution 1." Change tips to add sterile water. Vortex to mix well. Use a new tip to pipette into a new microcentrifuge tube labeled "cDNA dilution 2." Change tips to add sterile water. Vortex to mix well. Note that the RNA has been converted to more stable cDNA, so it is not necessary to use RNase-free water. Autoclaved, deionized water is fine.
2. Obtain 12 real-time compatible (optical) PCR tubes and label them 1–12 on the side of the tube if they are not already printed with numbers. Do not label the lids. It is a good idea to record in which wells of the thermal cycler your tubes are placed, in case the ink wears off the sides. Also label the side of each tube with your initials. Alternatively, your instructor may give you a multi-well plate to share with other groups. Place these samples in a PCR tube rack on ice.
3. Label two microcentrifuge tubes "*egfp* master mix" and "*23S* master mix." To ensure that multiple samples are treated equally with each reagent and for ease of pipetting, you will make a master mix of each of the components listed in Table 14.4. Each master mix will contain everything necessary for the PCR reaction except for the cDNA. One master mix will contain the primers specific for *egfp*. The other master mix will contain the primers specific for *23S* rRNA. *Note to instructor: Table 14.4 shows the volumes needed for six PCR reactions with each primer pair (three +RT and three −RT reactions), including an additional 20 μL volume to account for pipetting error. If you decided to use a larger or smaller sample number, you will need to adjust the volumes accordingly.*

TABLE 14.3 cDNA Dilutions

	Reverse Transcription Reaction	Sterile dH_2O
cDNA dilution 1 (1:500)	2 μL original reaction	998 μL
cDNA dilution 2 (1:25,000)	2 μL dilution 1	98 μL

No Permission Required.

TABLE 14.4 qPCR Master Mixes

Reagent	For 1 PCR Reaction (μL)	*egfp* Master Mix (μL)	*23S* Master Mix (μL)
Sterile dH_2O	6.75	47.25	47.25
Primer mix (10 μM each F and R primer)	0.75	5.25 μL *egfp* primer mix	5.25 μL *23S* primer mix
iQ SYBR Green Supermix	12.5	87.5	87.5
Total	20	140	140

No Permission Required.

TABLE 14.5 qPCR Setup With*egfp* Master Mix

Tube	1	2	3	4	5	6
Master mix (20 μL)	*egfp*	*egfp*	*egfp*	*egfp*	*egfp*	*egfp*
cDNA (5 μL)	−no	+no	−IPTG	+IPTG	−lac	+lac

No Permission Required.

TABLE 14.6 qPCR Setup With*23S* Master Mix

Tube	7	8	9	10	11	12
Master mix (20 μL)	*23S*	*23S*	*23S*	*23S*	*23S*	*23S*
cDNA (5 μL)	−no	+no	−IPTG	+IPTG	−lac	+lac

No Permission Required.

TABLE 14.7 qPCR Program

Cycle	Repeats	Step	Dwell Time	Set Point (°C)
1	1	1	3 min	95
2	40	1	15 s	95
		2[a]	30 s	60

[a]Note to instructor: Make sure that the thermal cycler is programmed to collect real-time data during Cycle 2, Step 2. Including a melt curve after the program is optional. The instrument will have a pre-programmed step for a melt curve that can be added.
No Permission Required.

4. Pipette each reagent in the order shown into a microcentrifuge tube on ice.
5. Vortex well to mix and return to ice.
6. Pipette as accurately as possible 20 μL of the *egfp* master mix into each PCR tube labeled 1–6. Change tips before switching to the next master mix. Pipette 20 μL of the *23S* master mix into each PCR tube labeled 7–12. Avoid creating bubbles that may interfere with the fluorescence reading.
7. Pipette as accurately as possible 5 μL of the appropriate diluted cDNA into the tubes as shown in Table 14.5 and Table 14.6. Be sure to take from the tube labeled "cDNA dilution 2." Also, change tips between samples. Avoid creating bubbles that may interfere with the fluorescence reading.
8. Keep PCR tubes on ice until all groups are done.
9. Your instructor will perform a quick centrifugation of all the samples or finger-flick to remove bubbles. Immediately place the tubes in a thermal cycler that has been programmed as shown in Table 14.7.

10. The program will take approximately 1.5 hours to complete. The data collected during the course of the experiment will be saved by your instructor, and you will analyze it during your next lab session.

REFERENCE

1. Thermo Fisher. Real-time PCR (qPCR) Learning Center.

DISCUSSION QUESTIONS

1. If you had primed reverse transcription of your *E. coli* RNA with an oligo-dT primer rather than a random hexamer, how would that affect the experiment?
2. You are amplifying both *gst::egfp* mRNA and *23S* rRNA in the PCR steps. These are two different types of RNA (messenger and ribosomal, respectively). Could an *E. coli* mRNA rather than an rRNA be amplified as a reference gene? Why? What are the main criteria for selecting a reference gene?

Lab Session 15

Analysis of *gst::egfp* mRNA Levels by RT-qPCR: Part II

Goal: You will analyze RT-qPCR data obtained in the previous session to determine the relative levels of *gst::egfp* mRNA in *Escherichia coli* induced with IPTG and lactose compared to those with no inducer.

I. INTRODUCTION

During the qPCR run, fluorescence readings were recorded during the annealing/extension step of every sample at every cycle (in "real-time"). By the end of the run, a complete amplification plot (fluorescence readings versus cycle number) was generated for each sample. Now, you must analyze these plots in order to quantify gene expression levels. Before these types of quantification methods are introduced, we will first take a closer look at the results generated in this type of experiment.

Recall from Lab Session 3 that PCR amplifies DNA exponentially, with the amount of DNA doubling each cycle. Depending on the amount of starting template DNA, it will take a number of cycles to produce detectable fluorescence. Quantification cannot take place this early in the experiment. Later, when reagents in the PCR become limiting, the amplification reaches a plateau; therefore, you cannot accurately quantify cDNA levels in this plateau phase. Accurate quantification can only occur during exponential amplification. It is easy to obtain data from the exponential phase using qPCR because data was collected at every cycle, not just at the end.

Typically, each sample's amplification plot is shown in a different color. Fluorescence versus cycle number can be displayed in either the linear view or the log view (Figure 15.1). In either case, you can clearly see that fluorescence intensity (indicating DNA abundance) increases as cycle number increases. An important concept to keep in mind is that the more input cDNA that went into the PCR, the sooner an increase in fluorescence will occur. In Figure 15.1A (linear view), all samples produce minimal fluorescence during the first 13 cycles. These cycles are considered the baseline. Then, the blue amplification plot begins to increase significantly, followed by the red and green plots at later cycle numbers. Since the blue plot showed an increase in fluorescence at the earliest cycle number, we can conclude that the sample it represents contains more input cDNA (in this case, *23S*) than the other samples. To assign numeric values to these observations, a threshold is set, intersecting all the plots during exponential amplification. The threshold in Figure 15.1 is indicated by an orange horizontal line. Notice that in the linear view, each plot crosses the threshold in the early cycle numbers. Keeping all settings the same but viewing the data on a logarithmic scale (Figure 15.1B), we see that the threshold is high enough to be above background noise, but low enough to stay out of the plateau phase. The cycle at which a plot crosses the threshold is called the threshold cycle (C_T). The lower the C_T, the more abundant the template is. During exponential amplification, DNA should in theory double with each cycle. So, after two cycles, there will be four times as much as you started with, and after about 3.3 cycles, there will be 10 times as much. Even though the C_Ts of the amplicons shown in Figure 15.1B differ by only about three cycles, these represent 10-fold differences in *23S* amplicon abundance. If the same reactions were performed using conventional PCR, we may have analyzed amplicon abundance after 30 or more cycles on an agarose gel. Based on the qPCR results shown (especially in the log view), it becomes clear that by this point in the reaction, the plateau phase has been reached and the reagents have become limiting. We would conclude that there is little difference in *23S* rRNA levels between samples. However, by examining amplicons early in the reaction, in the exponential phase, we are able to determine that there is actually a 10-fold difference between each sample.

There are two general methods for reporting quantification: absolute and relative. Absolute quantification calculates the quantity of a target gene in your unknown samples by interpolation from an absolute standard curve. This curve is commonly made by *in vitro* transcribed RNA or plasmid DNA with known copies of a target gene. This method is useful when you want to determine the exact copy number of a gene. The second type of quantification is called relative quantification. This aims to

Molecular Biology Techniques. DOI: https://doi.org/10.1016/B978-0-443-26714-7.00003-X

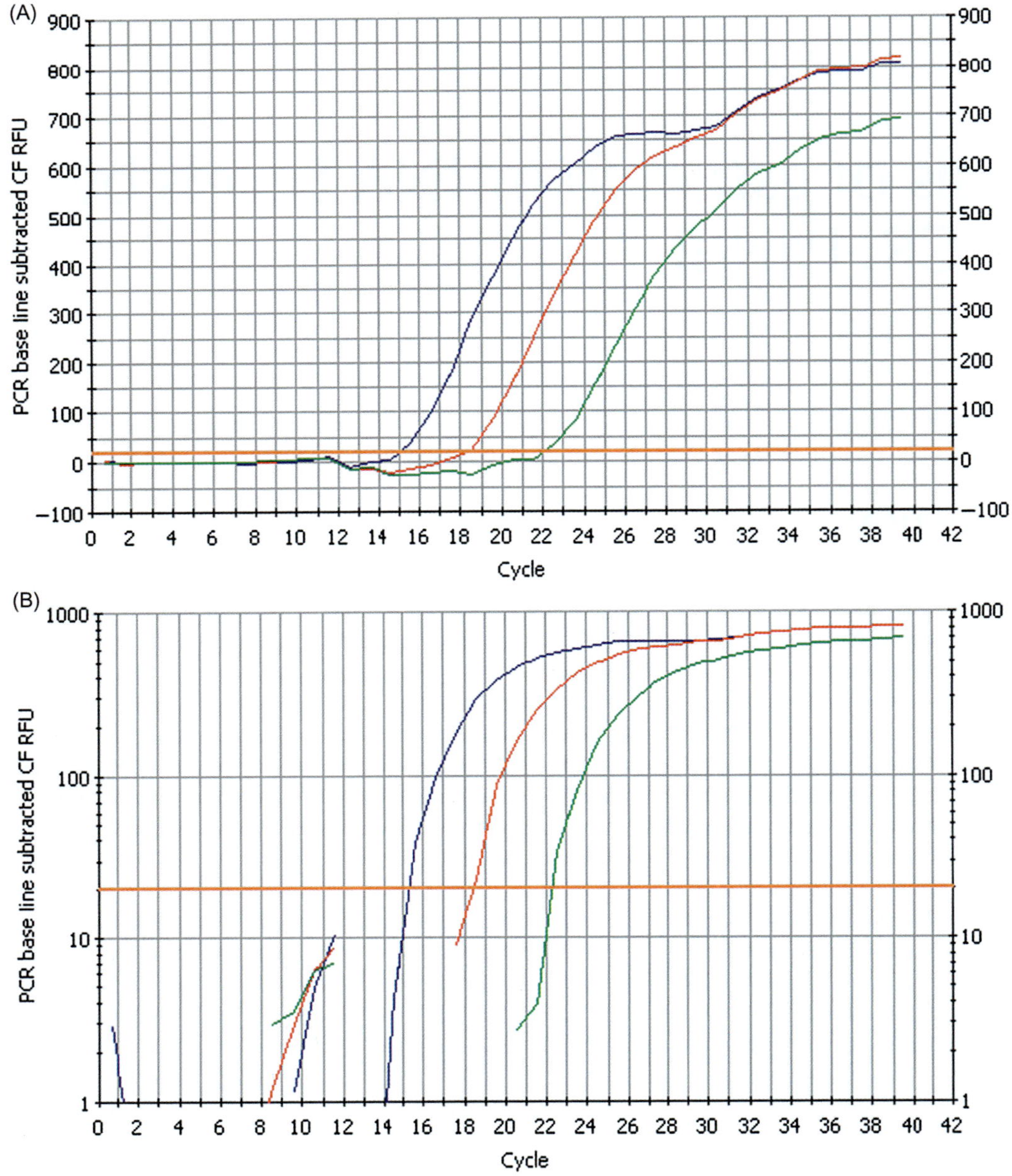

FIGURE 15.1 Amplification plots. Ten-fold serial dilutions of cDNA were used as templates for RT-qPCR reactions using 23S primers. The horizontal orange line represents the threshold. The x-axis displays cycle number and the y-axis displays relative fluorescence units (RFU). (A) Linear view. (B) Log view. *No Permission Required.*

determine the difference in expression of a target gene between two samples. Reporting relative quantities is often sufficient when studying changes in gene expression. Basically, it aims to answer the question "How much of gene X mRNA is present in sample 1 compared to sample 2?" Relative quantification reports *n*-fold changes of gene expression rather than exact quantities: "There is fivefold more of gene X mRNA in sample 1 versus sample 2." In this example, sample 2 serves as a calibrator (the sample used for the basis of a comparison). In our experiments, we will be comparing *gst::egfp* levels between no induction, induction with IPTG, and induction with lactose. Therefore, samples with no induction will serve as calibrators, and *gst::egfp* levels of samples induced with IPTG or lactose will be expressed relative to those with no induction.

Two methods exist for relative quantification: the relative standard curve method and the comparative C_T method (also called the $\Delta\Delta C_T$ method). The comparative C_T method is what we will use in this lab session. It has the advantage that no standard curve is required. Instead, mathematical formulas are used to calculate relative gene expression. It should be noted that the comparative C_T method assumes PCR amplification efficiencies are extremely similar between amplification of the test and reference genes. Not all reactions will have 100% amplification efficiency, where the DNA amount exactly doubles each cycle. Most reactions will be slightly lower or higher than 100%, and this can ultimately affect quantification. A validation experiment was performed previously to confirm that the primers have similar amplification efficiencies.

In the last experiment, you amplified not only *gst::egfp* (our test gene) but also *23S* rRNA (our reference gene). This reference gene is needed for the comparative C_T method. We will be comparing C_T values of *gst::egfp* reactions to *23S* rRNA reactions for normalization purposes. Then, these differences (ΔC_T) will be compared to the calibrator, which is no induction (two comparisons, so $\Delta\Delta C_T$). In this way, *gst::egfp* levels will be 1× in the no induction control and *gst::egfp* levels will have an *n*-fold difference with IPTG and lactose induction.

STEP 1: NORMALIZE TO AN ENDOGENOUS REFERENCE

$$\Delta C_T = C_{T\text{egfp}} - C_{T23\text{S}}$$

For each cDNA in our experiment (no induction, IPTG, and lactose), you will be taking the C_T of the *egfp* reaction minus the C_T of the *23S* reaction to get a ΔC_T.

STEP 2: COMPARE TO A CALIBRATOR

$$\Delta\Delta C_T = \Delta C_{T\text{ IPTG}} - \Delta C_{T\text{ no}}$$

Once a ΔC_T value is calculated for no induction, IPTG, and lactose cDNAs, you will report each value relative to the calibrator (no induction) to get a $\Delta\Delta C_T$. Notice that $\Delta\Delta C_T$ will always be zero for the calibrator sample itself.

STEP 3: CALCULATE RELATIVE QUANTITY

$$2^{-\Delta\Delta C_T}$$

Finally, the amount of the test gene (*gst::egfp* with IPTG or lactose), which is now normalized to an endogenous reference (*23s* with IPTG or lactose) and relative to a calibrator (no induction), is calculated to provide us with the relative quantity of *gst::egfp* mRNA.

With these values in hand, you can draw conclusions as to how much *gst::egfp* mRNA was expressed with IPTG and lactose in your positive clone. The references at the end of this chapter provide detailed guides on performing and analyzing RT-qPCR experiments.[1,2]

II. LABORATORY EXERCISE

Note: The following instructions are specific for analyzing data obtained on the Bio-Rad iCycler using MyiQ Software. If you are using equipment made by another manufacturer, your instructor will provide you with data analysis instructions. Your instructor may perform Steps 1–17 for you.

1. Open MyiQ software on the computer connected to the iCycler.
2. Click the "view post-run data" tab.
3. Double-click the data file (.odm) from your lab session to open it.
4. Click the "select wells" button and select the positions of your group's reactions. If your group had 12 reactions occupying the same row, you can click on the row's label (A, B, etc.) to automatically select that entire row. Click the "analyze selected wells" button.
5. Make sure that the "log view" button is displayed. That means that the current view is the normal view.
6. Verify that the amplification plots are running along 0 relative fluorescence units (RFU) (parallel to the *x*-axis) in the early cycles of the experiment. If the plots dip well below 0 or if RFU values vary greatly in the early cycles, you will need to adjust the baseline setting. To do this, determine the cycle at which your first (left-most) plot begins to rise above 0 RFU (amplification). Set the baseline cycles to between two and this cycle number.
7. Click the "log view" button.
8. Verify the auto-calculated threshold position (orange line) is high enough on the *y*-axis to be above any background noise (it should not intersect any sharp peaks seen in the early cycles) but is low enough to intersect the plots in the bottom third of the exponential phase.

9. If any adjustments are made in Steps 6 and/or 8, you will need to click the "recalculate threshold cycles" button.
10. Click the "Reports" button.
11. In the "Select Report" drop-down menu, select "PCR baseline."
12. Click "save to file" and save the report to a thumb drive.
13. Click the "print" button to print a copy for your lab notebook.
14. Open Microsoft Word. Open the report you saved (a Rich Text Format file).
15. Copy the well number and C_T values for each of your reactions.
16. Open Microsoft Excel.
17. Create a new spreadsheet with column and row headings shown in Figure 15.2. *Paste your C_T values from +RT reactions only* in the appropriate cells.
18. Enter the formulas shown in Figure 15.2 to calculate ΔC_T, $\Delta\Delta C_T$, and $2^{-\Delta\Delta C_T}$ values.
19. The last column (relative quantity) now shows the level of *gst::egfp* in each sample, relative to no inducer
20. Below the data you have entered, create the column and row headings shown in Figure 15.3. *Paste your C_T values for the six −RT reactions and six +RT reactions.*
21. For each RNA sample (no inducer, IPTG, lactose), locate the C_T values corresponding to its −RT and +RT reactions for each primer pair (you will analyze *gst::egfp* first). Remember that these −RT reactions were included to control for DNA contamination.
22. Amplification in the −RT reactions with the *egfp* primers will indicate plasmid DNA contamination. To calculate the amount of plasmid DNA that is attributable to amplification in your +RT samples, perform the ΔC_T calculation for each *gst::egfp* reaction. This time, you will use the +RT sample as a calibrator:

$$\Delta C_T = C_{T\text{ without RT}} - C_{T\text{ with RT}}$$

You are not interested in performing a second comparison between the C_T in one RNA sample relative to another in this step, so you will not use $\Delta\Delta C_T$ values. You stop at one comparison (ΔC_T). Finally, the equation $2^{-\Delta C_T}$ will calculate the quantity (in percent) of contaminating plasmid DNA. Multiply this value by 100 to obtain a percentage.

$$\text{DNA contamination}_{\text{IPTG}}(\%) = 2^{-[(C_T\text{ without RT})-(C_T\text{ with RT})]} \times 100$$

23. Amplification with the *23S* primers will indicate genomic DNA contamination. Perform the calculations described in Step 22 for each *23S* reaction. Refer to Figure 15.3.
24. To summarize, these results should be collected for your notebook:

	A	B	C	D	E	F
1		*egfp*	*23S*	CT egfp - CT 23S	ΔCT inducer - ΔCT no inducer	Relative Quantity (Fold Change)
2		CT	CT	ΔCT	ΔΔCT	2^-ΔΔCT
3	No inducer			=B3-C3	=D3-D3	=2^-E3
4	IPTG			=B4-C4	=D4-D3	=2^-E4
5	Lactose			=B5-C5	=D5-D3	=2^-E5

FIGURE 15.2 Excel example. Screenshot shows the layout and formulas used in the comparative C_T method to calculate relative quantity. *No Permission Required.*

	A	B	C	D	E	F
1		*egfp*	*23S*	CT egfp - CT 23S	ΔCT inducer - ΔCT no inducer	Relative Quantity (Fold Change)
2		CT	CT	ΔCT	ΔΔCT	2^-ΔΔCT
3	No inducer			=B3-C3	=D3-D3	=2^-E3
4	IPTG			=B4-C4	=D4-D3	=2^-E4
5	Lactose			=B5-C5	=D5-D3	=2^-E5
6						
7		*egfp*		CT without RT - CT with RT	Quantity of contaminating plasmid DNA (%)	
8		without RT	with RT	ΔCT	2^-ΔCT	
9	No inducer			=B9-C9	=(2^-D9)*100	
10	IPTG			=B10-C10	=(2^-D10)*100	
11	Lactose			=B11-C11	=(2^-D11)*100	
12						
13		*23S*		CT without RT - CT with RT	Quantity of contaminating genomic DNA (%)	
14		without RT	with RT	ΔCT	2^-ΔCT	
15	No inducer			=B15-C15	=(2^-D15)*100	
16	IPTG			=B16-C16	=(2^-D16)*100	
17	Lactose			=B17-C17	=(2^-D17)*100	

FIGURE 15.3 Excel example. Screenshot shows the layout and formulas used to calculate the percentage of DNA contamination. *No Permission Required.*

a. Relative quantity of *gst::egfp* mRNA in uninduced, IPTG-induced, and lactose-induced samples. This should be expressed as a fold-change (Steps 17–19).
b. Plasmid DNA contamination for each sample (Step 22)
c. Genomic DNA contamination for each sample (Step 23)

REFERENCES

1. Bio-Rad Real-Time PCR Applications Guide. 2006.
2. T.D. Schmittgen, K.J. Livak, Analyzing real-time PCR data by the comparative *CT* method, Nat Protoc 3 (2008) 1101–1108.

DISCUSSION QUESTIONS

1. Why do standard PCR programs often include 30 cycles or more, while qPCR amplification plots (like the one in Figure 15.1B) show that at these later cycles, the plateau phase is often reached?
2. You were able to determine the levels of *gst::egfp* mRNA in the induced cultures relative to the non-induced. Without an inducing molecule to derepress the lac operator of pET-41a, was any *gst::egfp* mRNA transcribed? Why?

Lab Session 16

Analysis of *gst::egfp* mRNA Levels by Semiquantitative RT-PCR: Part I

Goal: Total RNA that you purified will be converted to cDNA by a reverse transcription (RT) reaction. That cDNA will then be used as a template for semiquantitative polymerase chain reaction (PCR). You will amplify DNA with primers specific to *egfp* (your test gene) and *23S* rRNA (your reference gene).

I. INTRODUCTION

Read the Introduction in Lab Session 14 for information on RT. Semiquantitative RT-PCR is, as its name would suggest, less quantitative than RT-quantitative PCR (qPCR) (see Lab Sessions 14 and 15), but can be used to detect changes in gene expression if they are approximately 10-fold or more. More subtle changes may be detected; however, the magnitude may be dampened due to the limited sensitivity of the assay. RT-PCR is considered an "end-point" method because you examine cDNA levels, and by inference mRNA levels, at the end of the PCR steps by agarose gel electrophoresis. You do not monitor cDNA levels throughout the PCR as you do in RT-qPCR. Recall from Lab Session 3 that PCR amplifies DNA exponentially, with the amount of DNA doubling each cycle. Depending on the amount of starting template DNA, it will take a number of cycles to produce detectable fluorescence in qPCR. Quantification cannot take place this early in the experiment. Later, when reagents in the PCR become limiting, the amplification reaches a plateau; therefore, you cannot accurately quantify cDNA levels in this plateau phase. Accurate quantification can only occur during exponential amplification. Therefore, it is important to determine the cycles included in the exponential phase when using RT-PCR to quantify gene expression. In practice, the exponential phase should be determined with each new cDNA sample with each primer pair before performing the actual experiment. This is most critical when trying to detect small-fold changes, but in our case, we are examining large-fold changes. Furthermore, a range of cycles will generally fall within the exponential phase, so for the following experiment, it has been predetermined. There are several other reasons why semiquantitative RT-PCR is considered semiquantitative. First, detection of amplified cDNAs is based on the intensity of bands on an agarose gel. Ethidium bromide or fluorescent dyes like GelRed are reagents used to stain DNA on a gel; however, these are not as sensitive as measuring fluorescence with a real-time capable thermal cycler. Additionally, comparing intensities of different PCR products is not quantitative since a smaller product is typically amplified with higher efficiency than a larger product. To control for this limitation, the regions that you amplify for both genes (*egfp* and *23S* rRNA) will be the same length. Additionally, the amount of dye that is available to bind DNA will be kept constant by running all samples on the same gel. You should always avoid comparing band intensity from bands that are present on different gels. Finally, semiquantitative RT-PCR relies on several successive steps being performed carefully in order to prevent imprecise results from sample to sample. This method, as well as RT-qPCR, assumes that the RT efficiency will be the same in each sample. In other words, if you pipette 1 μg of total RNA into each tube, the same amount of cDNA should be produced in each tube. Therefore, loading equivalent amounts of total RNA into each RT reaction is vital. To limit the possibility that RT efficiencies vary from sample to sample, a master mix of all reagents except for RNA will be aliquoted to ensure the equal treatment of each RNA. After this step, you must accurately pipette equivalent volumes of cDNA template into each PCR reaction. Again, we will utilize a master mix. Semiquantitative RT-PCR includes a final step of agarose gel electrophoresis. You can imagine that if you underload a sample into the well, it will alter your quantification because this lane will have a lower intensity than if you had loaded it correctly. The *23S* products will serve as our loading control. Above all, sources of error can be minimized by careful pipetting. The major advantage of semiquantitative RT-PCR over RT-qPCR is that it does not require a special thermal cycler capable of detecting fluorescence (which can be quite expensive).

Molecular Biology Techniques. DOI: https://doi.org/10.1016/B978-0-443-26714-7.00048-X

II. LABORATORY EXERCISES

A. Reverse Transcription

Please follow the RT protocol in Lab Session 14a.

B. Semiquantitative Polymerase Chain Reaction

Note to instructor: A thermal cycler with 96 wells will hold PCR reactions for eight pairs of students (12 reactions per pair). If you have more than eight pairs of students, you will need to either use a second thermal cycler or eliminate reactions in order for one thermal cycler to accommodate all the reactions. Preferably, eliminate the lactose-induced samples to leave eight reactions instead of 12. The thermal cycler will then accommodate up to 12 pairs of students. Since both egfp and 23S need to be amplified for quantification and −RT reactions ensure contaminating DNA is not contributing to the quantification of mRNA, it is important to include these if at all possible.

The primers used to specifically amplify *gst::egfp* from your positive clone are:

- egfp 16F 5′-CCTGAAGTTCATCTGCACCA-3′
- egfp 16R 5′-TGCTCAGGTAGTGGTTGTCG-3′

The primers used to specifically amplify *23S* rRNA from your positive clone are:

- 23S 16F 5′-ACTGCGAATACCGGAGAATG-3′
- 23S 16R 5′-CCTGTTTCCCATCGACTACG-3′

The forward and reverse primers will be included together as a mix. Each primer pair should amplify a product 479 bp long.

1. The cDNA that you synthesized will now be used as a template in a semiquantitative PCR assay. An extremely small volume of cDNA is needed (0.5 µL for a 25 µL reaction), so you will dilute each sample 1:5 in order to pipette a larger volume. Add 80 µL of water to each cDNA reaction tube (total volume is now 100 µL). Vortex to mix well. Repeat (with a fresh tip) for each cDNA reaction.
2. Obtain 12 PCR tubes and label them 1–12 on both the cap and side of the tube. Also label each tube with your initials. Place these in a PCR tube rack on ice.
3. Label two microcentrifuge tubes "*egfp* master mix" and "*23S* master mix." To ensure that multiple samples are treated equally with each reagent and for ease of pipetting, you will make a master mix consisting of each of the components listed in Table 16.1. Each master mix will contain everything necessary for the PCR reaction except for the cDNA. One master mix will contain the primers specific for *egfp*. The second master mix will contain the primers specific for *23S* rRNA. *Note to instructor:* Table 16.1 *shows the volumes needed for six PCR reactions with each primer pair (three +RT and three −RT reactions), including an additional 22.5 µL volume to account for pipetting error. If you decided to use a larger or smaller sample number, you will need to adjust volumes accordingly.*
4. Pipette each reagent, in the order shown, into microcentrifuge tubes on ice.
5. Vortex well to mix and return to ice.

TABLE 16.1 PCR Master Mixes

Reagent	For 1 PCR Reaction (µL)	*egfp* Master Mix (µL)	*23S* Master Mix (µL)
1. Sterile dH_2O	18.8	131.6	131.6
2. 10× Taq buffer	2.5	17.5	17.5
3. dNTP mix (10 mM each dNTP)	0.5	3.5	3.5
4. Primer mix (10 µM each F and R primer)	0.5	3.5 µL *egfp* primer mix	3.5 µL *23S* primer mix
5. Taq polymerase (5 U/µL)	0.2	1.4	1.4
Total	22.5	157.5	157.5

TABLE 16.2 PCR Setup With *egfp* Master Mix

Tube	1	2	3	4	5	6
1. Master mix (22.5 µL)	*egfp*	*egfp*	*egfp*	*egfp*	*egfp*	*egfp*
2. cDNA (2.5 µL)	−no	+no	−IPTG	+IPTG	−lac	+lac

TABLE 16.3 PCR Setup With *23S* Master Mix

Tube	7	8	9	10	11	12
1. Master mix (22.5 µL)	*23S*	*23S*	*23S*	*23S*	*23S*	*23S*
2. cDNA (2.5 µL)	−no	+no	−IPTG	+IPTG	−lac	+lac

TABLE 16.4 PCR Program

Cycle	Repeats	Step	Dwell Time	Set Point (°C)
1	1	1	2 min	95
2	26	1	30 s	95
		2	30 s	60
		3	45 s	72
3	1	1	10 min	72
4	1	1	∞	4

6. Pipette as accurately as possible 22.5 µL of the *egfp* master mix into each PCR tube labeled 1–6. Change pipette tips in between master mixes. Pipette 22.5 µL of the *23S* master mix into each PCR tube labeled 7–12.
7. Pipette as accurately as possible 2.5 µL of the appropriate diluted cDNA into the tubes as shown in Table 16.2 and Table 16.3. Change pipette tips between each tube.
8. Keep the tubes on ice until all groups are complete.
9. Place the tubes in a thermal cycler that has been programmed as shown in Table 16.4.
10. The program will take approximately 1.5 hours to complete. Your instructor will remove the tubes once the program is finished and store them at −20°C until your next lab session.

DISCUSSION QUESTIONS

1. If you primed RT of *E. coli* RNA with an oligo-dT primer rather than a random hexamer, how would that affect the experiment? Why?
2. You are amplifying both *gst::egfp* mRNA and *23S* rRNA in the PCR steps. These are two different types of RNA (messenger and ribosomal, respectively). Could an *E. coli* mRNA rather than an rRNA be amplified as a reference gene? Why? What is the main criterion for selecting a reference gene?

Lab Session 17

Analysis of *gst::egfp* mRNA Levels by Semiquantitative RT-PCR: Part II

Goal: You will visualize semiquantitative RT-PCR reactions by agarose gel electrophoresis. Imaging software (if available) will be used to quantify the levels of *gst::egfp* normalized to *23S* rRNA. In this way, you will be able to compare the relative levels of induced *gst::egfp* mRNA in *E. coli* treated with IPTG and lactose.

I. INTRODUCTION

Review Lab Session 16 Introduction.

II. LABORATORY EXERCISES

A. Agarose Gel Electrophoresis

1. Pour a 1% agarose gel that includes GelRed, as described in Lab Session 3. If 12 PCR reactions were performed, you will need two rows of wells (eight wells per row) on your mini gel in order for all reactions and a molecular weight marker to fit. To do this, place one comb at the top of the gel casting tray as usual and an additional comb in the middle of the gel casting tray. Alternatively, you could run all samples on a larger gel that will accommodate at least 13 wells across.
2. Obtain your PCR samples from the last lab session and allow them to thaw at room temperature.
3. Add the appropriate amount of DNA loading buffer directly into each sample. Flick to mix.
4. Add an appropriate amount of 1 kb ladder to the first well in each row (see Table 17.1).
5. Load 5 μL of each sample in the order shown in Table 17.1. Numbers indicate the label on the PCR tube (1–6 are *egfp* reactions, 7–12 are *23S* reactions).
6. Close the apparatus and connect the leads to the voltage pack. Run the gel at approximately 85 V or at an appropriate voltage for your apparatus (10 V/cm). When the top row's bromophenol blue dye-front has migrated one quarter of the way down the gel (halfway to the second row of wells), turn the power off and place the gel on a UV transilluminator. Photograph your gel and save it before proceeding.

B. Quantification

Note to instructor: If imaging software is not available, instruct students to visually compare band intensities of the gst::egfp PCR products (+RT) between the following samples: no inducer, IPTG, and lactose. Also, have them observe and take into account any differences in band intensities of the 23S rRNA PCR product. The presence and intensity of products in (−RT) lanes should also be recorded. The following directions are written for ImageJ downloaded to a personal computer. Slight variations may exist for other operating systems and versions of ImageJ.

1. Download and open NIH ImageJ. ImageJ is freely available for a variety of different operating systems from the National Institutes of Health. System requirements and downloads are available here: (https://imagej.net/).
2. From file menu, select "Open," and choose your agarose gel image.
3. You will want to ensure your gel is completely straight. If not, from the Image menu select "Open," "Rotate," and "Arbitrarily." A new Rotate window will appear. Check the "Preview" box and adjust the angle of your gel image until it appears straight. Select "OK" when finished.
4. From the ImageJ toolbar at the top, select the "Rectangle" tool.
5. Using the rectangle tool and your mouse, carefully outline your gel image.

Molecular Biology Techniques. DOI: https://doi.org/10.1016/B978-0-443-26714-7.00040-5

TABLE 17.1 Agarose Gel Loading Order

Top row	1 kb ladder	1	2	3	4	5	6	Empty
Bottom row	1 kb ladder	7	8	9	10	11	12	Empty

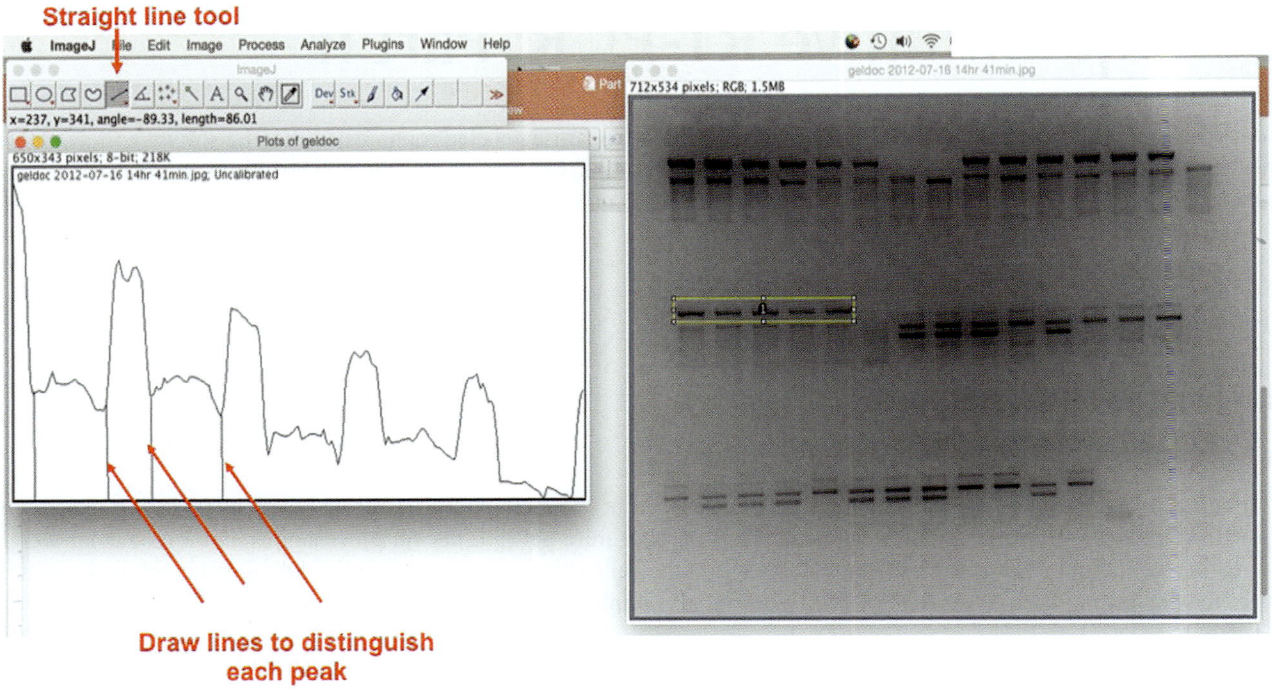

FIGURE 17.1 An example of a DNA agarose gel quantified using NIH ImageJ. Note placement of the straight line tool and results window illustrating band peak intensity. Note that this image does not correspond to the actual experiment you performed.

6. Once you have outlined your gel, under the "Image" menu select "Crop" to crop the region of interest.
7. Next, under the Image menu, select "Look up tables" and "Invert LUT." Inverting your background to white and bands to black will make it easier to view your bands.
8. Choosing the "Rectangle" tool again, draw a tight horizontal box around the gel bands of the appropriate molecular weight you want to analyze across all lanes. For example, draw a single box around all of the *egfp* bands. Next press "CTRL 1" to designate the first lanes. Using your cursor, touch the "1" in the center of the rectangle. Drag the box over to the next series of *23S* bands. Press "CTRL 2" to designate these bands. If a new window appears that asks if your lanes are horizontal, select "Yes."
9. Next, under the Analyze menu select the "Gels" and "Gel Analyzer" option. A new window will appear. Ensure the invert peaks boxes is *unchecked*. Everything else can be left as a default. Press "OK."
10. Going back to the Analyze menu, select "Gels" and then "Plot-lanes." You should see a new window appear with peaks of intensity from each band selected on your gel. Note for the background, there should not be tall peaks shown.
11. Using the "Straight line tool," isolate each peak by drawing a straight line down between each peak as shown in Figure 17.1.
12. To assign each peak/band a numerical value, select the "Wand tool" from the ImageJ menu. Using the wand tool, click in the center of each peak to assign it an area value. Do this for each peak/band. These area values will be shown in a table (Figure 17.2).
13. You can copy and paste your area data into Microsoft Excel. Then, calculate the relative *gst::egfp* mRNA levels in each sample using these two formulas:

   a. $$\text{Normalized } gst::egfp \text{ level} = \frac{\text{Band area}_{\text{egfp}}}{\text{Band area}_{23\text{S}}}$$

   b. $$\text{Relative } gst::egfp \text{ level} = \frac{\text{Normalized } gst::egfp \text{ level}_{\text{with inducer}}}{\text{Normalized } gst::egfp \text{ level}_{\text{no inducer}}}$$

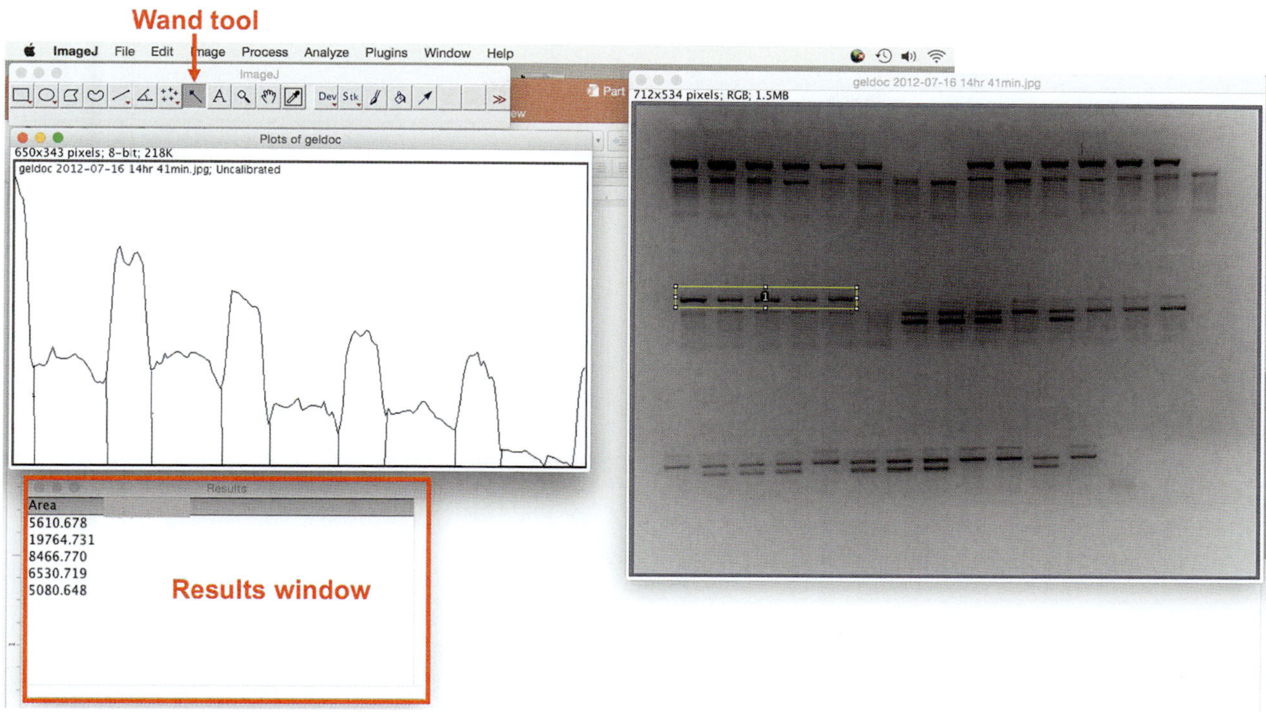

FIGURE 17.2 NIH ImageJ screenshot indicating the location of the wand tool and band area within the results window.

14. Determine if there is any product in the −RT reactions. Amplification with the *23S* primers will indicate genomic DNA contamination because it is an endogenous gene encoded by the *E. coli* genome. Amplification with the *egfp* primers will indicate plasmid DNA contamination, because it is not an endogenous gene; rather, it is encoded by a plasmid you engineered. Evaluate the level of DNA contamination (if any) by looking at each −RT reaction. Contamination levels for each sample can be estimated by using these formulas:

   a. $\%\ \text{Genomic DNA contamination} = \dfrac{\text{Band area}_{-\text{RT }23\text{S}}}{\text{Band area}_{+\text{RT }23\text{S}}} \times 100$

   b. $\%\ \text{Plasmid DNA contamination} = \dfrac{\text{Band area}_{-\text{RT egfp}}}{\text{Band area}_{+\text{RT egfp}}} \times 100$

DISCUSSION QUESTIONS

1. You ran the agarose gel but forgot to include the lactose-induced *23S* reaction containing reverse transcriptase (PCR tube 12). Without repeating the gel, what conclusions could you make about *gst::egfp* mRNA levels after lactose and IPTG induction? What would remain inconclusive? What if the sample you forgot to include was the lactose-induced *23S* reaction *without* reverse transcriptase (PCR tube 11)?
2. You were able to determine the levels of *gst::egfp* mRNA in the induced cultures relative to the non-induced. Without an inducing molecule to derepress the lac operator of pET-41a, were any *gst::egfp* mRNA transcribed? Why?

Part V

Modulation of Gene Expression

Congratulations! You have made tremendous headway in learning and successfully manipulating recombinant DNA. Thus far, you have created a recombinant pET-41a expression vector containing the *egfp* gene and transformed this plasmid into the model organism *E. coli* to express and purify the GST::EGFP protein. You may have also induced the overexpression of the protein in *E. coli* using IPTG, purified the protein, as well as quantified the amount of gene expression by using RT-qPCR to examine *gst::egfp* mRNA levels. The goal of the following advanced series of experiments is to take your scientific skills to the next level. You will transfect (introduce) the *egfp* gene in a different in vitro model system, HEK293 human embryonic kidney cells, a mammalian cell culture model. Once you verify through fluorescence screening that HEK293 cells express the EGFP protein, instead of *upregulating* (overexpressing) *egfp* as you did in *E. coli*, you will now *downregulate* expression of the *egfp* gene. To do this, you will utilize two cutting-edge and effective techniques: RNA interference (RNAi) and CRISPR-Cas9. Each method has its own unique advantages and disadvantages, which are explained in the introductions of the respective lab sessions. In the next series of lab sessions, you will create a stable HEK293 cell line that constitutively (always) expresses the EGFP protein using the mammalian expression vector pEGFP-N1. Next, you will screen HEK293 cells transfected with the pEGFP-N1 plasmid by fluorescence microscopy to determine successful transfection of the *egfp* transgene. You will then downregulate the expression of the EGFP protein in HEK293 cells by designing siRNA molecules to specifically target and knockdown the expression of *egfp* mRNA, thus considerably reducing the amount of EGFP protein. Finally, you will downregulate expression of *egfp* by designing guide RNAs to be used in the CRISPR-Cas9 system and compare and contrast your results.

All the experiments in Part V of this manual will require you to have a good understanding and knowledge of how to work with mammalian cells in culture. Therefore, mastering your aseptic technique is a crucial first step.

TIMELINE

The experiments proposed in Part V, Modulation of Gene Expression are designed to be completed as a complement to Parts I–IV, conducted during a separate semester. However, due to the complexity of experiments, we suggest those with time limitations commercially purchase HEK293 cells expressing GFP (*Note: not* EGFP) in lieu of creating them in Lab Session 19, Transient Transfection of Mammalian Cells. These cells are available from Cell BioLabs (catalog number AKR-200).

Furthermore, these labs are designed to be flexible to fit various curricular needs. Therefore, it may be suitable for students to only create stable cells expressing EGFP (Lab Sessions 18–19), while other student learning outcomes may require knockdown and/or knockout of gene expression utilizing RNAi (Lab Sessions 20–22) and/or the CRISPR-Cas9 system (Lab Sessions 23–30), respectively. Please refer to Figure V.1 to choose the experiments most suitable for your course.

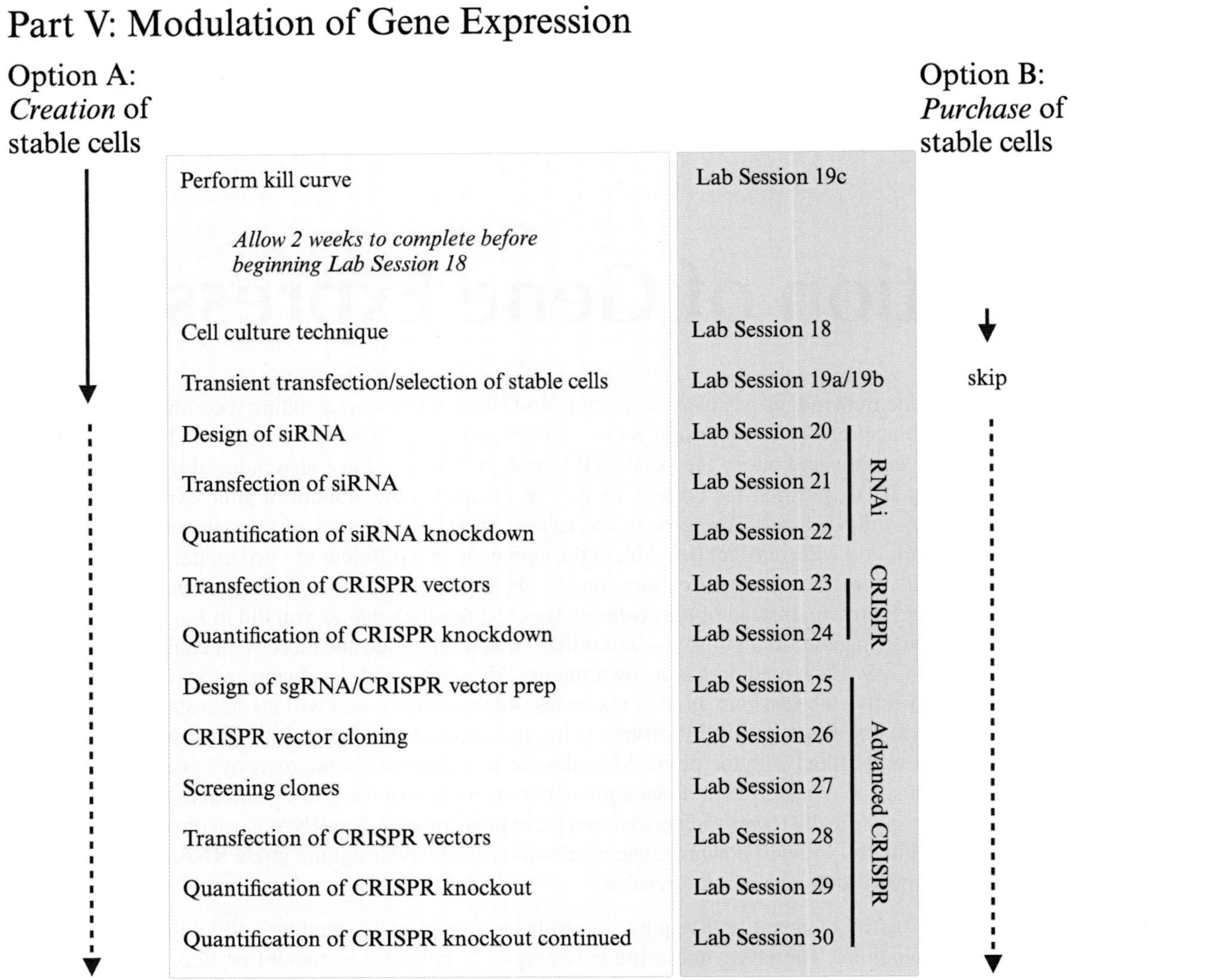

FIGURE V.1 Proposed path(s) for experiments outlined in Part V. If stable HEK293 cells expressing GFP are purchased (Option B), students will begin with Lab Session 18 and skip Lab Session 19. Both Option A and Option B should include Lab Session 18 to introduce cell culture technique. Solid arrows indicate required lab sessions and dotted arrows indicate optional lab sessions.

Lab Session 18

Culturing Mammalian Cells

Goal: To learn proper cell culture sterile technique and the basics of culturing mammalian cells in vitro, while preparing HEK293 cells for transfection with the pEGFP-N1 plasmid.

You will learn to culture the immortalized human embryonic cell line HEK293. However, the general rules for sterile technique can be applied to all in vitro cell cultures. Be aware, though, that culture conditions and medium will vary from cell line to cell line.

I. INTRODUCTION TO STERILE TECHNIQUE

Cell culture (aka tissue culture) is an important tool used to study a wide range of cellular and biochemical processes in both normal and pathophysiological states. For example, it can be used to study carcinogenesis and mutagenesis, to screen and develop pharmaceuticals, as well as for the manufacturing biological compounds.

Cell culture refers to the removal of cells from an animal or plant and their subsequent growth in an artificial environment. The cells may be removed from tissues directly or derived from a cell line or strain that has already been established.

Normal primary cells (those taken directly from a tissue) will only divide and proliferate for a limited amount of time before they senesce (metabolically active but no longer dividing), and some may eventually die. Senescent cells have irreversibly stopped dividing and enter a state of permanent growth arrest without dying. This amount of time varies based on a number of genetic factors. Some cell lines, on the other hand, can divide and proliferate indefinitely through a process known as transformation. Transformation can occur spontaneously, or be induced by chemicals or viral infections, usually by inactivation of genes modulating the cell cycle. Transformed cells are known as a continuous or immortal cell line; that is, under the proper growth conditions they can grow indefinitely in an artificial cell culture environment.

Each cell type requires its own unique set of culture conditions. In an artificial environment, however, cells require a certain set of factors:[1]

1. Culture vessel—typically a special plastic that has been sterilized and coated with poly-D-lysine to encourage cell adherence by the formation of cell junctions between adherent cells and the plastic. These cells are known as anchorage-dependent or adherent and will form monolayers in a culture. In contrast, suspension cells are anchorage-independent and still require a similar vessel, but do not form cell contacts with the plastic; they remain floating in solution. Many times companies will list these culture vessels as suitable for tissue culture or as tissue culture treated.
2. Growth medium—a liquid that supplies essential nutrients such as amino acids, carbohydrates, vitamins, and minerals.
3. Growth factors—contained in fetal bovine serum (FBS) or calf serum if required by the particular cell line. Any other special growth factor will be indicated by the company selling the cell line and will need to be added separately to the growth medium.
4. Hormones (cell type dependent).
5. Gases—(CO_2, O_2).
6. Regulated physio-chemical environment (pH, osmotic pressure, temperature). The regulation of gases and the physio-chemical environment is accomplished by growing cells in a specialized chamber known as a cell culture incubator. This is different from a bacterial incubator in that CO_2 (or other gases) can be regulated to enter the chamber and maintain a consistent and defined concentration (e.g., 5% or 10% CO_2) in addition to temperature. *It is extremely important to never place your cell cultures in a bacterial incubator or bacterial cultures in a cell culture incubator due to contamination issues, which we will discuss shortly.*

When culturing cells in vitro, it is important to observe their *morphology* (e.g., appearance) as well as their *confluency* (e.g., the percentage of surface area cells occupy in a culture vessel) under the microscope each time you handle your cells. This is crucial to maintaining the health of your cells, as well as to detect signs of contamination early before it spreads to other cells in the culture among commonly used equipment. Signs of unhealthy or contaminated adherent cells

Molecular Biology Techniques. DOI: https://doi.org/10.1016/B978-0-443-26714-7.00027-2

may include shrinking of the cell body, detachment, cloudy medium, a change in the growth medium's pH (most include a pH indicator), granularity around the nucleus, and apoptotic bodies and/or vacuolation.

Mammalian cells in culture can be segregated into three basic categories of morphology based on their shape and appearance (Figure 18.1).[1]

1. Fibroblastic (fibroblast-like) cells are bipolar or multipolar, have elongated shapes, and grow attached to a substrate.
2. Epithelial-like cells are polygonal in shape with more regular dimensions and grow attached to a substrate in discrete patches.
3. Lymphoblast-like cells are spherical in shape and usually grow in suspension without attaching to a surface.

Confluency of cells is usually described as the percentage of surface area occupied by adherent cells in a culture vessel. For example, if cells form a monolayer where there is no space between cells they are said to be 100% confluent. If cells occupy only half of the space observed they are said to be 50% confluent (Figure 18.2). These observations give us a rough estimate of the number of cells we have. When making visual observations of your cells it is important to look at multiple fields of view under the microscope to get an overall picture of the confluency and morphology of your cells in culture.

One of the largest challenges in maintaining in vitro cell lines is to keep them free of contamination. In order to maintain healthy cells in vitro, it is important to have proper aseptic (sterile) technique. This will be different than the sterile technique used when handling bacteria in previous lab sessions. Aseptic technique provides a barrier to microorganisms that can contaminate your cell cultures such as bacteria, yeast, mycoplasma, fungi, and viruses. Furthermore, it will help prevent cross-contamination between multiple cell lines. Any non-sterile supplies, media, reagents, work surfaces, equipment, plastic, and glass that come into contact with your cell cultures pose a contamination hazard. Aseptic technique is a set of standard operating procedures to drastically reduce contamination from these sources as well as keep you safe. There are three basic things to consider when practicing aseptic technique:

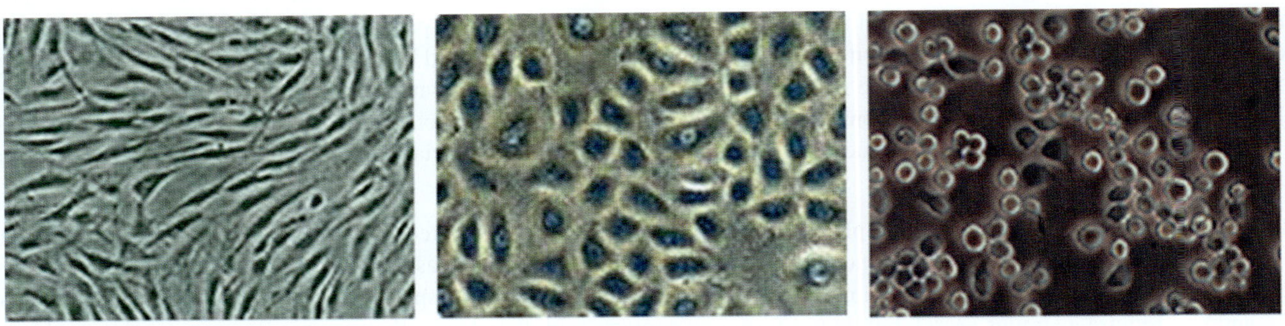

FIGURE 18.1 The basic morphologies of cells in culture. (Left panel) Fibroblastic cells are bipolar, adherent, and have elongated shapes. (Middle panel) Epithelial-like cells are adherent and appear in colonies with a polygonal shape, and (Right panel) Lymphoblast-like cells are non-adherent, grow in suspension, and have a circular shape. *Photo and caption taken with permission from Thermo Fisher Scientific.*

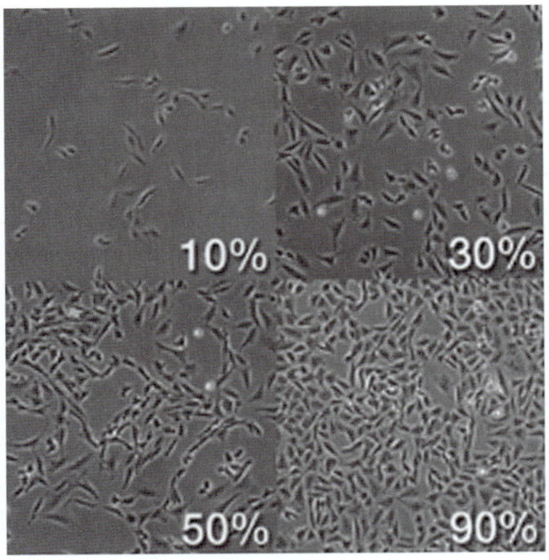

FIGURE 18.2 Visualization of various cell confluencies. *Photo taken with permission from Mirus Bio LLC.*

1. Personal hygiene: Always wash your hands before and after working with cell cultures. Wear personal protective equipment (PPE) such as lab glasses, gloves, and a lab coat to reduce the risk of dirt, dead skin cells, and other contaminants from entering your cell cultures. It is important to keep a dedicated set of PPE in the cell culture room that is not used for any other purpose.
2. Sterile reagents, media, glass, and plasticware: Tissue culture plastics, reagents, and media obtained directly from the manufacturer will come pre-sterilized; however, they can become contaminated from improper sterile technique. Always sterilize any solutions, media, reagents, or glassware prepared in the laboratory using the appropriate sterilization procedures (e.g., autoclave, sterile filtering with a 0.2 μm filter).
3. Sterile handling:
   a. Wipe your gloved hands and work area with 70% ethanol frequently and profusely.
   b. Wipe any object that enters the biological safety cabinet (BSC) cabinet with 70% ethanol.
   c. Do not pour liquids from one container to the next. Only use serological pipettes to transfer liquids.
   d. Use sterile plastic pipettes and sterile glass Pasteur pipettes only once and keep all items you are working with in close proximity to your work area.
   e. Never open a sterile item outside of the BSC cabinet.
   f. When opening any sterile container in the BSC cabinet, only leave them uncovered when you are actively removing or placing liquid in them. Promptly recap as soon as you are finished to prevent microorganisms and airborne particles from gaining entry.
   g. If you remove a cap or cover and must put it down on the work surface, place the cap with the opening face down or on its side with the opening facing the back.
   h. Perform your experiments as rapidly and carefully as possible.

II. LABORATORY EXERCISES

Note: Manufacturer's instructions for operating BSCs may vary and it is important to double-check the instructions prior to using cells in culture. The BSCs used in this protocol have a vacuum trap within each cabinet for removing hazardous biological liquid waste.

It is important to obtain biosafety approval from your institutional biosafety office as well as train your students in appropriate safety for Biosafety Level 2 (BSL-2) organisms prior to implementation of this exercise.

Our biggest foe for cells in culture is contamination! These steps will help lower the risk of contamination as well as the spread of contaminants between cultures, especially in shared research spaces.

A. Operating and Using a Biological Safety Cabinet (BSC)

1. Open the sash to the biological safety cabinet (BSC) to the line indicated on the cabinet. Turn on the blower *and* germicidal (UV) light at least 15 minutes prior to using the BSC. This allows sufficient time for the air inside the cabinet to be filtered. Before working in the BSC, be sure to switch from germicidal light to incandescent light.
2. Turn on the vacuum and spray the interior of the aspirator line with 70% EtOH.
3. Wipe the interior of the cabinet and *any objects* that are placed in the cabinet (including your gloved hands!) with 70% EtOH and a lint-free wipe (e.g., Kimwipe or paper towel). Be liberal.
4. When placing objects in the cabinet, avoid placing them in areas where they block the air flow.
5. Periodically spray gloves with 70% EtOH to minimize contamination. Always do this when taking hands out of the cabinet and putting them back in.
6. In case of a spill, spray the area with 70% EtOH and use a Kimwipe or paper towel to clean it up immediately. This includes if you drop medium on the top of a tissue culture container. Even spilled PBS should be cleaned immediately. The salts can damage the metal of the cabinet.
7. Any object that will come in contact with your cells must be sterile: sterile flasks, sterile wrapped pipettes, autoclaved glass bottle, Pasteur pipettes, etc. If any of these items have been compromised (e.g., broken/cracked flasks, autoclaved bottle opened outside of the cabinet) DO NOT use them.
8. Use glass Pasteur pipettes and sterile plastic pipettes once, then throw them away.
9. When handling sterile containers with caps or lids, place the cap on its side or face-down toward the back of the cabinet if it must be laid on the work surface. Recap the container as soon as possible.
10. Make sure not to touch the tip of the pipette to the rim of any flask or sterile bottle. If you do so, throw out the pipette and use a new one.

11. Try not to reach over open containers in the cabinet. Always work around them unless they are capped or covered.
12. Discard gloves after use and do not wear them when entering any other lab area.
13. Avoid working too closely to the front of the cabinet, or the non-sterile area, of the cabinet. Keep the working area at the center or toward the back. Keep objects needed for the current procedure within reach; keep the others in the back.
14. The indicator stripes on the autoclave tape should turn black if an object has been properly autoclaved.
15. Never pour from one sterile container to another. Pouring will generate a liquid path to introduce contamination from the outside to the inside. Always pipette or use filters when transferring from one bottle to another.
16. Withdraw the pipette from its wrapper at the center of the work area. Unwrap the top portion, where it hooks into the pipette aid, attach it to the pipette aid, and then slide it out of its wrapper into the cabinet. Do not allow the tip of the pipette to contact the cabinet itself.
17. When working with glass Pasteur pipettes, do not remove pipettes directly from the box with your fingers. Use the end of the aspirator hose to pick them up if possible.
18. Avoid having clutter in the biological safety cabinet. If there are items in the cabinet that you are not using, remove them. Make sure that when you remove these items the lids are securely closed so the contents remain sterile.
19. Clean the work area when finished by wiping with 70% alcohol, lower the glass sash, and turn off the blower and light.
20. Spray 70% EtOH into the aspirator tube, and then turn off the vacuum.

B. Other Considerations: Using Incubators and Microscopes Within a Cell Culture Facility

Commonly, rooms that are used to culture cells will have additional pieces of equipment in addition to biological safety cabinets. This additional equipment is necessary to maintain cell culture lines. Many times, these are shared pieces of equipment with other classes and research laboratories, so they present a great vehicle with which to spread contamination if poor aseptic technique is used!

Provided is a list of procedures that can help minimize contamination when using these additional pieces of equipment:

1. Before using the microscope, spray a Kimwipe (or paper towel) with 70% EtOH and wipe down the stage. Do this when you are done to prevent media spills from spreading between plates. Be sure to not get EtOH on any objective lens.
2. If you need counter space to set plates down on by the microscope, clean the area with 70% EtOH first.
3. The incubators are not technically sterile. However, every effort must be made to maintain their cleanliness and to prevent contamination from spreading. If media is spilled in the incubator, clean the spill with 70% EtOH. Spills into the water basin at the bottom of the incubator, which is used to maintain humidity, MUST be immediately taken care of—notify your TA or instructor.
4. Place tissue culture flasks and/or plates carefully into the incubator. Be sure not to bump other people's plates and minimize stacking as much as possible. Always clearly label your cell culture vessel with your initials, date of passage, passage number (the number of times cells were passaged from one vessel to another, indicating their age), and the name of the cell line. Keep your own plates set up to allow easy access.
5. If you notice that the medium for your cell cultures is extremely purple, this indicates a rise in pH and you should notify your instructor. The CO_2 may have run out. If you notice the medium for your cell culture is extremely yellow, the medium is acidic, and the cells need to be passaged or fed.

For helpful videos and more information on cell culture sterile technique, please refer to Thermo Fisher Scientific's web-based resources: https://www.thermofisher.com/us/en/home/references/gibco-cell-culture-basics.html

C. Counting and Plating Mammalian Cells

In this exercise, you will trypsinize and count your HEK293 cells in preparation for transfection with the pEGFP-N1 plasmid. HEK293 cells are human-derived embryonic kidney cells that have an epithelial morphology and are adherent in nature. This procedure should be done 1–2 days prior to transfection to allow ample time for the cells to re-adhere to the culture vessel.

Always practice aseptic technique when working with cells in culture. If needed, refresh your memory by reading the introductory material in this lab session again or ask your instructor if you have questions BEFORE you begin.

1. Obtain a 10 cm^2 dish or flask of HEK293 cells provided by your instructor. Observe their confluency and morphology and write the details in your lab notebook. If you observe any signs of contamination, notify your instructor immediately. Also observe the color of your cell culture *complete growth medium*. The complete medium for HEK293 cells is high glucose (4500 mg/L) Dulbecco's Modified Eagle's Medium supplemented with 10% FBS, 0.1 mM non-essential

amino acids, 2 mM L-glutamine, 1 mM sodium pyruvate, 1500 mg/L sodium bicarbonate, and 1% penicillin-streptomycin. It contains phenol red, a pH indicator. If the medium drastically changes color to yellow (acidic) or purple (basic), notify your instructor.

2. Using a sterile Pasteur pipette, aspirate the medium from the dish. Do this by gently, tipping the dish toward you and placing the tip of the Pasteur pipette to the bottom of the vessel until all the medium is aspirated. Remember the spot where you touched the dish. You will always aspirate from this spot to avoid aspirating cells from your dish.
3. Gently add ~2 mL of sterile 1x PBS to the dish and swirl carefully. Whenever you remove or add liquid to your cells, be sure to do so from the same location to avoid dislodging cells.
4. Aspirate the 1x PBS from the dish.
5. Add 2.5 mL of pre-warmed 0.25% Trypsin-EDTA solution to the dish and swirl to ensure adequate coverage on the bottom. Place the dish back in the 37°C cell culture incubator and set your timer for 2 to 3 minutes. Trypsin is a serine protease that will break up cell–cell junctions as well as contacts made between the cells and the plastic dish by cleaving the C-terminal side of lysine or arginine. Over-trypsinization can hurt your cells, so be sure not to leave it on beyond the specified time. EDTA is a divalent metal chelator. Adhesion molecules that comprise the cell–cell and cell–matrix interactions require Ca^{2+} and Mg^{2+}; therefore, EDTA can weaken these interactions and help facilitate detachment by trypsin.
6. Observe your cells under the microscope; detached cells will float in the medium and appear rounded. You may want to gently but firmly tap your dish on a sterile surface to help further detach your cells. If you still have 50% or greater cells attached to the dish, return it to the 37°C cell culture incubator for another 2 minutes.
7. Once cells appear detached, add 5 mL of pre-warmed complete growth medium to inactivate the trypsin. Gently pipette the medium over the cell layer surface several times to ensure detachment and to break up cell clumps. This process is also known as trituration. Vigorous pipetting can cause cell damage, so be careful.
8. Gently swirl the dish to ensure the cells are evenly distributed. You are now ready to count your cells using a hemacytometer.

   Time-Saving Tip: Alternatively, an automated cell counter such as a Cellometer from Nexcelom can be used.

   Hemacytometers are devices used to count cells. The following procedure is appropriate for glass hemacytometers, although plastic disposable ones can be used as well.
9. Clean the chamber and cover slip with 70% EtOH. Dry and put the cover slip in position.
10. Pipette cells up and down to mix. Using a sterile pipette tip, carefully remove 20 μL of cell suspension. Add the 20 μL to each chamber of the hemacytometer. Do this by slowly pipetting the cell suspension between the cover slip and hemacytometer; capillary action will spread the solution in an even layer between the two surfaces. Be sure not to overfill.
11. Place the chamber in the microscope under a 10× objective using phase contrast and course/fine focus to distinguish the cells and grid lines of the hemacytometer.
12. Using a hand counter, count the cells in one set of 16 squares (Figure 18.3). When counting, include cells that fall in the middle, top, and right-hand boundaries.

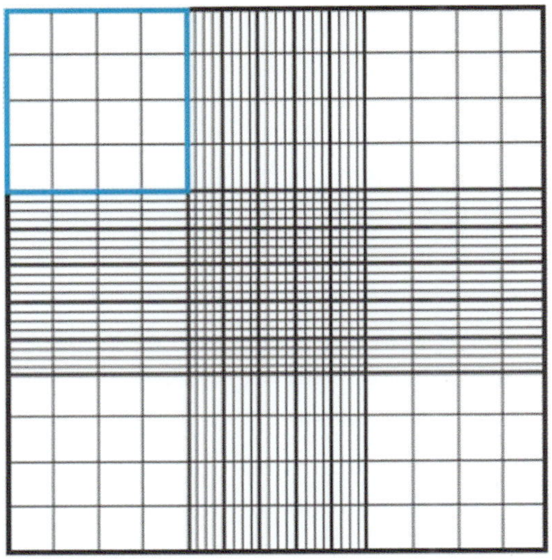

FIGURE 18.3 Hemacytometer grid lines indicating one of the sets of 16 squares that should be included for cell counting. *Zephyris at the English-language Wikipedia.*

13. Move the hemacytometer to the next set of 16 squares and count until all four sets of 16 squares are counted. Be sure to record your cell numbers from each square in your laboratory notebook.
14. Take the average cell count from each of the sets of 16 corner squares and multiply by 10,000 (10^4). The final number is the total amount of cells/mL in your cell suspension.
15. Calculate the volume of cells needed to seed a density of 5×10^5 cells in one well of a 6-well plate.

$$\text{Using } C_1V_1 = C_2V_2$$

where

C_1 is the concentration of cells counted (cells/mL):_______________

V_1 is the amount of cell suspension (mL) required (unknown):_______________

C_2 is the desired concentration of cells: 2.5×10^5 cells/mL

V_2 is the desired total volume: 14 mL

Add the appropriate volume of cell suspension to a sterile 50 mL conical tube. Bring the total volume up to 14 mL with complete medium. Mix well, by carefully pipetting up and down (avoid forming bubbles), and immediately add 2 mL of cell suspension per well of your 6-well plate.

16. Gently rock the plate back and forth to disperse the medium evenly on the bottom. Place the lid back on and leave in the 37°C cell culture incubator for at least 24 hours to allow the cells to re-adhere to the bottom of the plate.

REFERENCE

1. Thermo Fisher. *Cell culture basics handbook.* <thermofisher.com/gibcoeducation>; 2024.

DISCUSSION QUESTIONS

1. What are some of the biggest challenges when culturing cell lines in vitro? What are three steps that can prevent these issues?
2. By what mechanisms does trypsin-EDTA work to dissociate adherent cells from the tissue culture dish as well as from other cells?
3. If you counted a total of 1.5×10^4 cells/mL, how much volume of cell suspension would be required to make a 3×10^3 cells/mL solution in 100 mL of medium?

Lab Session 19

Transient Transfection of Mammalian Cells

This week, you will transiently transfect the pEGFP-N1 plasmid into HEK293 cells. Next, you will determine the success of your transfection by qualitatively assessing the amount of EGFP protein expressed using fluorescence microscopy. Finally, you will have the option of creating HEK293 cells that stably express EGFP using antibiotic selection.

Molecular Biology Techniques. DOI: https://doi.org/10.1016/B978-0-443-26714-7.00039-9

Lab Session 19a

Interim Laboratory Session

TRANSIENT TRANSFECTION OF PEGFP-N1 IN HEK293 CELLS USING FUGENE HD

Goal: To perform transient transfection of pEGFP-N1 plasmid in the mammalian HEK293 cell line. This will allow for the expression of EGFP protein in your cells, which will, in turn, allow for its genetic manipulation in later lab sessions.

I. INTRODUCTION

Plasmids do not naturally exist in mammals as they do in bacteria. However, scientists have been able to engineer plasmids to be used in cultured mammalian cells, thus enabling the study of gene function or protein expression in a variety of mammalian cell types. Transfection is the process of introducing nucleic acids into eukaryotic cells by various chemical (cationic lipid) or physical methods. Mammalian transfection is similar to bacterial transformation (introduction of foreign DNA into bacteria). However, the techniques and reagents vary. Recall that nucleic acids are negatively charged molecules due to their phosphodiester backbones, and, similarly, lipid bilayers are also negatively charged due to their biochemical structure. Since negative charges repel one another, neutralizing the phosphate backbones of nucleic acids is a crucial step in facilitating their entry through the lipid bilayer of mammalian cell membranes. There are a variety of ways to coat nucleic acids to allow for their passage across the membrane. For the purposes of this lab exercise, we will focus on the method you will be using, lipid-based transfection, which shields negatively charged nucleic acid molecules (e.g., DNA, RNA) so they may enter cells with a negatively charged membrane.

The process of lipid-mediated transfection can vary depending on the agent used. In many techniques, it is based on creating an artificial liposome (lipid bilayers that form a spherical vesicle in aqueous medium) composed of synthetic cationic lipids.[1] Cationic lipids are amphiphilic molecules that have a positively charged head group linked to a nonpolar hydrophobic domain typically consisting of two alkyl chains (Figure 19.1). Cationic lipids can vary in structure, especially within their hydrophobic domain according to length and degree of saturation. The positively charged head group associates with high affinity to the negatively charged phosphates on the nucleic acid backbone, condensing the nucleic acid and forming a complex wherein the nucleic acid becomes encased within the liposome.[2] Thus, the negative charge of the nucleic acid is shielded and the overall net charge of the liposome is positive at physiological pH. This thereby facilitates a high rate of transfer through the lipid bilayer via endocytosis or membrane fusion. Once inside cells, the transfection complex must evade degradation and deliver the nucleic acid to the cytoplasm (for siRNA, mRNA, miRNA) or the nucleus (for plasmid DNA or shRNA encoding vectors). In the cytoplasm, mRNA is in proximity to the ribosomes where it can be translated, and siRNA or miRNA can associate with the RNA-induced silencing complex, which we will discuss in later chapters. DNA, on the other hand, must enter the nucleus for the subcellular machinery to facilitate transcription; however, the exact mechanism remains unknown (Figure 19.2).

The advantages of lipid-mediated transfection techniques are that DNA can be expressed as a protein for both transient and/or stable, long-term experiments. For transient transfections, the DNA is only expressed for a short time (~24–96 hours) and is not integrated into the genome. Stable transfection, on the other hand, requires the foreign DNA to be

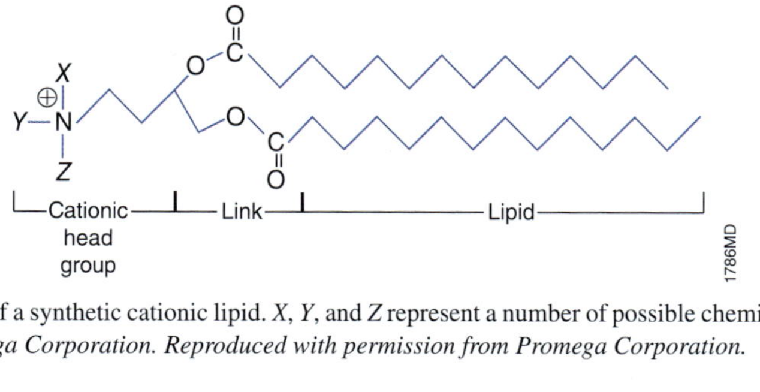

FIGURE 19.1 The general structure of a synthetic cationic lipid. *X*, *Y*, and *Z* represent a number of possible chemical moieties, which can differ, depending on the specific lipid. *©2024 Promega Corporation. Reproduced with permission from Promega Corporation.*

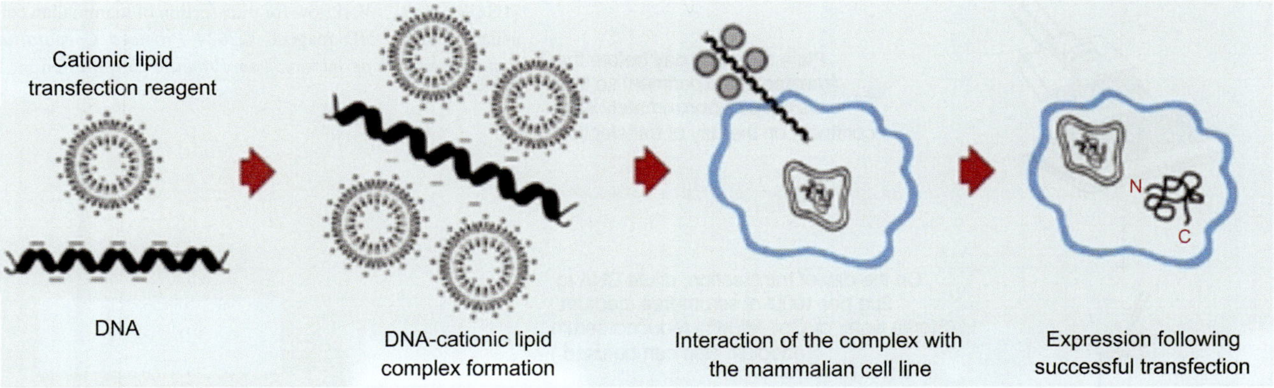

FIGURE 19.2 The process of cationic lipid–based transfection. *Figure used with permission from Thermo Fisher Scientific.*

fully integrated into the host genome, therefore allowing indefinite expression of the foreign DNA. Another advantage of lipid-mediated transfection is that it occurs at a relatively high efficiency with minimal experimental steps required. However, lipid-based transfection is not suitable for all cell types, especially primary cells.

For the purposes of this experiment, you will create a HEK293 cell line that transiently expresses the EGFP protein. To do this, you will use the transfection reagent FuGENE HD. It is a reagent with wide applications in different cell types and low toxicity. It uses a proprietary mixture of lipids and other components to deliver nucleic acids into eukaryotic cells and can be used in the presence of serum.[3]

II. LABORATORY EXERCISE

Refer to Lab Session 18, Culturing Mammalian Cells, before beginning. It is important to follow proper sterile technique in all experiments involving the manipulation of mammalian cells in culture. Furthermore, it is necessary to always wear the appropriate personal protective equipment.

Please refer to Figure 19.3 to familiarize yourself with the overall workflow of the transfection experiment. The FuGENE HD transfection reagent protocol was modified with permission from Promega Corporation (©2024 Promega Corporation):

1. Obtain your group's six-well plate of HEK293 cells that were plated from Lab Session 18, Culturing Mammalian Cells. Using the microscope and 10× objective lens, observe and record the morphology and confluency of your cells under white light. Your cells should be 60%–80% confluent with few floating cells in the medium. If you notice any signs of contamination or your cells are outside the 60%–80% confluent range, notify your instructor. Return your plate to the cell culture incubator.
2. Remove the FuGENE HD reagent from the 4°C refrigerator and allow it to reach room temperature. Mix gently by inversion before use.
3. In the biological safety cabinet, label three sterile microcentrifuge tubes as: No DNA, pcDNA3 (empty vector/negative control), and pEGFP-N1. Both pEGFP-N1 and pcDNA3 are mammalian expression vectors driven by a cytomegalovirus promoter; however, pcDNA3 does not contain *egfp*.
4. Pipette 150 μL room temperature Opti-MEM into each tube.
   $ Money-Saving Tip: Serum-free medium can be used instead of Opti-MEM.
5. Create a table like Table 19.1 to calculate the volume of each plasmid vector that will equal 3 μg. Add 3 μg of the correct plasmid DNA to each labeled tube now containing the Opti-MEM. Pipette up and down to mix.

TABLE 19.1 Plasmid Vectors Used in Transfection

DNA	Conc. (μg/μL)	Vol. to Add (μL)
No DNA	0	0
pcDNA3 (negative control)		
pEGFP-N1		

No Permission Required.

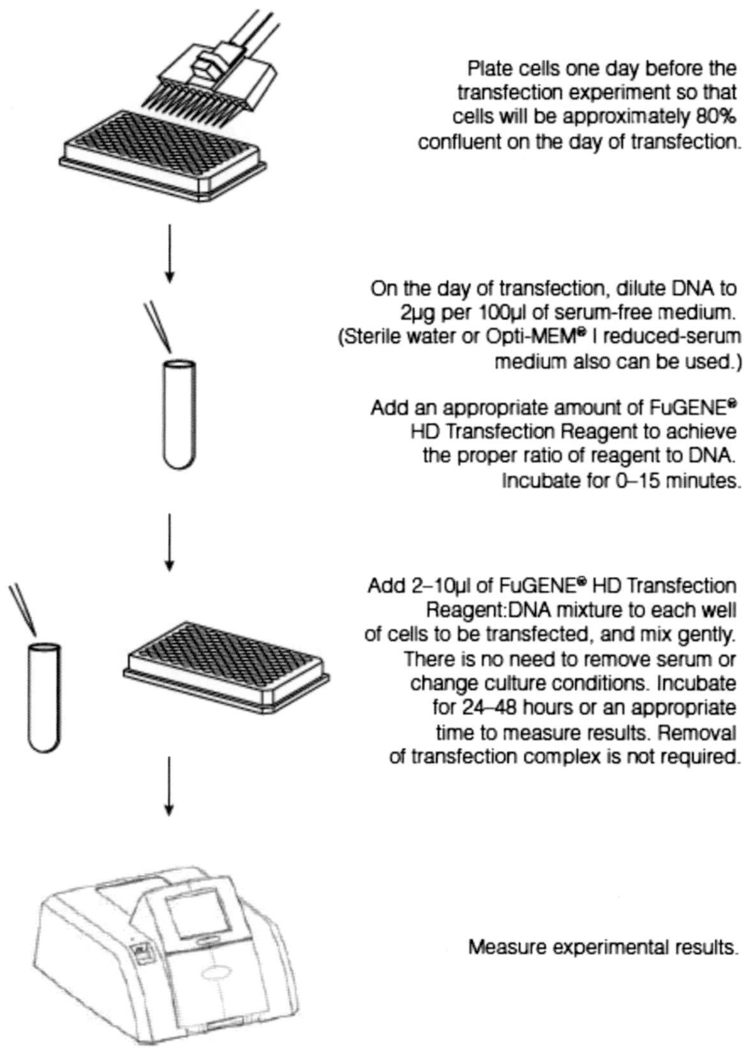

FIGURE 19.3 Workflow for transfection of mammalian cells using FuGENE HD reagent. ©*2024 Promega Corporation. Reproduced with permission from Promega Corporation.*

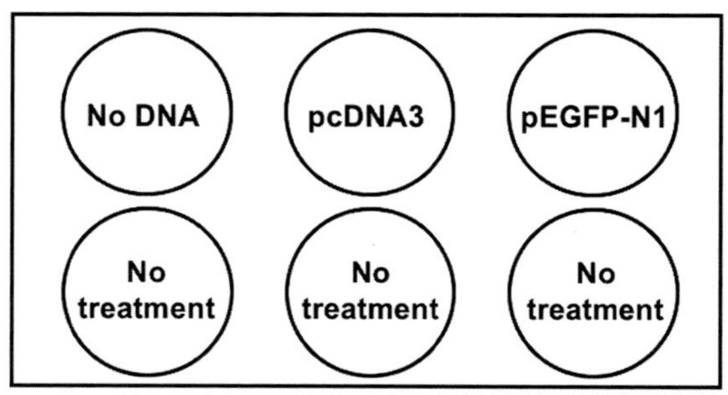

FIGURE 19.4 Six-well plate set-up for transient transfection of HEK293 cells.

6. Pipette 10 µL FuGENE HD reagent (using 3:1 ratio) DIRECTLY into the Opti-MEM diluted DNA in each tube. *DO NOT allow the FuGENE HD reagent to directly contact the plastic walls of the tube*, as this will drastically reduce transfection efficiency. Change tips for each new aliquot.
7. Carefully mix the contents in each tube by pipetting up and down or by vortexing briefly. Leave in the biological safety cabinet for 15 minutes. This time will allow the formation of DNA:FuGENE HD complexes, thus enabling the DNA to enter the cells.
8. Remove your six-well plate of HEK293 cells from the incubator. Label (on the lid) what each well will contain using the names given in Figure 19.4. You will have three wells that remain untreated.

9. Add the entire ~170 μL of the complex mixture drop-wise to each of its corresponding wells of the six-well plate. Rock the plate back and forth *gently* to ensure complete mixing of the DNA:FuGENE HD complexes throughout the medium and return the plate to the 37°C incubator.
10. Qualitatively assay for transient expression of EGFP 48 hours after transfection using fluorescence microscopy as described in Lab Session 19b. *Note: After 48 hours, it is recommended to aspirate and replace the medium with fresh complete medium, which will remove floating, dead cells prior to fluorescence imaging.*

Lab Session 19b

Analysis of EGFP Expression Using Fluorescence Microscopy

Goal: To qualitatively verify the transient transfection of the pEGFP-N1 plasmid in the mammalian HEK293 cell line using fluorescence microscopy.

I. INTRODUCTION

Fluorescence microscopy is a commonly employed tool in the molecular biology/biochemistry laboratory. Many specimens and various organic compounds/proteins have unique fluorescence properties, which can be exploited for visualization, quantitation, and other purposes. A vast spectrum of fluorophores has made it feasible to tag proteins with high specificity to identify subcellular machinery, protein-protein interactions, and even single molecule detection. Examples of a few common fluorophores used in molecular biology imaging are listed in Table 19.2.

The basics of fluorescence microscopy are to shine light of a particular wavelength (called excitation) onto the specimen, then visualize emitted light at another wavelength (emission). On a subatomic level, fluorescence occurs when electrons in the molecule are excited to a higher and more unstable energy state, then relax and fall back to its ground state, resulting in the release of a photon of light. Excitation light, as the name implies, moves electrons to that higher energy state. Excitation light has a shorter wavelength and higher energy than fluorescence emission, which is characterized by a higher wavelength, lower energy, and different color. For example, we are using EGFP, which is a naturally fluorescing protein. It has an excitation wavelength of 488 nm (short wavelength, high energy, blue color) and an emission wavelength of 509 nm (longer wavelength, lower energy, green color).

Using a fluorescence microscope enables scientists to visualize only the emission wavelength by eye or by a detector such as a camera against a dark background, so the structure stands out in high contrast. The type of fluorescence microscope utilized will determine the exact limits of detection for the specimen under investigation. The main components of the fluorescence microscope are extremely similar to light microscopes, with the exception of the light source and specialized filter sets, which selectively block and direct light of different wavelengths.

The light sources utilized by fluorescence microscopes are much more powerful compared to traditional white light microscopes. They may include xenon or mercury arc lamps which are ~10–100× brighter than incandescent lamps and provide a wide spectrum of light wavelengths spanning from ultraviolet through to infrared. When light leaves the arc lamp, it travels through an excitation filter which will select the appropriate excitation wavelength for the sample

TABLE 19.2 A List of Common Fluorophores and Their Excitation/Emission Range

Fluorophore	Excitation Max (nm)	Emission Max (nm)	Chemical Property	Visible Color
Alexa Fluor 488	494	517	Organic dye	Green
Enhanced Green Fluorescent Protein (EGFP)	488	507	Fluorescent protein	Green
Cyan Fluorescent Protein (CFP)	435	485	Fluorescent protein	Cyan
Alexa Fluor 568	579	603	Organic Dye	Orange
Cy5	647	665	Organic Dye	Red
Alexa Fluor 647	652	670	Organic Dye	Red
Tetramethylrhodamine (TRITC)	552	578	Organic Dye	Yellow
Hoechst 33342	343	483	Organic Dye	Blue
DAPI	345	455	Organic Dye	Blue

No Permission Required.

(Table 19.2). LED light sources are another popular option. They provide a greater intensity of light specifically in the blue and red excitation regions compared to mercury bulbs, have a longer lifespan, emit less heat, and are more energy efficient. Excitation filters vary depending on the wavelength of light required to excite the specimen/fluorophore. Wavelengths passed through the excitation filter will reflect off the dichromatic (aka dichroic) mirror/beamsplitter through the microscope objective and to the specimen. The fluorescent emissions from the specimen are passed back through the objective lens and next through the dichroic mirror. Finally, the light is filtered by the emission filter, which blocks unwanted emission wavelengths and filters out any light from the light source or microscope components. The filtered fluorescent emission is sent through to the eyepiece for viewing or to a detector where the image can be digitized. Many times the excitation filter, dichroic mirror, and barrier filter are housed within one component, often referred to as a filter cube (Figure 19.5). A number of fluorescence microscopes can house up to six separate filter cubes, allowing for a wide range of fluorophores to be visualized.

Remember, when working with fluorescent molecules, the emission wavelength has a peak absorbance that is typically lower in wavelength and magnitude than the excitation peak. Therefore, the emitted light has a longer wavelength and thus lower energy compared to the exciting light. The difference in energy between the absorbed and emitted light is known as the Stokes shift. The greater the distance in this shift, the easier it is to separate excitation and emission light by utilizing the appropriate filter combinations. This is an important consideration when using multiple fluorophores in your specimen, as bleed-through of light between different and overlapping spectrums can occur.

Importantly, continuous exposure to excitation light will cause a fluorophore to gradually decrease or lose fluorescence, also known as photobleaching. This occurs when a photoexcited fluorophore interacts with molecular oxygen, resulting in destruction of the fluorophore and production of the highly reactive singlet ($^1O_2$) and triplet oxygen ($^3O_2$).[3] These species can react and degrade neighboring fluorophores. Furthermore, in live specimens, such as the ones you will be using in subsequent lab sessions, photobleaching can also modify cellular homeostasis as photobleached cells are damaged cells. Therefore, to reduce photobleaching, you should always optimize excitation filter sets and choose the lowest reasonable light intensity and exposure times needed to visualize your specimen, whether fixed or live. You should reduce continuous light exposure by only turning on the arc lamp when you are ready to view your specimen. Fixed samples should be prepared with anti-fade mounting media and sealed with clear nail polish (if the mounting media is non-setting). Lastly, slides should be kept in the dark when not being imaged and live cells should be kept in the tissue culture incubator when not being used.

II. LABORATORY EXERCISE

Viewing EGFP fluorescence and acquiring images for later analysis using a fluorescence microscope

EGFP is a fluorescent molecule that can be excited by 488 nm wavelength of light and emit in the green spectrum at 509 nm. We will exploit this property of EGFP to verify its expression in our HEK293 cells using fluorescence microscopy 48 hours after transient transfection (Lab Session 19a). If cells express EGFP, fluorescence will be visualized throughout the cellular cytoplasm. Since you will be using these cells for subsequent experiments, be sure to limit their exposure time on the microscope and return them to the cell culture incubator when not in use.

Note: Fluorescent microscopes can have different types of light sources such as mercury-vapor lamps, xenon arc lamps, lasers, or high-powered LEDs. It is important to identify and understand which microscope you are using as they

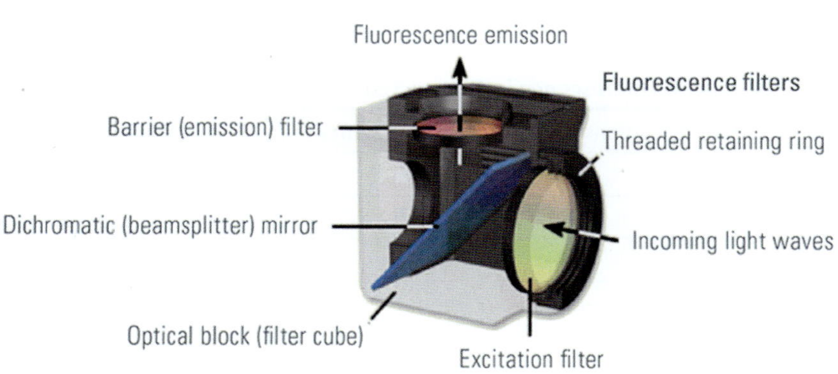

FIGURE 19.5 Filter cube image. *Photo courtesy of Leica Science Lab, author Wymke Ockenga.*

each require different operational procedures. The steps outlined are for use with an Olympus LED powered inverted fluorescence microscope equipped with a charged coupled device (CCD) camera and cellSens life science imaging software. Refer to your manufacturer's instructions for specific equipment guidelines.

For your safety, it is important to never look directly into unfiltered light from your fluorescent microscope because it can result in serious eye damage.

1. Remove the protective cover from your fluorescence microscope. Lower the stage to the lowest baseline position and the light intensity to low before turning on the microscope.
2. Turn on the computer and CCD camera in this order.
3. Ensure the collar of the objective is placed at the appropriate numerical aperture corresponding to the thickness of the plastic dish you are using. Next rotate the lowest power objective lens into the light path.
4. Secure your six-well plate (from Lab Session 19a) on the stage using the appropriate stage adaptor. Do not remove the lid to the plate.
5. Turn on the white light source (brightfield) and look through the eyepiece. Make the necessary adjustments to bring the cells into focus. First use the course adjustment and move the stage carefully upwards to bring cells into view. Find a field of view of healthy, adherent cells with an epithelial-like morphology. Avoid fields of view with excessive rounded or floating cells, as this is an indication of dead cells. You may also need to move your stage adjustment knobs to find the areas of interest. Once you see healthy cells, use the fine adjustment for final clarity. See your instructor if you are having difficulty finding your cells.
6. Rotate to the next power of objective lens and focus the cells again. Continue to do this until you have focused the cells using the 20× (or 40× if available) objective lens.
7. Switch the eyepiece output light path from "eye" to "camera." Using the cellSens software, view a live image of your cells on the computer screen. You may need to use the fine adjustment once more. When your cells are focused and in the appropriate field of view, acquire the image.

   $ Money-Saving Tip: If your microscope does not come equipped with a camera, careful placement and focus of a cell phone camera against the eyepiece has been used with success to take a picture in the absence of other equipment.
8. Save your image as a.TIFF or .JPG file. Be sure to also record visual qualitative observations in your notebook.
9. Turn off the white light source and all other lights in the room to reduce background light. Do not move the plate, as you want to next image the same field of cells for fluorescence.
10. Switch the eyepiece output light path back from "camera" to "eye." Turn on the LED light source (you may need to adjust the light intensity depending on the degree of EGFP expression) and rotate the appropriate excitation/emission filter into place. For maximum visualization of EGFP, you want an emission wavelength between 502 and 538 nm and excitation wavelength between 457 and 487 nm. *For your safety do not look directly at the fluorescent light!*
11. View the same sample through the eyepiece and make any necessary adjustments to bring your sample into focus using the fine focus knob only. Cells that have successfully been transfected with the pEGFP-N1 plasmid will fluoresce green due to expressed EGFP protein. Note that cells will have varying levels of green fluorescence, which corresponds to the amount of EGFP expressed, and some cells may not fluoresce at all (untransfected cells). Denote these qualitative observations in your lab notebook according to Table 19.3.
12. Switch the eyepiece output light path from "eye" to "camera." Using the cellSens software, view a live image of your cells on the computer screen. You may need to use the fine adjustment once more to get your cells into focus. When your cells are focused and in the appropriate field of view, acquire the image by adjusting the exposure time on the camera to maximize the brightness of EGFP in your pEGFP-N1 transfected cells. Note the exposure time and keep it constant for each and every sample you image in your experiment.

TABLE 19.3 Experimental Observations of EGFP-Mediated Fluorescence in Transfected HEK293 Cells

Experiment	Fluorescence Intensity	Confluency	Morphology
No DNA			
pcDNA3 (empty vector; negative control)			
pEGFP-N1			

Note: Fluorescence intensity should be scored as: +++, high intensity; ++, medium intensity; +, low intensity; –, no fluorescence detected. Confluence should be expressed as a percentage. Morphology as epithelial- or fibroblast-like, adherent, and/or floating.
No Permission Required.

13. Turn off the fluorescent light when you are not actively viewing/imaging your cells.
14. Save your images as a .TIFF or .JPG file with a file name that distinguishes it from the corresponding brightfield image.
15. Repeat Steps 5–14 for at least 2–3 fields of view for each sample. You should have white light (brightfield) and fluorescence images for each of your samples and respective fields of view.
16. After you have imaged all your samples, qualitatively score the average level and percent of cells expressing the EGFP protein. Write these observations in your lab notebook (Table 19.3).

Note: If you wish to quantify the level of EGFP fluorescence after transient transfection, see Lab Session 22, RNAi-Mediated Knockdown of EGFP: Part III.

Lab Session 19c

Creating Stable EGFP Expressing HEK293 Cells

Goal: To create HEK293 cells that stably express the pEGFP-N1 plasmid using antibiotic selection.

I. INTRODUCTION

Recall that stable transfection involves foreign DNA to be fully integrated into the host genome, therefore allowing indefinite expression of the foreign DNA. In order to create stable cell lines (in our case, HEK293 cells stably expressing EGFP), there needs to be a selective pressure to force cells to incorporate the plasmid DNA into the genome. The process of selection in mammalian cells is also identical to what you learned using bacterial cells in previous chapters. However, selection in this case is not used for general plasmid selection, but to generate cells that stably express integrated recombinant DNA. The selective agent you will use will be specific for mammalian cells and is identified on the plasmid. Typical mammalian selection includes resistance to antibiotics such as Geneticin (aka G418), bleomycin, puromycin, zeocin, or hygromycin B.

Prior to transfection, it is recommended you create a "kill curve" with the mammalian selection agent (determined by the plasmid used for transfection) for your particular cell line. In the case of pEGFP-N1, the selection for mammalian cells is G418. A kill curve allows you to empirically determine the lowest concentration of selection required to kill nontransfected cells. Since this will vary from cell line to cell line and selection agent to selection agent, it is recommended that you perform a kill curve for each particular cell line/selection you will use. Typically, cells that have not been transfected with plasmid should begin to die within 3–7 days. Once a kill curve has been completed, the appropriate concentration of selective medium is added to transiently transfected cells ~48 hours after transfection. Over time, cells that did not stably integrate the foreign DNA will die, leaving only those that have integrated the DNA. The remaining stable cells can then be expanded, stored, and/or used for subsequent experiments.

II. LABORATORY EXERCISES

If you wish to make stable cells with pEGFP-N1 plasmid integrated into the HEK293 genome, follow the given instructions *after* you have successfully confirmed EGFP expression after transient transfection (Lab Session 19b).

A. Performing a Kill Curve

It is recommended you create a kill curve prior to the initial transfection as it may take 1–2 weeks to achieve the desired results.

Plate your *nontransfected* cells at ~20%–30% confluency in a six-well plate. Of note, 24–48 hours later, add varying doses of G418 to complete medium in each well of your plate for a total volume of 2 mL per well. Typically starting around 100–200 μg/mL of G418 is a good place to begin. For example, you may create a kill curve of the concentrations of G418 (in μg/mL): 0, 100, 200, 300, 400, and 500.

After adding the selective medium, observe the cells under light microscopy for the next 1–2 weeks. Cells should start to die in 3–7 days. The dose of G418 you will want to use is the lowest dose that kills 100% of your nontransfected cells. This will be the concentration you will use to select for stable clones after transfection of pEGFP-N1 or another plasmid of interest.

B. Selecting for Stables

1. Forty-eight hours after your initial transfection, add the appropriate selective medium (in this case, DMEM supplemented with the appropriate G418 determined from the kill curve) to the cells to create stable clones.

2. You will observe your cells by light microscopy when possible in the next 7–10 days. All cells not containing the pEGFP-N1 plasmid will be killed in the presence of G418. Cells that took up the pEGFP-N1 plasmid will be alive in the presence of G418. Cells should start to die in 3–7 days and resistant colonies (those that integrated the DNA into the genome) will remain and proliferate in 10–14 days depending on how fast the cells divide.
3. Replace medium with the appropriate concentration of G418 every 2–3 days for up to 1 week.
   Note: It is normal for cationic lipid-based transfection efficiencies to vary between 20% and 60%, so do not be alarmed if you have a high number of dead cells during the selection process. The pooled population of cells that remain which have stably integrated egfp can be expanded through multiple rounds of cell division.
4. To maintain your stable clones, you will need to continue to grow them in the presence of complete medium and G418. However, it is recommended for any subsequent experiments to replace the medium with antibiotic-free medium. Selective antibiotics may influence your experimental results.
5. When you have a sufficient number of cells, verify their EGFP expression by fluorescence microscopy (Lab Session 19b), and it is recommended to freeze and store aliquots in liquid nitrogen for future use. The stable cells created are from a genetically heterogenous pooled population of transfectants, which will have varying levels of EGFP expression. Other methods for creating stables will consist of isolating single cell clones to generate a genetically homogenous and clonal population.

For more information, see the FuGENE HD Transfection Reagent manual (http://bit.ly/1GaElqy)[1].

REFERENCES

1. P.L. Felgner, et al., Lipofection: a highly efficient, lipid-mediated DNA-transfection procedure, Proc Natl Acad Sci 84 (21) (1987) 7413–7417.
2. A.V. Kabanov, V.A. Kabanov, DNA complexes with polycations for the delivery of genetic material into cells, Bioconj Chem 6 (1995) 7–20.
3. V. Magidson, A. Khojakov, Circumventing photodamage in live-cell microscopy, Methods Cell Biol 114 (10) (2013) 545–560.

DISCUSSION QUESTIONS

1. What are some challenges of transfecting nucleic acids into mammalian cells? How can these barriers be overcome?
2. What is the difference between transient versus stable transfection? Which method would be best to create a cell line that would be used to examine long-term changes in gene expression after treatment with various pharmacological agents?
3. How can you confirm if your HEK293 cells were successfully transfected with pEGFP-N1?
4. Is it possible to have successfully transfected the pEGFP-N1 plasmid into HEK293 cells, but not see any fluorescence using the fluorescence microscope? What other ways could you confirm the transfection was successful?

REFERENCE

1. Promega Corporation, FuGENE® HD Transfection Reagent Technical Manual, Promega Corporation, Madison, WI, 2013.

Lab Session 20

RNAi-Mediated Knockdown of EGFP: Part I

Goal: For this exercise, you will use a web-based small (short)-interfering RNA (siRNA) design tool, BLOCK-iT, from Thermo Fisher Scientific to design siRNA oligonucleotides targeted to *egfp*. Once the appropriate siRNAs are chosen, you will have them commercially synthesized to test in vitro in the next lab session. Note that many other companies (e.g., Dharmacon, Ambion) offer similar free software on their websites as well.

I. INTRODUCTION

RNA interference (RNAi) or post-transcriptional gene silencing is a powerful biotechnological tool that can be exploited to study the effects of genes in biological systems both in vitro and in vivo. RNAi is an endogenous and highly conserved pathway present in many organisms that involves the post-transcriptional downregulation of gene expression. Double-stranded RNAs (dsRNAs) prevent the expression of endogenous mRNA via complementary interactions with their target mRNA, thus preventing the production of proteins (aka silencing). Silencing occurs in a sequence-specific manner, such that only the gene whose mRNA matches the dsRNA is affected. Suppression of gene expression by RNAi occurs by two main mechanisms: (1) mRNA is targeted for degradation; or (2) mRNA is blocked from translation via sequestration into processing bodies (p-bodies) or interference of ribosome action. It is believed that RNAi is an ancient "immune system" that protects organisms from both endogenous parasitic and exogenous pathogenic nucleic acids as well as the modulation of the expression of protein-coding genes. Several types of small nonprotein-coding RNAs have been discovered that perform silencing functions. To date, they include classes of siRNAs, microRNAs (miRNAs), repeat-associated RNAs (rasiRNAs), and piwi-interacting RNAs (piRNAs).

The mechanism of RNAi is complicated; however, the basic pathway found in mammals is summarized here:

RNAi can be initiated by a number of exogenous triggers such as foreign DNA or dsRNA from a virus. These "foreign" nucleic acids are recognized by a RNase III ribonuclease enzyme called Dicer or Drosha. The Dicer enzyme acts like a molecular ruler, binding a dsRNA end and cleaving both strands into ~21–23 nt siRNA fragments. The siRNA is loaded onto the RNA-induced silencing complex (RISC). RISC is a multi-subunit complex containing a key catalytic component known as Argonaute (AGO). The siRNA is separated into single-stranded RNAs: the guide strand (antisense to the target mRNA) stays loaded onto RISC, and the passenger strand (sense to the target mRNA) is released and degraded. Next, the siRNA-RISC complex hybridizes with its mRNA target. The hybridization is 100% complementary, and gene silencing is the result of nucleolytic degradation of the targeted mRNA by the RNase H enzyme AGO (Figure 20.1; siRNA pathway on the right). Since the gene encoding the target protein is still present, repression is typically not 100%; therefore, we refer to this as protein knockdown (rather than knockout) (Figure 20.1).

RNA silencing by miRNAs is used by many organisms to regulate the expression of endogenous genes. The main difference between the various classes of silencing molecules is their origin. siRNAs are derived from complementary dsRNA which are typically introduced from outside the cell, such as from viruses. siRNAs hybridize 100% to their mRNA target. miRNAs may form from nonprotein-coding regions in endogenous genomic DNA or as part of a transcript for protein-coding RNAs.[1,2] The miRNA transcript has a region that folds back on itself, forming a hairpin and a large unpaired loop. In the nucleus, the primary miRNA transcript is then processed to a ~22 mer by the ribonuclease Drosha. The now precursor miRNA is shuttled to the cytoplasm by the exportin-5 protein, where it is a substrate for Dicer cleavage. Dicer processing and loading onto the RISC complex is analogous to that for siRNAs. Once the miRNA is loaded on the RISC complex, it targets mRNAs in the cell to regulate gene expression. Unlike siRNAs, however, the mRNA target is typically not a perfect match to the guide miRNA, which results in translational inhibition of the mRNA through sequestration or by preventing ribosome action[3] (Figure 20.1; miRNA pathway on the left).

Molecular Biology Techniques. DOI: https://doi.org/10.1016/B978-0-443-26714-7.00028-4

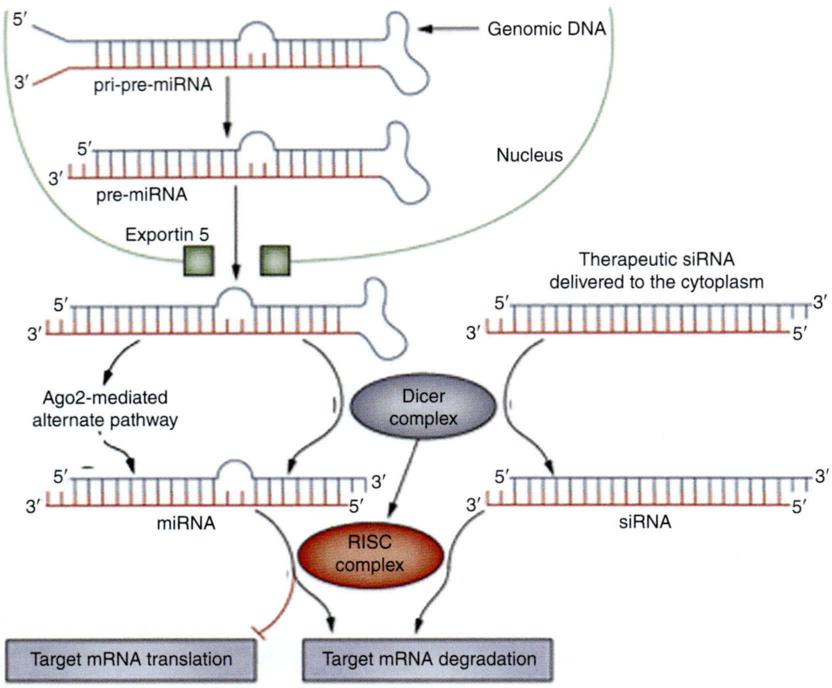

FIGURE 20.1 Mechanisms of RNA interference (RNAi). Reprinted by permission from Czech MP, et al. RNAi-based therapeutic strategies for metabolic disease. Nat Rev Endocrinol 2011;7(3):473–84.

In the lab, RNAi can be triggered in a number of different model organisms ranging from mammalian cell culture,[4] plants,[5] *Caenorhabditis elegans*,[3] and others by exogenous introduction of dsRNA (for short-term knockdown) or constructs which express short hairpin RNA (shRNA) plasmids (for long-term knockdown). The high specificity and efficacy (how many mRNA target molecules can be inhibited by the siRNA) of gene repression make RNAi an advantageous tool to study gene function. The success of RNAi in this context depends on how well the siRNA binds the target mRNA, as each siRNA will vary in silencing ability as well as off-target effects. Off-target effects can include the initiation of cells' natural interferon immune response (in mammals) or targeting unintended mRNAs. This may occur through sequence similarity of either strand of the duplex siRNA with an mRNA that is not the intended target. Furthermore, extremely high concentrations of siRNA (>100 nM) can produce nonspecific off-target activity. The higher the concentration of an individual siRNA, the increased likelihood it will anneal to nonspecific mRNA targets. Therefore, designing efficient and specific siRNA oligonucleotides is a key step in its successful application.

Numerous companies offer custom siRNA design, as well as specialized web-based tools to facilitate the design of your siRNAs. These tools take into consideration various published selection criteria. Furthermore, many companies incorporate proprietary chemical modifications into their synthesized siRNA molecules, which can increase the stability and specificity of siRNAs. Regardless of the method for siRNA design, it is helpful to know the general guidelines for what constitutes a good siRNA effector molecule. A list of general rules is provided next. Note that this list is not comprehensive and does not guarantee a successful effector molecule. All in silico designed siRNAs should be tested in vitro and alongside the appropriate controls.

The list of general considerations when designing siRNAs includes:

1. *Select target region of mRNA*: The target region is typically ~21 nt long. Do not design siRNAs greater than 30 nt to avoid nonspecific silencing.[6] It is helpful to first find your mRNA of interest using the RefSeq database from the NCBI (https://www.ncbi.nlm.nih.gov/refseq/). Scan the sequence starting with the AUG start codon until you reach an AA dinucleotide sequence. Record the placement of each AA and the 3′ adjacent 19 nt as potential siRNA target sites. Elbashir et al.[7] suggest that siRNAs with 3′ UU overhangs are the most effective silencing siRNAs, although other dinucleotide overhangs can be utilized. It is recommended to avoid 3′ GG dinucleotide overhangs, as it may become a target for cleavage by RNAase at single-stranded G residues. You will want to choose two to four target sequences, all of which meet these guidelines.
2. *Starting nucleotide*: The first nucleotide of the siRNA sequence can be an A or a G.
3. *Avoid multiple nucleotide repeat regions*: Avoid areas with three or more G's or C's. This will help avoid intra-molecular secondary structures from forming, which will prevent proper hybridization to the target.

4. *GC content*: Choose sequences with low GC content (30%–50%), with a higher GC content at the 5′ end. However, some effective siRNAs are known to have high GC content; it is therefore important to validate all your silencing constructs in vitro.
5. *BLAST for homology to the mRNA database*: Compare your target sites to the appropriate genome database (human, mouse, etc.). Eliminate any putative target siRNA sequence that shares more than 13 continuous base pairs of homologies to other coding sequences. The BLAST website can be found here: https://www.ncbi.nlm.nih.gov/BLAST.
6. *Use the appropriate siRNA controls*: Any siRNA experiment should include a number of controls to ensure specificity and lack of off-target effects.
   a. *Scrambled siRNA*. This negative control is a scrambled nucleotide sequence of the gene-specific siRNA. An additional BLAST search of the scrambled siRNA is needed to verify that it does not have homology to any other gene.
   b. *Additional siRNA sequences* that target the same mRNA and achieve similar biochemical effects.
   c. *A nonspecific siRNA* that is not known to target any other genes in the genome of your organism.
   d. *Rescue experiment* to ensure the knockdown effect is due to the gene of interest. This can be performed by transfecting a mammalian expression vector encoding your gene of interest back into your cells after siRNA-mediated knockdown to "rescue" the phenotype.

For this lab session, you will be designing siRNAs specific for *egfp* mRNA that will then be synthesized commercially as individual complementary RNA strands that are annealed together. Some researchers, however, may choose to use pooled siRNAs, which are multiple siRNAs to the same target (siRNA pool). The advantage of using pooled siRNAs is that they can reduce off-target effects because the initial concentration of each siRNA is low, but the overall silencing effectiveness is maintained.

II. LABORATORY EXERCISE

In silico design of siRNAs targeting *egfp*

This protocol was adapted with permission from Thermo Fisher Scientific:

1. On the BLOCK-iT RNAi site (https://rnaidesigner.thermofisher.com/rnaiexpress/), leave "Stealth RNAi™ siRNA" selected as the default.
2. Input target sequence information either by accession number or nucleotide sequence (Figure 20.2). For the *egfp* sequence, use the NCBI database nucleotide search (https://www.ncbi.nlm.nih.gov/nuccore) to find the synthetic

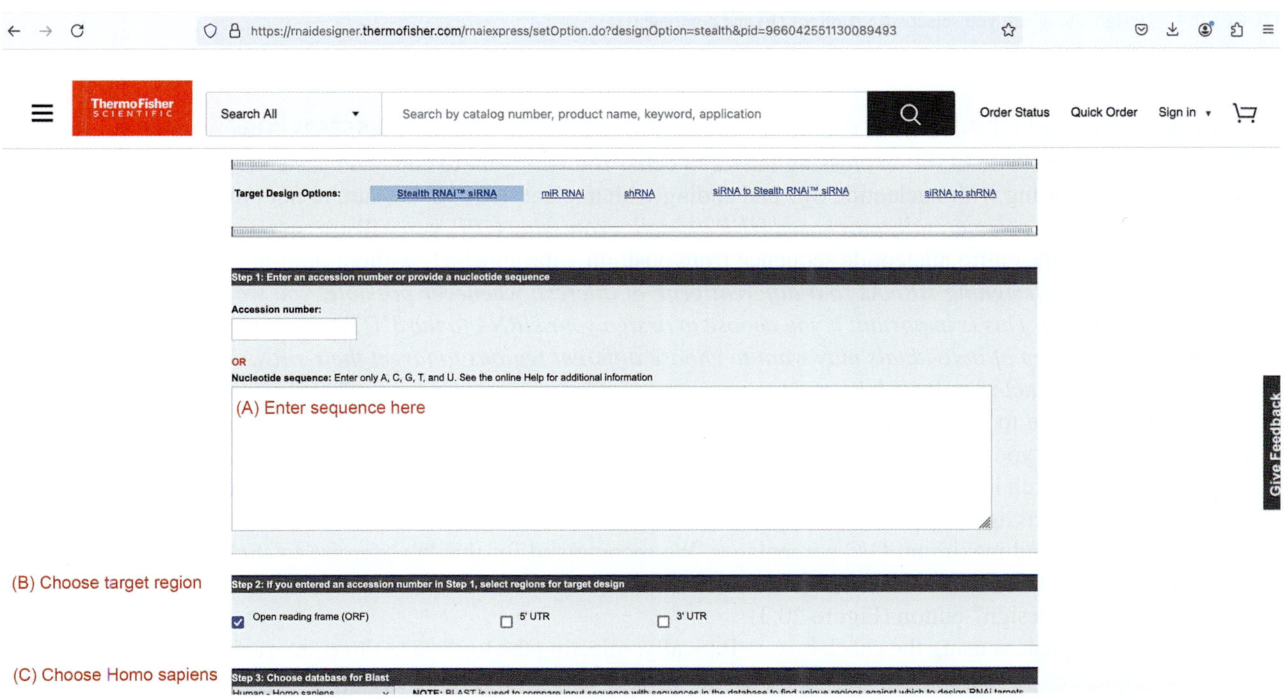

FIGURE 20.2 Screenshot of the BLOCK-iT RNAi Designer Interface. (A) Box where you enter egfp coding sequence in FASTA format. (B) Check boxes where you may select regions for siRNA target design. (C) Drop-down box to select organism database to BLAST. This will identify unique regions on the target mRNA.

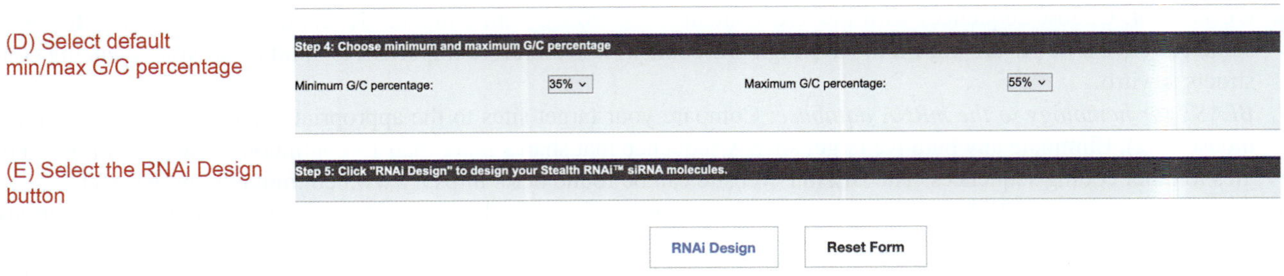

FIGURE 20.3 Design site where you select default (D) min/max G/C percentages and then (E) design your siRNAs.

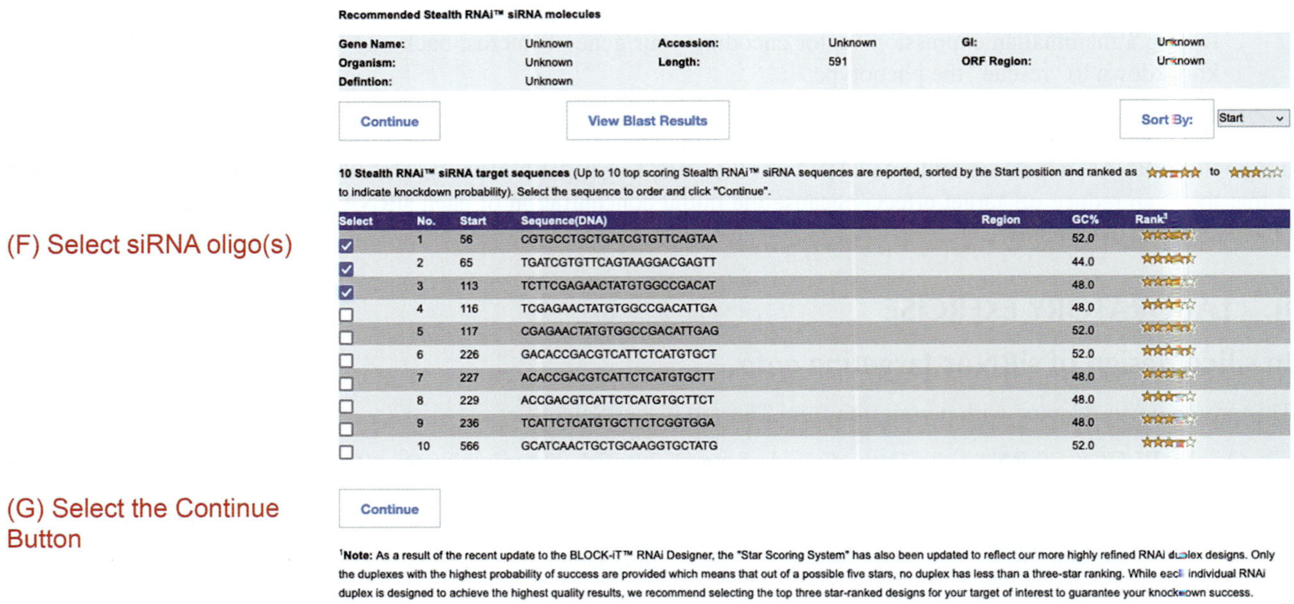

FIGURE 20.4 Design site where you select siRNA oligos (F) and continue (G).

construct plasmid pEGFP-N1 from Clontech Laboratories (accession number: U55762). This plasmid is what you used to create your stable EGFP-expressing HEK293 cells. You will want to copy and paste only the coding sequence (CDS) for *egfp* starting with nucleotide 679 and ending with nucleotide 1398. Be sure to not omit any nucleotides.

If you are using the commercially purchased HEK293 cells expressing GFP, you will use accession number: M62653.1. Copy and paste the entire nucleotide sequence from position 1 through 961, without omitting any numbers.

Note: If you are designing siRNAs to a different gene of interest, whenever possible, you want to enter the RefSeq accession number. This is important if you choose to design your siRNA to the 3'UTR of your mRNA transcript. Each individual or group of individuals may want to choose different regions to target their siRNA and compare/contrast the efficacy of silencing in later lab sessions.

3. Choose the database to perform a BLAST search of your gene (e.g., *egfp*) against all other genes in the organism (Figure 20.2). Since you want to knockdown EGFP in human embryonic kidney cells, you will choose the human database. A BLAST search is important in the design of your siRNA. It is used to compare input sequences in the database to find unique regions against which to design RNAi targets.
4. Choose minimum and maximum G/C percentages. We recommend for this lab exercise to stick with the default settings of 35%–55% (Figure 20.3).
5. Select the "RNAi Design" button (Figure 20.3).
6. Select siRNA sequences using the check boxes. Typically, selecting the top two to three-star rankings will be effective. Note the sequences of your chosen siRNA molecules in your notebook (Figure 20.4).
7. Click "Continue."

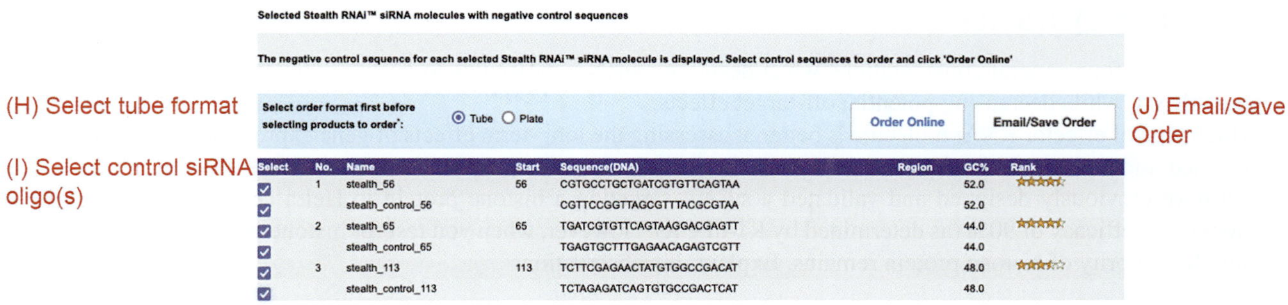

FIGURE 20.5 (H) Select the tube format and (I) the matching control scrambled siRNAs which correspond to the siRNA oligos previously chosen. Click (J) to email/save the order when finished.

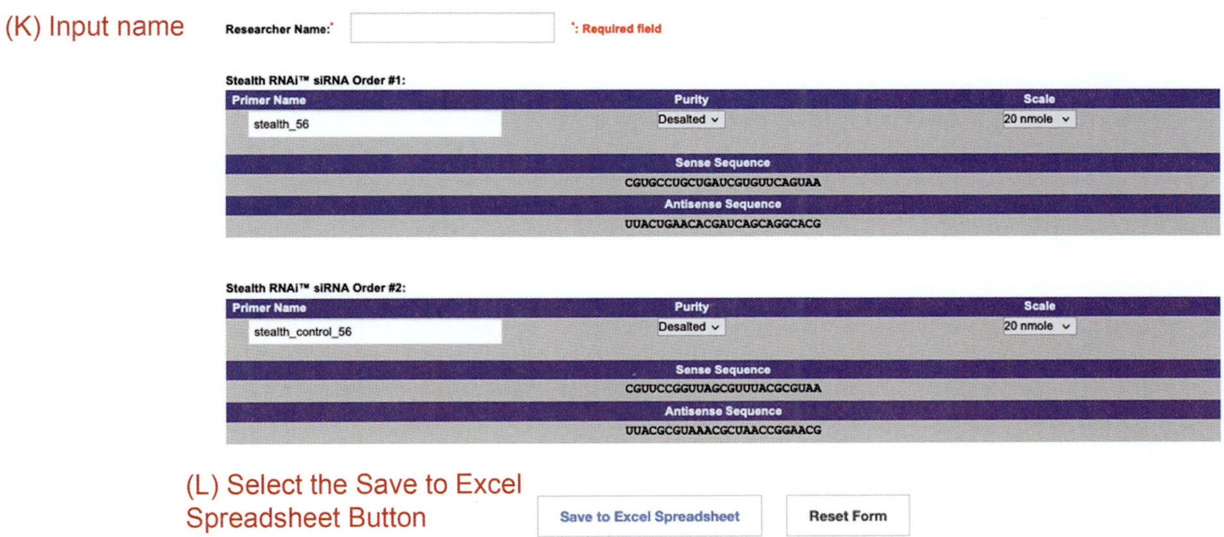

FIGURE 20.6 (K) Enter your name as instructed and (L) select save to Excel Spreadsheet.

8. You will want to also select a negative control. Your negative control can be a scrambled siRNA or a siRNA targeting a gene not present in your organism of interest (Figure 20.5).
9. Click "Tube" format and "Email/Save Order" (Figure 20.5)
10. You may review your choices here. Fill out "Researcher Name" and save the designed siRNAs to a Microsoft Excel spreadsheet by clicking "Save to Excel Spreadsheet" at the bottom of the screen (Figure 20.6). Submit a copy to your instructor, who will choose the siRNAs you will test in subsequent experiments.

Note to instructors: You may use the default settings when you are ready to order your siRNA oligos.

REFERENCES

1. T. Du, P.D. Zamore, MicroPrimer: the biogenesis and function of microRNA, Development 132 (21) (2005) 4645–4652.
2. V.N. Kim, MicroRNA biogenesis: coordinated cropping and dicing, Nat Rev Mol Cell Biol 6 (2005) 376–385.
3. Editors of Nature Cell Biology, Whither RNAi? Nat Cell Biol, 5 (2003) 489–490.
4. J.Y. Yu, et al., RNA interference by expression of short-interfering RNAs and hairpin RNAs in mammalian cells, PNAS 99 (9) (2002) 6047–6052.
5. C. Napoli, C. Lemieux, R. Jorgensen, Introduction of a chimeric chalcone synthase gene into petunia results in reversible co-suppression of homologous genes in trans, Plant Cell 2 (1990) 279–289.
6. S.M. Hammond, E. Bernstein, D. Beach, G.J. Hannon, An RNA-directed nuclease mediates post-transcriptional gene silencing in Drosophila cells, Nature 404 (2000) 293–296.
7. Elbashir, et al., Functional anatomy of siRNA for mediating efficient RNAi in Drosophila melanogaster embryo lysate, EMBO J 20 (2001) 6877–6888.

DISCUSSION QUESTIONS

1. List three to four general considerations when designing siRNA molecules to ensure high specificity and efficacy of your siRNA, while decreasing potential off-target effects.
2. Which type of effector RNAi molecule is better at assessing the long-term effects of gene expression: siRNA or shRNA designed into a plasmid? Justify your answer.
3. You have previously designed and validated a siRNA targeting a histone protein in HeLa cells to achieve a mRNA knockdown efficacy of 90% (as determined by RT-qPCR). However, when you test for histone protein levels by western blot, the majority of histone protein remains. Explain this observation.

Lab Session 21

RNAi-Mediated Knockdown of EGFP: Part II

Goal: You will count and plate HEK293 cells expressing EGFP (created in Lab Session 19: Transient Transfection of Mammalian Cells) in preparation for transient transfection of siRNA EGFP oligos to knockdown the EGFP protein later this week.

I. INTRODUCTION

Review the Introduction from Lab Sessions 18–20.

Molecular Biology Techniques. DOI: https://doi.org/10.1016/B978-0-443-26714-7.00041-7

Lab Session 21a

Counting and Plating EGFP Expressing Mammalian Cells

I. LABORATORY EXERCISE

Before beginning, review how to culture, count, and plate mammalian cells in Lab Session 18, Culturing Mammalian Cells.

Approximately 24 hours before Lab Session 21b, plate 3×10^5 HEK293 cells expressing the pEGFP-N1 plasmid (created in Lab Session 19, Transient Transfection of Mammalian Cells) per well in a six-well plate. The cells should be plated in 2 mL of complete growth medium *without any antibiotics*.

Lab Session 21b

Interim Laboratory Session

TRANSIENT TRANSFECTION OF siRNAS

Goal: To knockdown expression of the EGFP protein in HEK293 cells. You will transiently transfect your designed silencing, positive, and negative control siRNA oligonucleotides into the HEK293 EGFP expressing cells using the transfection reagent FuGENE HD. An overview of the transfection procedure is illustrated in Lab Session 19, Transient Transfection of Mammalian Cells (Figure 19.3).

I. LABORATORY EXERCISE

When working with siRNAs, adhere to the protocol "Challenges of working with RNA" found in Part IV of the manual. Ensure that in addition to using sterile items, they are also RNase free. This includes water, barrier pipette tips, microcentrifuge tubes, and serological pipettes. The use of RNase inactivating spray can reduce the risk of contaminating RNases in your experiment. An overview of the siRNA transfection procedure, which was modified with permission from Promega Corporation (©2024 Promega Corporation), is described below.

1. Remove the FuGENE HD reagent from the 4°C refrigerator and allow to reach room temperature. Mix gently by inversion before use.
2. Observe the morphology, confluency, and fluorescence of each well of your HEK293 EGFP expressing cells, plated from the previous lab session, under a 20× objective of a fluorescence microscope. Record your observations in your lab notebook similar to Lab Session 19b, Table 19.2 from Chapter 19: Transient Transfection of Mammalian Cells. Return your plate back to the 37°C cell culture incubator.
3. In the biological safety cabinet, label five microcentrifuge tubes as: No siRNA (Opti-MEM only), non-silencing or scrambled siRNA (negative control), siRNA eGFP (positive control), siEGFP #1, siEGFP #2 (siEGFP #1 and #2 refer to the experimental siRNA oligos previously designed in Lab Session 20: RNAi-Mediated Knockdown of EGFP: Part I), and Opti-MEM only.

   $ Money-Saving Tip: You may use serum-free medium in lieu of Opti-MEM.
4. Pipette 100 µL Opti-MEM into each tube.

TABLE 21.1 Experimental Set-up for siRNA-Mediated Knockdown of EGFP

Tube	Conc. of siRNA (nM)	Vol. to add of siRNA (µL)[a]	Opti-MEM (µL)	FuGENE HD (µL)
No siRNA (Opti-MEM only)	0	0	100	10
Scrambled siRNA (negative control)			100	10
siRNA eGFP (positive control)			100	10
siEGFP #1			100	10
siEGFP #2			100	10

[a]The total volume of siRNA should not exceed 15 µL.
No Permission Required.

5. Add each siRNA oligonucleotide to its respective labeled tube for a final concentration of 40 nM in a total of 2 mL medium. In your lab notebook, draw the table provided in Table 21.1 to record the amount of siRNA you need to add for each treatment. Pipette gently up and down to mix. *Note: For experimental siRNAs, you may need to adjust the final total concentration. We recommend starting within the range of 20–100 nM of siRNA.*
6. Pipette 10 µL FuGENE HD reagent DIRECTLY into each specified tube. *DO NOT allow the FuGENE HD reagent to directly contact the plastic walls of the tube*, as this will drastically reduce transfection efficiency. Be sure to change tips for each new aliquot.
7. Pipette up and down to mix and leave in the biological safety cabinet for 15 minutes. This time will allow the formation of siRNA:transfection reagent complexes, enabling the siRNA to enter the cells.
8. Remove your group's plate from the incubator. Label (on the lid) what each well will contain using the names given in Table 21.1. Label one well as no siRNA. Cells in that well will not receive any siRNA but will have transfection reagent.
9. Add the entire volume of the FuGENE HD:siRNA complexes drop-wise to the corresponding well of the six-well plate. Rock the plate back and forth gently to ensure complete mixing of the FuGENE HD:siRNA complexes throughout the medium and return the plate to the 37°C incubator.
10. Observe the qualitative knockdown of the EGFP protein by fluorescence microscopy 72 hours after siRNA transfection. *Note: For self-designed siRNAs, it is recommended to monitor for knockdown at 48, 72, and 96 hours.*

DISCUSSION QUESTIONS

1. When designing your experimental setup for siRNA-mediated knockdown of EGFP, why is it important to treat some samples with transfection reagent alone (with no siRNA oligos)?
2. To introduce our siRNAs into our target cells, we are using a cationic lipid-based transfection method. How does this method work? What are other methods that could be employed to introduce siRNA (or other nucleic acids) into our cells in culture? *Note: The last part of the question will require outside resources.*

Lab Session 22

RNAi-Mediated Knockdown of EGFP: Part III

Goal: To quantify the efficiency of RNAi-mediated knockdown of the EGFP protein in HEK293 cells using fluorescence microscopy and NIH Image J.

I. INTRODUCTION

Refer to Lab Session 19, Transient Transfection of Mammalian Cells, Introduction.

II. LABORATORY EXERCISE

Seventy-two hours after transfection with siRNA oligonucleotides targeted to EGFP in HEK293 cells (Lab Session 21b), acquire and save bright-field and fluorescence images of each well as previously described in Lab Session 19b. Save your images as either .JPG or .TIFF files. Be sure to also record qualitative visual observations in your lab notebook according to Table 22.1.

1. Download and Open NIH ImageJ on your personal computer. ImageJ is freely available for a variety of different operating systems from the National Institutes of Health. (https://imagej.net/software/imagej/)
2. From file menu, select "Open," and choose the fluorescence image you wish to analyze first. It is best to start with your control HEK293 EGFP expressing cells that were not transfected with any siRNA (e.g., Opti-MEM only treated). You would anticipate these cells to have the highest levels of EGFP-mediated fluorescence.
3. From the analyze menu, select "Set measurements." From this display, be sure to check box the following items: area, integrated density, and mean gray value and then select "OK" (Figure 22.1).
4. From the ImageJ toolbar at the top, select the "Freehand tool" (Figure 22.1).
   Using the freehand tool and your mouse, carefully outline one healthy cell of interest in your fluorescence image (Figure 22.2). *Note: It is useful to also open the accompanying bright-field image to identify cells that may not express considerable EGFP protein but that you will still need to select for analysis in the fluorescence image.*
5. Once you have outlined your first cell with the freehand tool, select "Measure" (or Ctrl-M; PC computers or Command-M; Mac computers) under the analyze menu (Figure 22.2).
6. A new window will appear that will list all the measurement results for that particular cell (Figure 22.2).
7. Select a non-cellular, small region (exact size does not matter) next to your cell of interest that has no fluorescence. Record the measurement by selecting "Measure" under the analyze menu. This will serve as your background fluorescence reading.
8. Repeat Steps 4–7 for all your cells in your control experiment. Count a minimum of 10 cells from each treatment. You may need to analyze additional images from other fields of view for your sample if you do not have at least 10 cells.
9. Once you have taken measurements of all your cells, select all the data in the results window and copy and paste into a Microsoft Excel spreadsheet (Figure 22.3).
10. Calculate the corrected total cell fluorescence (CTCF) for each cell using the equation:

$$\textit{Corrected total cell fluorescence (CTCF)} = \textit{Integrated density} - (\textit{area of selected cell} \times \textit{mean background fluorescence})$$

Molecular Biology Techniques. DOI: https://doi.org/10.1016/B978-0-443-26714-7.00045-4

TABLE 22.1 Experimental Observations of siRNA-Mediated Knockdown of EGFP Protein

Experiment	Fluorescence Intensity	Confluency	Morphology
No siRNA (Opti-MEM only)			
Scrambled siRNA (negative control)			
siRNA eGFP (positive control)			
siRNA EGFP#1 (experimental)			
siRNA EGFP#2 (experimental)			

Note: Fluorescence intensity should be scored as: +++: high intensity; ++: medium intensity; +: low intensity; −: no fluorescence detected. Confluence should be expressed as a percentage. Morphology as epithelial- or fibroblast-like, adherent, and/or floating.

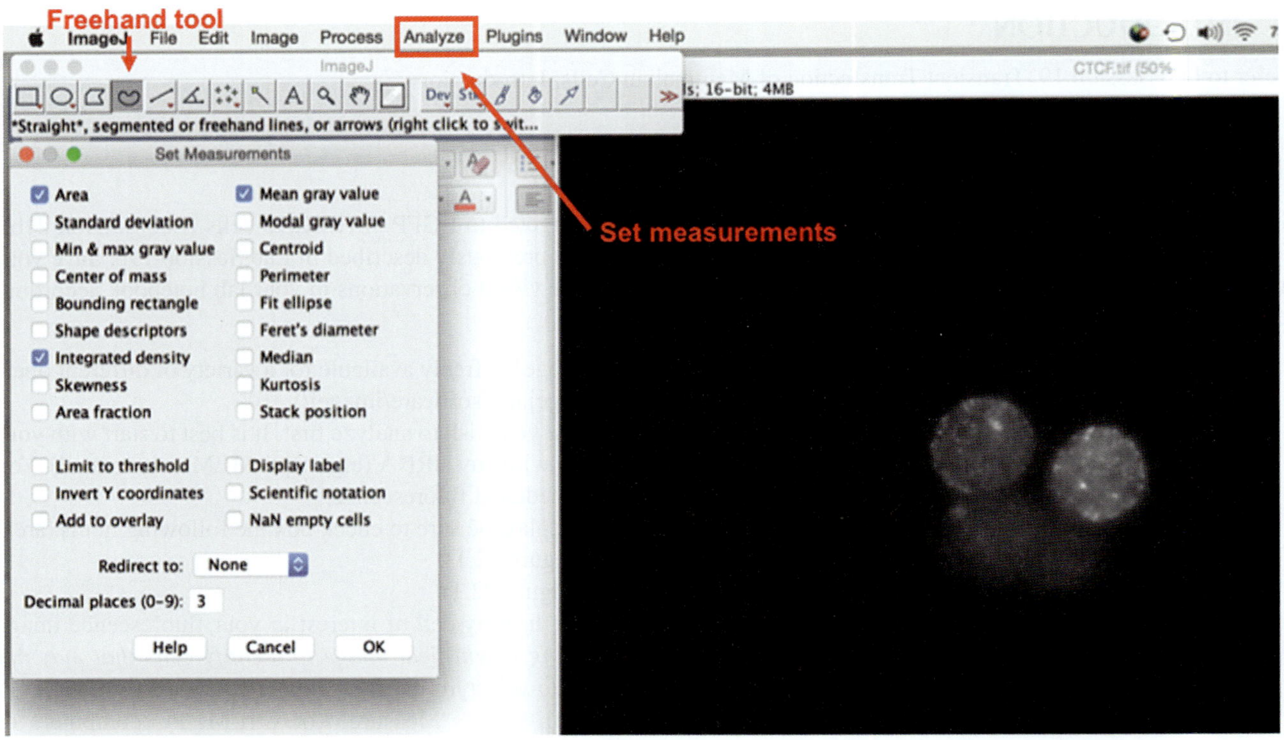

FIGURE 22.1 NIH ImageJ window indicating the location of the freehand tool and set measurements window. *No Permission Required.*

We use the integrated density since it represents the region of interest (ROI) × mean of the ROI. If you only took the mean fluorescence intensity, it will calculate the average fluorescence, which does not take the size of your cell into account. Therefore, if you have a large cell compared to a small cell, you will likely acquire a higher fluorescence value in the small cell compared to the large cell. However, with the greater size of the large cells more EGFP expression could be present even though the average may be lower at any one point in the cell. We will use integrated density so we do not have to worry about size issues between different samples, since that is already accounted for.

11. Determine the *average* value in a new cell in Microsoft Excel. To do this, in the formula bar type "= AVERAGE (select all cells containing the CTCF)."
12. Creating a new column, calculate the *standard deviation* of the corrected total cell fluorescence by typing in the formula bar "= STDEV.P (select all cells containing CTCF values)."
13. Repeat Steps 2–13 with each experimental group for a minimum of 10 cells per treatment: siRNA eGFP (positive control), scrambled siRNA (negative control), siRNA EGFP #1 (experimental), and siRNA EGFP #2 (experimental).

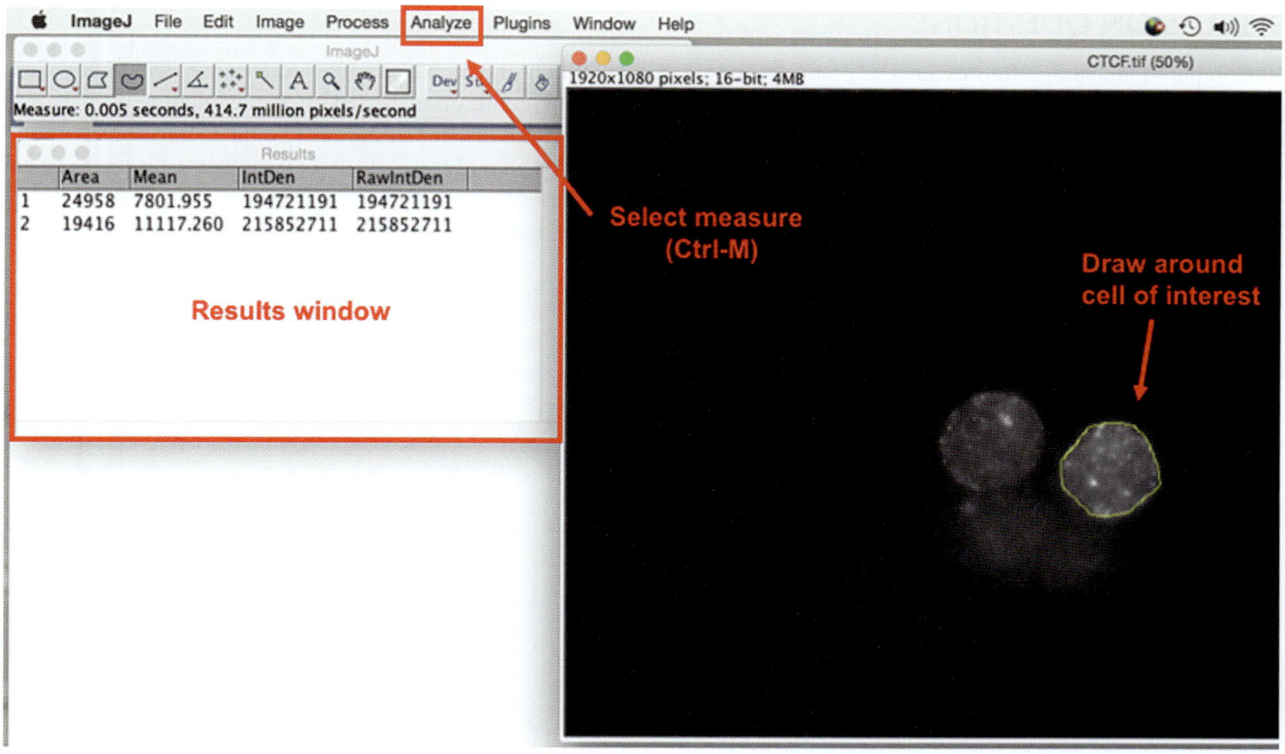

FIGURE 22.2 NIH ImageJ screenshot indicating outlined cell, measurement function, and results window. *No Permission Required.*

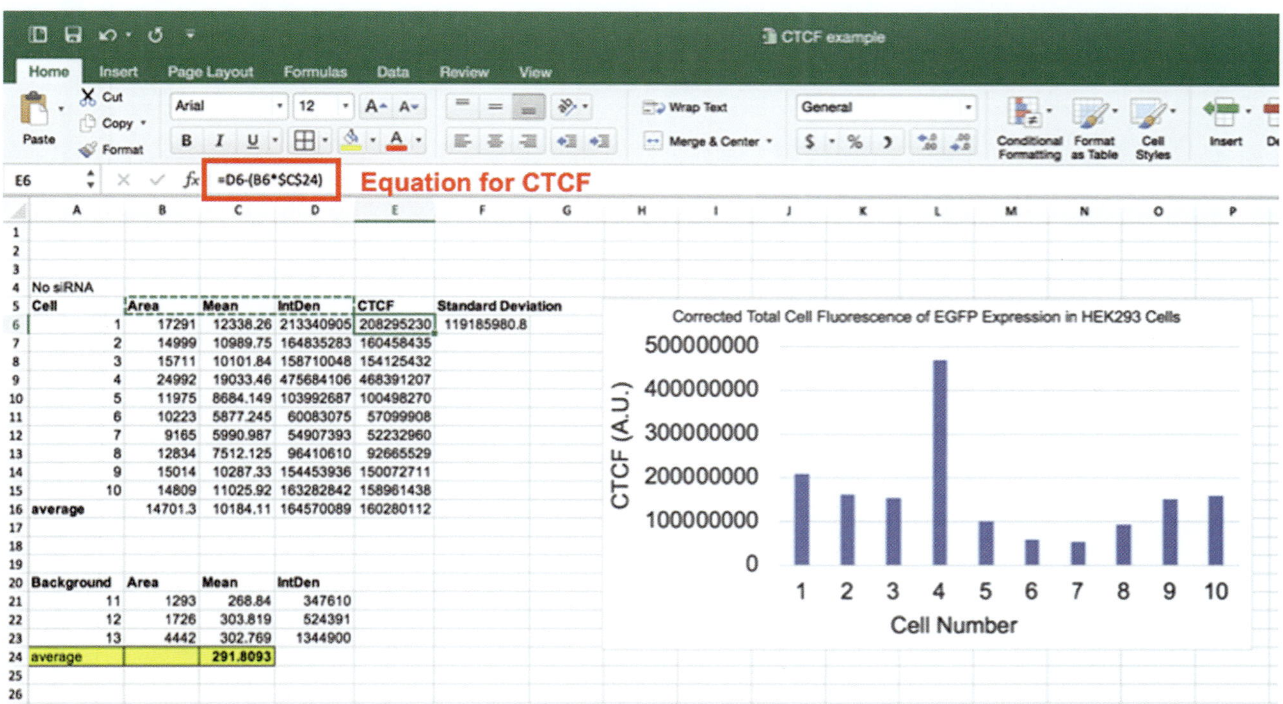

FIGURE 22.3 Example Microsoft Excel workbook calculating CTCF. *CTCF*, Corrected total cell fluorescence. *No Permission Required.*

14. When you are finished, create a bar graph comparing the *average CTCF fluorescence* (in Arbitrary Units) on the *y*-axis and the *cell treatment* (no siRNA, siRNA eGFP, etc.) on the *x*-axis for each of your experimental and control groups. An example is shown illustrating CTCF values per individual cell (Figure 22.3).
15. Be sure to label your graph appropriately, including your calculated standard deviation, labeling *x*- and *y*-axes with units (where appropriate), and a descriptive title.

DISCUSSION QUESTIONS

1. Compare the average fluorescence for each treatment group. Which groups have more fluorescence, which have less? Does the data agree with your expectations for the siRNA oligonucleotides transfected? If not, explain why.
2. Why is it important when measuring overall fluorescence intensity that the size of the cells be taken into consideration?
3. List two off-target effects of siRNAs. What are specific controls you should run in every experiment to ensure your data are not a result of off-target effects of the siRNA?
4. What is another technique that could be performed to quantify the extent of knockdown in your cells? How would this experiment be performed?

Lab Session 23

CRISPR-Mediated Knockout of EGFP: Part I

In the next several laboratory sessions, you will apply a genome engineering technology called CRISPR-Cas9 (clustered regularly interspaced short palindromic repeats—CRISPR associated protein 9) or CRISPR for short. Unlike RNAi, where levels of a gene product are knocked down, CRISPR can completely knockout a gene of interest. In essence, it creates a functional knockout where the information at the DNA level is disrupted. This is in contrast to a knockout mouse, for example, where the DNA sequence is removed. The versatility of CRISPR enables it to be used in other applications as well, including gene knockin, gene editing, or even gene activation and repression. The flexibility, relative ease of use, and low cost of this technology make it attractive to both basic and applied scientists. As our knowledge of CRISPR expands, so do its applications.

Your goal in the next two weeks will be to knockout *egfp* from a mammalian cell line using CRISPR. The same cell line that you may have used in RNAi lab sessions (the adherent human kidney cell line, HEK293, that stably expresses EGFP) will be used again. This cell line is diploid, meaning there should be two copies of each gene in the human genome. This is more desirable in a CRISPR experiment than, say, a cancer cell line that has abnormal copy numbers of each gene. HEK293 cells stably expressing EGFP will be transfected with plasmid DNA that encodes the two main components of the CRISPR system: Cas9 and a single guide RNA (sgRNA or gRNA) that is complementary to *egfp*.

Later, in Laboratory Sessions 25–30, you will have the option to design, create, and test sgRNAs specific to your own gene of interest.

This week will require at least one interim lab, as the cells must be transfected one day after plating them. Once transfected, you or your instructor/TA will need to expand the cells one time before analyzing for knockout next week.

Molecular Biology Techniques. DOI: https://doi.org/10.1016/B978-0-443-26714-7.00031-4

Lab Session 23a

CRISPR-Mediated Knockout of *egfp*, Part I

Goal: Today you will count and plate HEK293 cells stably expressing EGFP.

I. INTRODUCTION

Throughout the lab sessions in this book, you have cloned, expressed, and purified EGFP. You may have also expressed this gene in mammalian cells and performed RNAi experiments to knock it down. To gain additional experience in mammalian cell culture and introduce a revolutionary technology, you will next attempt gene editing of *egfp* via the CRISPR-Cas9 system, or just CRISPR for short.

CRISPR sequences (loci) were first described in archaea and bacteria in 1987 as short repetitive DNA fragments with spacer sequences, but their purpose was unclear.[1] In 2002, the term "CRISPR" was coined and Cas genes (CRISPR-associated genes) were identified.[2] The spacers were found to be homologous to bacteriophage DNA, revealing that the foreign genetic material was a remnant in previously infected bacteria. Subsequently, Cas proteins were identified that used transcribed CRISPR RNAs to recognize homologous sequences, cleave, and effectively destroy phage genes and transferred plasmids. Like our own adaptive immune systems, CRISPR can basically keep a record of invading pathogens and target those sequences for destruction upon reinfection.[2–8] This is similar in function to the RNAi pathway that occurs naturally in many organisms and helps combat foreign RNAs.

As more details about this phenomenon were revealed, scientists began to utilize CRISPR as a new type of genome engineering tool. Although gene editing technologies already existed, CRISPR was remarkable in its relative ease. Harnessing CRISPR's power now allows a researcher to target a specific gene, induce double-strand breaks (DSBs) in that DNA, and either delete it, modify it, or introduce a donor gene at that location. By targeting DNA, not RNA, the edits can be inherited. Additional applications of this technology continue to be discovered each year, including clinical trials for gene therapy using this editing technology. In 2020, Drs. Jennifer Doudna and Emmanuelle Charpentier were awarded the Nobel Prize in Chemistry for their discovery of this mechanism.[9] In 2023, the first CRISPR-based therapy was approved for use in the United States, among other countries worldwide. CASGEVY® is a therapy to cure sickle cell disease and transfusion-dependent beta thalassemia.[10] Before we discuss how you will use CRISPR to knockout *egfp*, we will first describe its mechanism.

There are two main components of the CRISPR system: Cas9, an endonuclease that catalyzes DSBs in the target DNA, and a gRNA that forms a complex with Cas9 to direct it to the specific sequence of DNA.[4,5] An overview of this process is shown in Figure 23.1. Once the Cas9:gRNA complex forms, it can randomly associate and dissociate with the genome. To accomplish this, Cas9 directly binds a protospacer adjacent motif (PAM). This 3-nucleotide sequence varies between organisms, but due to its short length, is highly abundant throughout the genome. Next, Cas9 unwinds a stretch of genomic/target DNA into single-stranded DNA. If the gRNA component is complementary with the now unwound target DNA, it will form hydrogen bonds, thus stabilizing the hybrid gRNA:DNA. This triggers Cas9 to undergo a conformational change, activating its nuclease activity. This results in cleavage of both strands of the DNA a few base pairs upstream of the PAM site. The PAM site is shown in purple in Figure 23.1.

Cells repair DSBs through one of two main mechanisms: non-homologous end joining (NHEJ) or homology-directed repair (homologous recombination repair). For the purposes of our experiment, we will focus on NHEJ, as that is how we will use CRISPR in the following laboratory sessions. NHEJ is illustrated in Figure 23.2 and shows that once a DSB occurs, a Ku heterodimer binds the break and recruits DNA polymerases, nucleases, and ligases involved in NHEJ. These enzymes catalyze reactions that can resect and/or add nucleotides in a template-dependent or -independent manner at the site of the break. NHEJ has great flexibility in the mechanisms of repair and can also result in the direct ligation of two DNA ends without any nucleotide addition or loss. Alternatively, extremely short sequences of homology (microhomology) can occur between the nucleotides on the end of both single strands. These either already exist (dashed boxes) or are the result of the polymerase activities during repair. This microhomology is often used to guide end joining.[11] This process is naturally error prone and will result in either an insertion or deletion of base pairs, which are known as "indels." These "indels" are evidence that genome editing has occurred at the intended DNA sequence. The presence of indels within a gene sequence can alter the gene's reading frame, inactivate it, and thus create a knockout.

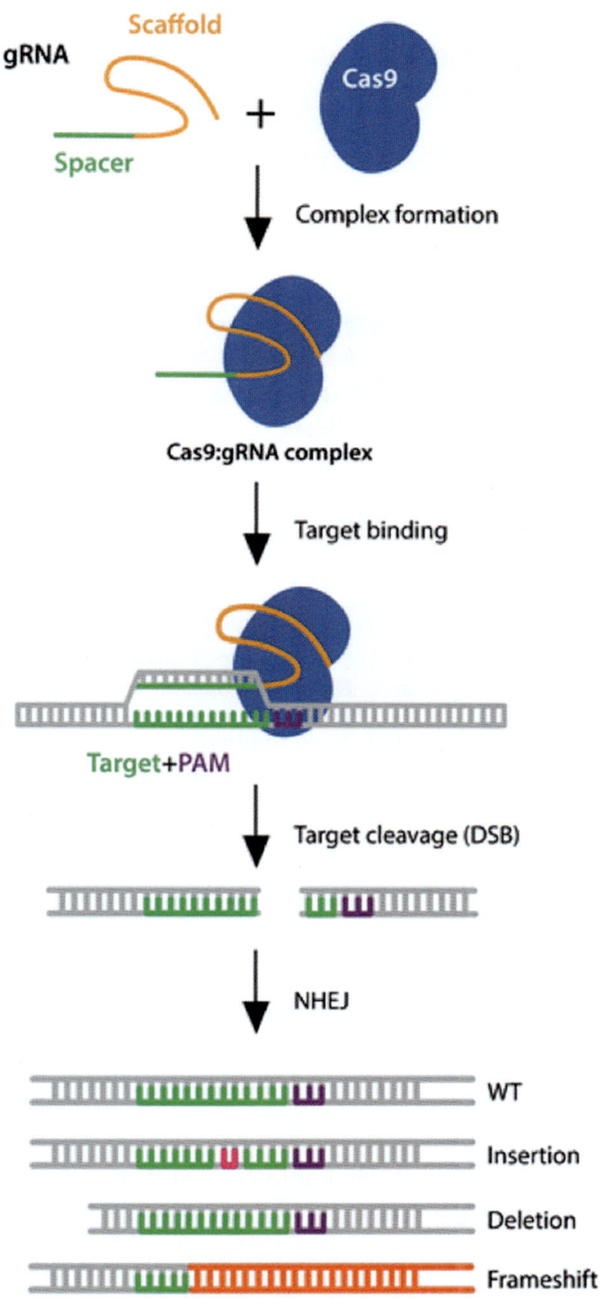

FIGURE 23.1 An overview of creating gene knockouts with CRISPR-Cas9gRNA, PAM, DSB, NHEJ, WT. *gRNA*, Guide RNA; *PAM*, protospacer adjacent motif; *DSB*, double-strand break; *NHEJ*, non-homologous end joining; *WT*, wild-type. *Used with permission from Addgene.*

By using CRISPR as a genome engineering tool, scientists are able to knockout, knockin, edit, activate, and repress individual genes in a wide variety of organisms. Unlike RNAi, which reduces or knocks down a gene by preventing translation, CRISPR is capable of completely inactivating the gene in the genome. Furthermore, these techniques are relatively inexpensive and quick to implement. The off-target effects in RNAi experiments have been well-documented,[12] and there is no way to completely eliminate them. Therefore, when CRISPR began being widely used, scientists were extremely interested in determining if the system's specificity outperformed that of RNAi. However, data suggests that significant off-target effects can occur during certain CRISPR experiments[13] (see Lab Session 25). The frequency of these off-target effects seems to vary widely.[14,15] This is a concern for researchers, as targeting an unintended gene (or multiple genes) could have detrimental effects in the experimental system.

CRISPR will most likely continue to revolutionize not only molecular biology research, but biotechnology applications, gene therapy, and other clinical applications as well. Ethical questions are continually asked about the future of

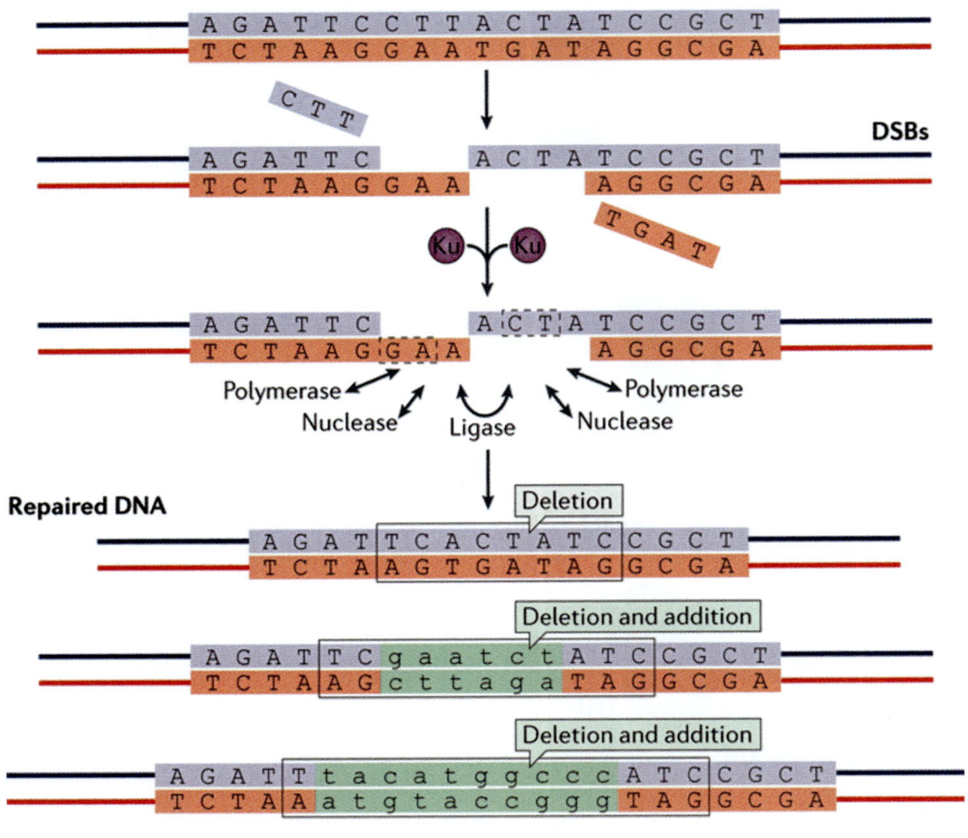

FIGURE 23.2 An overview of NHEJ. *NHEJ*, Nonhomologous end joining.[11] *Used with permission from Springer Nature.*

CRISPR and exactly how human genomes could be edited. Clearly, these important topics deserve attention from scientists around the world.

For an excellent interactive overview of CRISPR, please visit: https://www.hhmi.org/biointeractive/crispr-cas-9-mechanism-applications for The *CRISPR-Cas9: Mechanism and Applications* interactive (accessed 5 December 2024).

II. LABORATORY EXERCISE

Your instructor or TA will provide you with HEK293 cells stably expressing EGFP. Please use aseptic technique when working with these cells (for a review, see Lab Session 18: Culturing Mammalian Cells).

1. As performed in Lab Session 18, you will turn on, clean, and prepare the biological safety cabinet before you begin working with the cells. Collect all of the consumables you will need for these steps: sterile pipette tips, sterile Pasteur pipettes, sterile serological pipettes, and a sterile 48-well plate. Micropipettes and pipette guns should remain in the biological safety cabinet.
2. Obtain a dish or flask of cells from your instructor. Observe their confluency and morphology and record the details in your lab notebook, similar to Table 19.3. If you observe any signs of contamination, notify your instructor immediately. Also observe the color of your cell culture complete growth medium.
3. Using a sterile Pasteur pipette, aspirate the medium from the dish. Do this by gently tipping the dish toward you and with the lid/cap off, and place the tip of the Pasteur pipette to the bottom of the vessel until all the medium is aspirated. Remember the spot where you touched the dish. You will always aspirate from this spot to avoid aspirating cells from your dish.
4. Gently add ~2 mL of sterile 1× PBS to the dish and swirl carefully. Whenever you remove or add liquid to your cells, be sure to do so from the same location to avoid dislodging cells.
5. Aspirate the 1× PBS from the dish.

6. Add 2.5 mL of pre-warmed 0.25% Trypsin-EDTA solution to the dish and swirl to ensure adequate coverage on the bottom. Place dish back in the incubator and set your timer for 2–3 minutes.
7. Observe your cells under the microscope; detached cells will float in the medium and appear rounded. You may want to gently, but firmly, tap your dish on a sterile surface to help further detach your cells. If you still have 50% or greater cells attached to the dish, return it to the 37°C incubator for another 2 minutes.
8. Once cells appear detached, add 5 mL of pre-warmed complete growth medium (DMEM with 10% FBS) *without any antibiotics* to inactivate the trypsin. Gently pipette the medium over the cell layer surface several times to ensure detachment. Vigorous pipetting can cause cell damage, so be careful.
9. You are now ready to count your cells using a hemacytometer.

   Time-Saving Tip: An automated cell counter such as a Cellometer from Nexcelom can be used.
10. Clean the chamber and cover slip with 70% EtOH. Dry and put the cover slip in position.
11. Pipette cells up and down to mix thoroughly, then add 20 µL of your cell suspension to each chamber of the hemacytometer. Do this by slowly pipetting the cell suspension between the cover slip and hemacytometer; capillary action will spread the solution in an even layer between the two surfaces. Be sure not to overfill.
12. Place the chamber on the microscope under a 10× objective using phase contrast and course/fine focus to distinguish the cells and grid lines of the hemacytometer.
13. Using a hand counter, count the cells in one set of 16 squares (see Figure 23.3). When counting, include cells that fall in the middle, top, and right-hand boundaries.
14. Move the hemacytometer to the next set of 16 corner squares and count until all four sets of 16 squares are counted. Be sure to record your cell numbers from each square in your laboratory notebook.
15. Take the average cell count from each of the four sets of squares and multiply by 10,000 (10^4). The final number is the total amount of cells/mL in your cell suspension.
16. Calculate the volume of cells needed to seed a density of 4×10^4 cells in one well of a 48-well plate. An appropriate volume of solution in one well of a 48-well plate is 0.25 mL (250 µL). To add 4×10^4 cells in a volume of 0.25 mL, the concentration of the cell suspension needs to be 1.6×10^5 cells/mL.

    Using $C_1V_1 = C_2V_2$ where:

    C_1 is the concentration of cells counted (cells/mL):______________
    V_1 is the amount of cell suspension (mL) required (unknown):______________
    C_2 is desired concentration of cells: 1.6×10^5 cells/mL
    V_2 is desired total volume: 1.5 mL
17. Add the appropriate volume of cell suspension to a sterile microcentrifuge tube. Bring the total volume up to 1.5 mL with complete medium (DMEM) *without any antibiotics*. Mix well by carefully pipetting up and down (avoid forming bubbles) and immediately add 250 µL of cell suspension to three wells of your 48-well plate. You may be asked to share this plate with other student groups.
18. In addition, plate 125 µL of cell suspension to three different wells.
19. Add 125 µL of complete medium (DMEM) *without any antibiotics* to those wells from step 18 for a total of 250 µL per well. You should now have three wells with 4×10^4 cells each and three wells with half that amount (2×10^4 cells each) for a total of six wells containing cells.
20. Write your initials, as well as the date, on the lid of your plate. Return your plate to the cell culture incubator. Your instructor will tell you how to dispose of your extra cells.

Lab Session 23b

Interim Laboratory Session

TRANSFECTION OF CRISPR VECTORS

Goal: Today you will transfect mammalian expression vectors into the EGFP-expressing HEK293 cells that were previously plated. These vectors will introduce the required CRISPR components into the cells.

I. INTRODUCTION

The expression vectors you are using will each encode Cas9 driven by a mammalian promoter called EF-1α. This promoter drives expression of a human gene encoding the alpha subunit of elongation factor-1. The EF-1α promoter is commonly used in mammalian expression vectors. On the vector map of pLentiCRISPR-E (Figure 23.3), you will see that Cas9 is written as eSpCas9. The "Sp" stands for the organism *Streptococcus pyogenes*, from which this enzyme's sequence was obtained. Therefore, the Cas9 that will be produced in the human cells comes from a bacterium. The letter "e" in eSpCas9

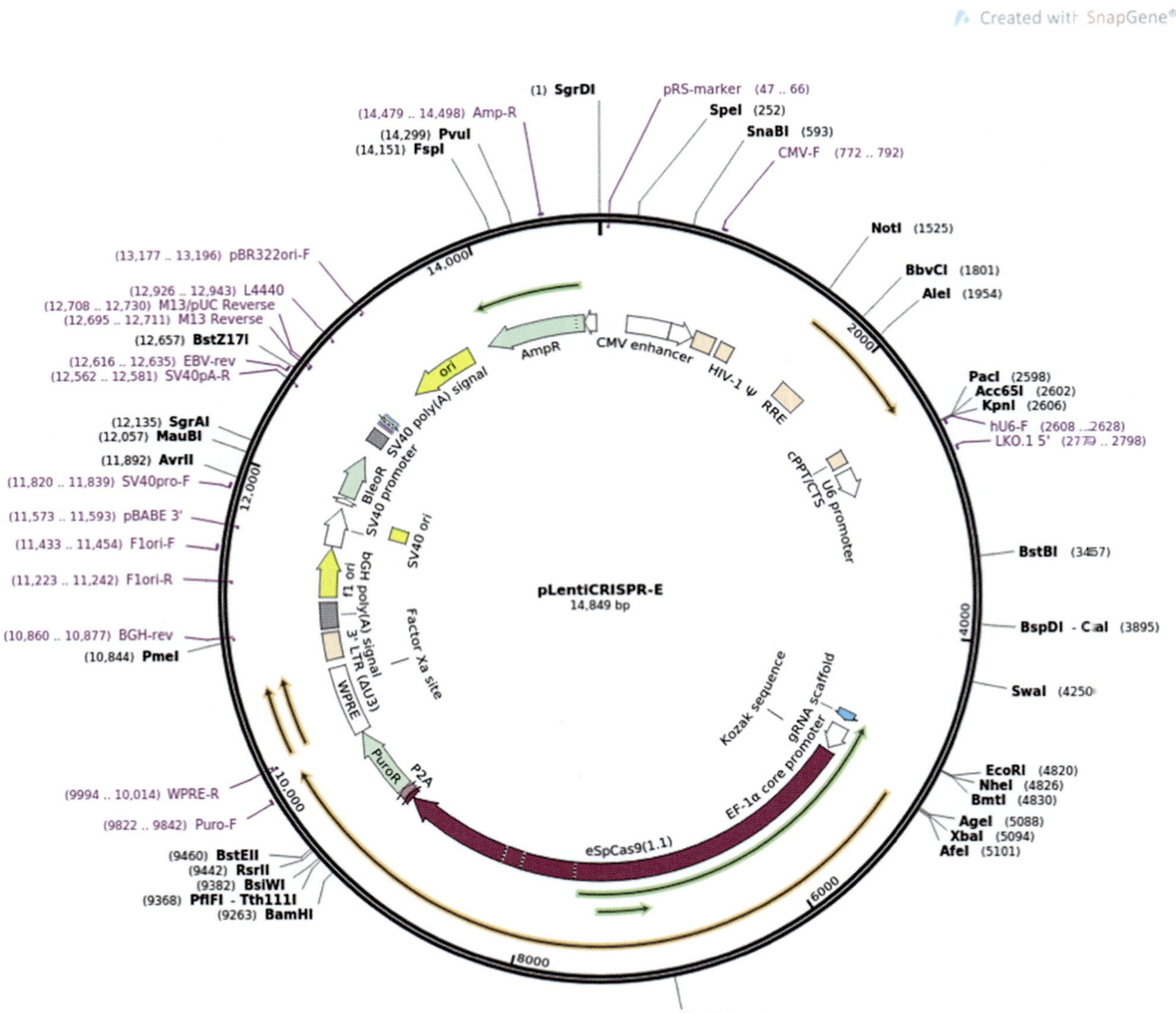

FIGURE 23.3 Vector map of pLentiCRISPR-E. *Used with permission from Dr. Philip Abbosh.*

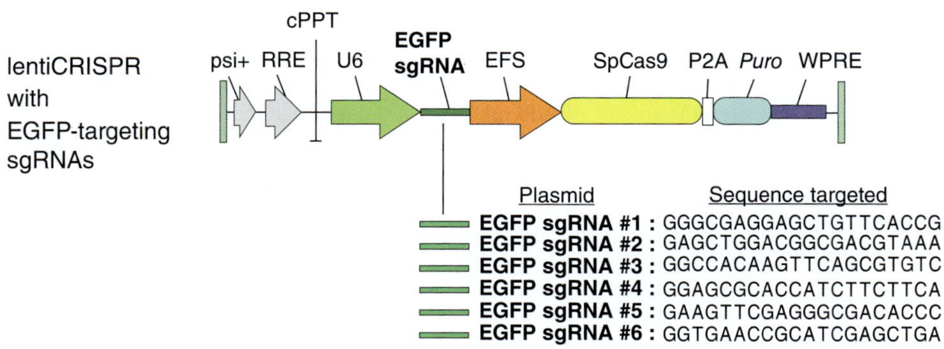

FIGURE 23.4 Validated sgRNAs against *egfp*. *sgRNAs*, Single guide RNAs. *Used with permission from Dr. Feng Zhang.*

stands for "enhanced." This variant of Cas9 was engineered to have enhanced specificity for the target DNA, therefore minimizing off-target effects (cleaving target DNA sequences other than the one you intended to).[16] pLentiCRISPR-E will serve as a negative control, since it contains no sgRNA.

Your experimental expression vector will be based on pLentiCRISPR, but it WILL contain an sgRNA specific to *egfp*. The Zhang lab has validated and made available six different CRISPR vectors that target *egfp*.[17] You will test one of them. On the restriction map of pLentiCRISPR-E, you will see another promoter called U6. This human promoter is immediately upstream of where the sgRNA is located. The relevant portion of the pLentiCRISPR vectors is shown in Figure 23.4. Note that downstream of the *egfp* sgRNA is the EFS promoter driving the expression of Cas9. EFS is a shorter version of the EF-1α promoter described earlier in this introduction. These *egfp*-targeting vectors, therefore, encode both Cas9 and an sgRNA which are necessary for CRISPR to function in the EGFP expressing human cell line.

You will notice that these vectors all have the word "lenti" in their names. In addition to being mammalian expression vectors, they can also be used for the production of lentivirus that will contain viral particles housing the encoded CRISPR components. In some applications, using viruses to infect host target cells (known as transduction) is a more efficient means of introducing foreign nucleic acids in typically hard to transfect cell lines. For the purposes of these lab exercises, you will not create lentivirus from the vectors, but will transiently transfect the vectors using a cationic lipid.

Please review Lab Session 19a for information on transient transfection of mammalian cells.

II. LABORATORY EXERCISE

Note: Carefully follow aseptic technique guidelines when completing today's interim laboratory session.

Please refer to Figure 19.3 from Lab Session 19: Transient Transfection of Mammalian Cells to familiarize yourself with the overall workflow of the transfection experiment. This procedure was modified with permission from Promega's FuGENE HD transfection reagent protocol:

1. Remove the FuGENE HD reagent from the 4°C refrigerator and allow to reach room temperature. Mix gently by inversion before use.
2. Prepare the biological safety cabinet as described in Lab Session 18: Culturing Mammalian Cells.
3. Obtain your group's 48-well plate of EGFP-expressing HEK293 cells that were plated in Lab Session 23a.
4. Using the microscope and 10× objective lens, observe and record the morphology and confluency of your cells under white light. You plated cells at two different densities. You will need to choose the density that resulted in 60%–80% confluency, as you will only transfect those three wells. There should be few floating cells in the medium. If you notice any signs of contamination or if none of your cells are in the 60%–80% confluent range, notify your instructor. Return your plate to the cell culture incubator.
5. In the biological safety cabinet, label three sterile microcentrifuge tubes as follows: Opti-MEM only, empty, and egfp.
6. In your lab notebook, draw a table as shown (Table 23.1) to record the amount of plasmid you need to add for each treatment to achieve 0.85 μg and the amount of Opti-MEM to bring the final volume to 36 μL.

   $ Money-Saving Tip: Serum-free media can be used in place of Opti-MEM.
7. Add 0.85 μg of the appropriate plasmid DNA to the correctly labeled tube.
8. Add the calculated volume of Opti-MEM to the correctly labeled tube.
9. You will use a 5:1 FuGENE HD:DNA ratio. Carefully add 4.2 μL of FuGENE HD reagent directly to the liquid in the tube labeled empty. *DO NOT allow the FuGENE HD reagent to directly contact the plastic walls of the tube*, as this will drastically reduce transfection efficiency. Using a new pipette tip, add the same volume of FuGENE HD to the tube labeled egfp.

TABLE 23.1 Experimental Setup for CRISPR-Mediated Knockout of *egfp* in HEK293 Cells

Tube	Vector	Plasmid Conc. (μg/μL)	Plasmid Vol. to Add (μL)	Opti-MEM (μL)	FuGene HD (μL)
Opti-MEM only (negative control)	No vector	0	0	36	0
Empty (negative control)	pLentiCRISPR-E			Up to 36 μL total	4.2
egfp (experimental)	pLentiCRISPR EGFP-2			Up to 36 μL total	4.2

No Permission Required.

10. Pipette up and down to mix and leave in the biological safety cabinet for 15 minutes.
11. Remove your group's 48-well plate from the incubator. Label (on the lid) what each well will contain using the names given in Table 23.1.
12. Add the entire volume of the transfection complexes drop-wise to the corresponding well of the 48-well plate. Rock the plate back and forth gently to ensure complete mixing of the FuGENE HD:DNA complexes throughout the medium and return the plate to the 37°C incubator.

DISCUSSION QUESTIONS

1. If a mammalian cell line was used that overexpressed green fluorescent protein (GFP), but the expression vector you transfected targets enhanced GFP (*egfp*), how could this potentially affect the knockout experiment?
2. After transfection, a number of biochemical processes must occur even before the target DNA gets cleaved. Describe the events and the order in which they occur.

Lab Session 23c

Interim Laboratory Session

EXPANSION OF TRANSFECTED CELLS

Goal: You or your instructor will expand the transfected cells, allowing room for growth over the next few days.

I. INTRODUCTION

At this point (2–3 days post-transfection), the cells you are working with are extremely confluent (>90%) and need to be passaged to a six-well plate. This will allow room for growth while the cells express the CRISPR-Cas9 components.

II. LABORATORY EXERCISE

1. Prepare the biological safety cabinet.
2. Obtain your 48-well plate of transfected cells from the incubator. Observe their confluency and morphology and write the details in your lab notebook. If you observe any signs of contamination, notify your instructor immediately. Also observe the color of your cell culture medium.
3. Using a sterile Pasteur pipette, aspirate the medium from the dish. Do this by gently tipping the dish toward you and placing the tip of the Pasteur pipette to the bottom of the vessel until all the medium is aspirated. Remember the spot where you touched the dish. You will always aspirate from this spot to avoid aspirating cells from your dish.
4. Gently add 100 μL of sterile 1× PBS onto each well and swirl carefully.
5. Aspirate the 1× PBS from each well.
6. Pipette 100 μL of 0.25% Trypsin-EDTA solution onto each well and swirl to ensure adequate coverage on the bottom. Place dish back in the incubator and set your timer for 2–3 minutes.
7. Observe your cells under the microscope; detached cells will float in the medium and appear rounded. You may want to gently but firmly tap your dish on a sterile surface to help further detach your cells. If you still have 50% or greater cells attached to the dish, return it to the 37°C incubator for another 2 minutes.
8. Once cells appear detached, add 100 μL of pre-warmed fresh complete medium containing G418 to each well, pipetting up and down to create a cell suspension. Remember that G418 is included to select for pEGFP in the cells.
9. Transfer half of this volume (100 μL) to one well of your labeled six-well plate. Be sure to change pipette tips between samples.
10. Add an additional 2 mL of pre-warmed complete medium containing G418 to each well.
11. Place the lid on the six-well plate, rock it gently, and place it in the cell culture incubator. The cells will remain there until you are ready to view fluorescence ~1 week post-transfection (4–5 days from today).

REFERENCES

1. Y. Ishino, H. Shinagawa, K. Makino, M. Amemura, A. Nakata, Nucleotide sequence of the iap gene, responsible for alkaline phosphatase isozyme conversion in *Escherichia coli*, and identification of the gene product, J Bacteriol 169 (12) (1987) 5429–5433.
2. R. Jansen, J.D.A. Van Embden, W. Gaastra, L.M. Schouls, Identification of genes that are associated with DNA repeats in prokaryotes, Mol Microbiol 43 (2002) 1565–1575.
3. R. Barrangou, C. Fremaux, H. Deveau, M. Richards, P. Boyaval, S. Moineau, et al., CRISPR provides acquired resistance against viruses in prokaryotes, Science 315 (5819) (2007) 1709–1712.
4. M. Jinek, et al., A programmable dual-RNA-guided DNA endonuclease in adaptive bacterial immunity, Science 337 (2012) 816–821.
5. E. Deltcheva, et al., CRISPR RNA maturation by trans-encoded small RNA and host factor RNase III, Nature 471 (2011) 602–607.
6. K.S. Makarova, N.V. Grishin, S.A. Shabalina, Y.I. Wolf, E.V. Koonin, A putative RNA-interference-based immune system in prokaryotes: computational analysis of the predicted enzymatic machinery, functional analogies with eukaryotic RNAi, and hypothetical mechanisms of action, Biol Direct 1 (2006) 7.
7. A. Bolotin, B. Quinquis, A. Sorokin, S.D. Ehrlich, Clustered regularly interspaced short palindrome repeats (CRISPRs) have spacers of extrachromosomal origin, Microbiology 151 (2005) 2551–2561.
8. J.E. Garneau, et al., The CRISPR/Cas bacterial immune system cleaves bacteriophage and plasmid DNA, Nature 468 (2010) 67–71.
9. Chang, et al., Non-homologous DNA end joining and alternative pathways to double-strand break repair, Nat Rev Mol Cell Biol 18 (2017) 495–506.

10. M. Kosicki, K. Tomberg, A. Bradley, Repair of CRISPR-Cas9–induced double-stranded breaks leads to large deletions and complex rearrangements, Nat Biotechnol 36 (2018) 765–771 (7/16/18 online).
11. A. Veres, et al., Low incidence of off-target mutations in individual CRISPR-Cas9 and TALEN targeted human stem cell clones detected by whole-genome sequencing, Cell Stem Cell 15 (1) (2014) 27–30.
12. Y. Fu, et al., High-frequency off-target mutagenesis induced by CRISPR-Cas nucleases in human cells, Nat Biotechnol 31 (9) (2013) 822–826.
13. I. Slaymaker, et al., Rationally engineered Cas9 nucleases with improved specificity, Science 351 (6268) (2016) 84-88.
14. O. Shalem, et al., Genome-scale CRISPR-Cas9 knockout screening in human cells, Science 343 (6166) (21014) 84-87.

Lab Session 24

CRISPR-Mediated Knockout of EGFP: Part II

Goal: To quantify the efficiency of CRISPR-mediated knockout of the EGFP protein in HEK293 cells using fluorescence microscopy and NIH ImageJ.

I. INTRODUCTION

Please review Lab Session 19b for background information on fluorescence microscopy. In that exercise, you transfected a mammalian expression vector (pEGFP-N1) that led to transcription and translation of EGFP. You examined the cells with a fluorescence microscope to look for evidence of overexpression of EGFP. In the RNAi exercises, you started with mammalian cells overexpressing EGFP and attempted to *knock it down*. Now, you have introduced an additional mammalian expression vector that should trigger the CRISPR-Cas9 pathway to *knockout* EGFP.

II. LABORATORY EXERCISE

Approximately 1 week post-transfection with CRISPR vectors (Lab Session 23b), acquire and save fluorescence and brightfield images of each well as previously described in Lab Session 19, Transient Transfection of Mammalian Cells. Save your files as either .JPG or .TIFF files. Be sure to also record qualitative visual observations in your lab notebook according to Table 24.1.

1. Download and Open NIH ImageJ on your personal computer. ImageJ is freely available for a variety of different operating systems from the National Institutes of Health (https://imagej.net/).
2. From file menu, select "Open," and choose the fluorescence image you wish to analyze first. It is best to start with your control HEK293 EGFP expressing cells that were not transfected with any plasmid vector (e.g., Opti-MEM only treated). You would anticipate these cells to have the highest levels of EGFP fluorescence.
3. From the analyze menu, select "Set measurements." From this display, be sure to check box the items: area, integrated density, and mean gray value and then select "OK" (Fig. 22.1 from Lab Session 22: RNAi-Mediated Knockdown of EGFP: Part III).
4. From the ImageJ toolbar at the top, select the "Freehand tool" (Fig. 22.1 from Lab Session 22: RNAi-Mediated Knockdown of EGFP: Part III).

   Using the freehand tool and your mouse, carefully outline one healthy cell of interest in your fluorescence image (Fig. 22.2 from Lab Session 22: RNAi-Mediated Knockdown of EGFP: Part III). *Note: It is useful to also open the accompanying brightfield image to identify cells that may not express considerable EGFP protein but that you will still need to select for analysis in the fluorescence image.*
5. Once you have outlined your first cell with the freehand tool, select "Measure" (or Ctrl-M; PC computers or Command-M; Mac computers) under the analyze menu (Fig. 22.2 from Lab Session 22: RNAi-Mediated Knockdown of EGFP: Part III).
6. A new window will appear that will list all the measurement results for that particular cell (Fig. 22.2 from Lab Session 22: RNAi-Mediated Knockdown of EGFP: Part III).

Molecular Biology Techniques. DOI: https://doi.org/10.1016/B978-0-443-26714-7.00032-6

TABLE 24.1 Experimental Observations of CRISPR-Mediated Knockout of EGFP Protein

Experiment	Fluorescence Intensity	Confluency	Morphology
Opti-MEM only (negative control)			
pLentiCRISPR-E (negative control)			
pLentiCRISPR-EGFP (experimental)			

7. Select a non-cellular, small region (exact size does not matter) next to your cell of interest that has no fluorescence. Record the measurement by selecting "Measure" under the analyze menu. This will serve as your background fluorescence reading.
8. Repeat Steps 4–7 for additional cells in your control experiment. Count a minimum of 10 cells from each treatment. You may need to analyze additional images from other fields of view for your sample if you do not have at least 10 cells.
9. Once you have taken measurements of your cells, select all the data in the results window and copy and paste into a Microsoft Excel spreadsheet (Fig. 22.3 from Lab Session 22: RNAi-Mediated Knockdown of EGFP: Part III).
10. Calculate the corrected total cell fluorescence (CTCF) for each cell using the equation:

$$CTCF = Integrated\ density - (area\ of\ selected\ cell \times mean\ background\ fluorescence)$$

We use the integrated density since it represents the region of interest (ROI) × mean of the ROI. If you only took the mean fluorescence intensity, it will calculate the average fluorescence and does not take the size of your cell into account. Therefore, if you have a large cell compared to a small cell, you will likely acquire a higher fluorescence value in the small cell compared to the large cell. However, with the greater size of the large cells more EGFP expression could be present even though the average may be lower at any one point in the cell. We will use integrated density, so we do not have to worry about size issues between different samples, since that is already accounted for.

11. Then determine the *average* value in a new cell in Microsoft Excel. To do this, in the formula bar type "=AVERAGE (select all cells containing the CTCF)."
12. Creating a new column, calculate the *standard deviation* of the CTCF by typing in the formula bar "=STDEV.P (select all cells containing CTCF values)."
13. Repeat Steps 2–12 within each well for a minimum of 10 cells per treatment: pLentiCRISPR-E (negative control) and pLentiCRISPR-EGFP (experimental).
14. When you are finished, create a bar graph comparing the average CTCF fluorescence (in Arbitrary Units) on the *y*-axis and the cell treatment on the *x*-axis. An example is shown illustrating CTCF values per individual cell (Fig. 22.3 from Lab Session 22: RNAi-Mediated Knockdown of EGFP: Part III).
15. Be sure to label your graph appropriately, including your calculated standard deviation, labeling *x*- and *y*-axes with units (where appropriate), and a descriptive title.

REFERENCE

1. Ophir Shalem, et al., Genome-scale CRISPR-Cas9 knockout screening in human cells, Science 343 (6166) (2014) 84–87.

DISCUSSION QUESTIONS

1. You observed a population of cells for a decrease in EGFP fluorescence. Based on what you know about transfection efficiency, why would you NOT expect a complete loss of fluorescence?
2. How is this CRISPR experiment different than attempting to knockout your favorite human gene?
3. Assuming you created stable cell lines that overexpress EGFP, how do they differ from those used in the paper that validated the same sgRNAs you used? Could this have an impact on your final experimental results and why? To answer this question, refer to: Shalem et al.[1]

Lab Session 25

Advanced CRISPR: Part I

Our previous experiments aimed to knockout *egfp* in a human cell line. The following experiments are entitled "Advanced CRISPR" because they are designed for the user to knockout a human gene of their choice. That process will involve experimental design, annealed oligo cloning, and screening by Sanger sequencing before the reagents can be introduced into cells. Because the gene editing will not result in an easily identifiable readout like loss of fluorescence, you will be instructed to perform a commonly used assay to detect DSBs in the last two lab sessions. The aim is to provide you with a solid foundation on which additional work with CRISPR technology can build.

You will still utilize HEK293 cells that stably express EGFP. This cell line is diploid, meaning there should be two copies of each gene in the human genome. This is more desirable in a CRISPR experiment than, say, a cancer cell line that has abnormal numbers of copies of each gene. Although your goal is no longer to target *egfp*, using this cell line allows for the inclusion of a control for the CRISPR pathway, since validated sgRNAs are available against *egfp*. This can be visualized through fluorescence microscopy. Furthermore, targeting *egfp* serves as a negative control, as those cells should not show targeting of your gene of interest. If you will be utilizing these cells outside of the context of this book (see Lab Session 30 for a complete overview of a CRISPR experiment workflow), it would be advised to confirm that your gene of interest is not affected by overexpression of EGFP.

Goal: Today, you will design sgRNAs to target a gene of interest. While these reagents are being ordered and synthesized, you will also prepare your CRISPR plasmid for cloning.

I. INTRODUCTION

Before beginning this session, please refer to the introduction of Lab Session 23, CRISPR-Mediated Knockout of EGFP: Part I for CRISPR mechanism.

A. sgRNA Design

One of the most exciting features of CRISPR technology is the ability to edit genes relatively quickly and easily. This allows researchers to either target a single genomic locus or perform large-scale screens of an entire genome. Although knocking out, knocking in, and editing genes in multiple cell lines and organisms is now possible, it requires careful experimental design. First, if your favorite gene is essential to survival, your cells will die if you successfully knock it out. Similarly, knocking in lethal genes or making mutations to the genome can result in cell death. Ultimately, your aim is to achieve the desired, specific effect without non-desirable off-target effects. For example, if you wish to knockout your favorite gene to study its normal function, you want an sgRNA that *only* targets your gene of interest. However, if the sgRNA sequence is similar enough to other sequences in the genome, it is possible for it to still recruit Cas9. The enzyme can tolerate five base mismatches in the protospacer motif[1] and even one base mismatch in the PAM sequence.[2] In this scenario, double-strand breaks and indels would be created in other unintended genes. The only truly comprehensive way to detect off-target effects is to perform whole-genome sequencing. As this is not always possible, it is extremely important to minimize these off-target effects from the beginning of the experimental design process. Emerging data suggest that significant off-target effects can occur during certain CRISPR experiments.[3] As data continues to be published, we are still learning the extent to which CRISPR-mediated gene editing can cause unintended genetic modifications. This is an important concern for both basic and applied research efforts. For example, attempting to correct a gene's sequence that causes disease but unintentionally editing some other region of the genome could have severe consequences. This is similar to the concern over off-target

Molecular Biology Techniques. DOI: https://doi.org/10.1016/B978-0-443-26714-7.00026-0

effects in RNAi. To this end, there has been a great amount of effort put into algorithms designed to identify specific sgRNAs for a particular sequence.

In summary, the primary criteria for designing a sgRNA are:

1. The sequence is unique compared to the rest of the genome to avoid off-target effects.
2. The target is present immediately upstream of a PAM.

Multiple online tools are available to help design sgRNAs. We will use the CRISPR tool within the free electronic lab notebook Benchling (https://benchling.com), although others are also based on the same design algorithms. It is necessary before beginning your CRISPR experiment to determine if you are using the techniques to knockout, knockin, or edit your gene of interest. The following lab sessions will walk you through knockout only.

This online tool will allow you to view your gene of interest and sgRNAs (CRISPR guides) with a built-in genome browser, choose your Cas9 variant, view PAM sites, view sgRNA sequences, and filter them based on corresponding on-target activity scores and off-target scores. An overview of the scores that the Benchling algorithm calculates is shown below.

On-Target Score—Ranges from 0 to 100. This number is a score that predicts the on-target activity of a particular guide RNA, assuming spCas9 is also used. An on-target score increases as a function of on-target activity, so the higher the score, the better. Although there is no guarantee that a gRNA of a certain minimum on-target score will perform effectively, you will be directed to use a minimum on-target score of 60.

Off-Target Score—Ranges from 0 to 100. This number is a score that predicts the specificity of the guide RNA. The higher the off-target score, the better. This high score means that the guide is less likely to target an unintended sequence.

Guide RNAs are generally designed to target the first few exons of a gene (promoter sequences are fine, too). The goal in CRISPR-mediated gene knockout is to create double-strand breaks, indels, and, ultimately, a non-functional transcript (mRNA). So, it is advisable to have this occur early in the DNA sequence, resulting in a shorter transcript being produced that will be degraded in the cell.

It is highly recommended that you design 2–3 sgRNAs per gene for several reasons. First, chromatin structure can influence the ability of an sgRNA to hybridize to a gene of interest. Therefore, designing 2–3 sgRNAs to different locations in your gene increases the chance that at least one will effectively have access to the genome in order to induce your desired knockout. Second, sometimes sgRNAs that fit all of the design criteria fail to produce the desired outcome. Testing additional sgRNAs to the same gene will increase the likelihood that your CRISPR experiment will be successful.

B. Annealed Oligo Cloning

In Lab Sessions 2–5, you performed cloning by PCR. An additional technique commonly used to create recombinant DNA is called annealed oligo cloning. Oligos (short, ssDNA molecules, also known as primers) can be designed and ordered from a number of manufacturers. Recall that dsDNA is both antiparallel and has complementary base pairs. So, if the reverse complement of a single oligo is also ordered, the two oligos can be annealed together to create a double-stranded DNA molecule. These molecules serve as insert DNA in annealed oligo cloning. The bold type in oligos 1 and 2 in Figure 25.1 represent an example of annealed oligos that are the reverse complement of each other. Keep in mind that by convention, only one strand of DNA is written and in the 5′–3′ direction. When you design your sgRNAs, the software will provide you with just one strand's sequence. You can use that as "oligo 1" (5′-CAGTCTGATCAGTTTTCCTT-3′ in Figure 25.1). Then, you would write out the complementary bases, keeping them 5′–3′ from left to right (5′-AAGGAAAACTGATCAGACTG-3′ shown as "oligo 2").

Notice that the oligos in Figure 25.1 also have nucleotides that are not in bold. Those are additional nucleotides designed to serve as overhangs. They do not contain sequences that will be transcribed into your sgRNA. Rather, they create compatible cohesive overhangs that will be ligated to your restriction enzyme digested vector. For oligo 1, you will add "CACCG" to the 5′ end when ordering your oligo to create an appropriate overhang that will hydrogen bond to an overhang on your BsmBI-digested CRISPR vector. For oligo 2, you will add "AAAC" to the 5′ end and a single "C" to the 3′ end of oligo 2, as well. When the oligos arrive, you will anneal them, creating hydrogen bonds. This dsDNA can then be ligated together with BsmBI-digested vector backbone to create your recombinant DNA (Figure 25.2). Note that your insert in this case is formed by two oligos that *do not get digested*. This is in contrast to the PCR cloning you may have performed earlier in the book. However, both methods result in insert DNA with compatible cohesive overhangs to vector DNA.

To summarize, the bold nucleotides in Figure 25.1 illustrate that the oligos you order match the target genomic sequence you wish to cleave. The overhangs that are not bold are there only to facilitate ligation to the overhangs in the

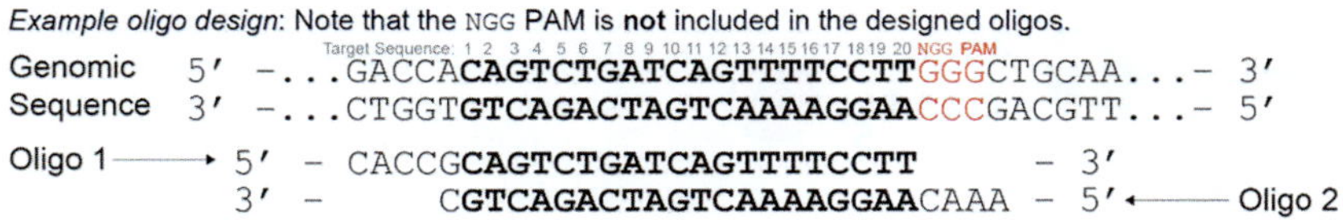

FIGURE 25.1 Example of oligo design for inserts cloned into pLentiCRISPR-E. *Used with permission from Dr. Feng Zhang.*

digested vector. So, even if the sgRNAs you aim to design are different in sequence or different from another group's, the overhangs you are instructed to add will be identical in each case.

The plasmid vector you will use is pLentiCRISPR-E, the same vector described in Lab Session 23. There are two BsmBI sites in this vector (Figure 25.3) ~2 kb apart. Therefore, digesting the vector with BsmBI will create two fragments: the ~2 kb stuffer or filler fragment between the sites, and the ~13 kb vector backbone. You will gel-purify the ~13 kb vector backbone from an agarose gel, which will later be ligated to your annealed oligos. This will position the insert DNA (annealed oligos) immediately downstream of the U6 promoter and will result in transcription of the sgRNA once introduced into mammalian cells.

Unlike the restriction enzymes you used previously (NcoI and NotI), BsmBI does not cut within its recognition sequence. It recognizes a specific sequence, but cleaves outside of it, which classifies it as a type IIS restriction enzyme. DNA cut with BsmBI should have four nucleotide long incompatible overhangs. However, it is common to take an extra measure when cloning with one restriction enzyme to prevent plasmids from reannealing with themselves. This will ultimately enrich for transformants that include both vector and insert. To prevent vector religation unto itself without insert being present, the BsmBI-digested vector will be treated with a phosphatase first. Phosphatases catalyze the removal of phosphoryl groups. Without these groups, phosphodiester bonds cannot form, preventing pLentiCRISPR from recircularizing. However, if insert DNA (the annealed oligos with BsmBI overhangs) is present and has phosphoryl groups, it can be ligated to the vector. In this way, you are working with a *dephosphorylated* vector, but *phosphorylated* insert to ensure that the insert is properly ligated to the vector (Figure 25.2).

C. PCR Primer Design Basics

Please see Lab Session 3, Completion of Vector Preparation and PCR Amplification of *egfp* for introductory information on PCR and the role of primers.

Oligos (or primers) can be used for many purposes. In addition to annealed oligo cloning, primers are more often used in PCR reactions. You may have used primers already in cloning by PCR, screening transformants by PCR, and RT-PCR in this book. There are some additional details about primer design that you should be aware of. These will be summarized, as you will require primers in order to confirm your CRISPR-mediated gene editing. Please keep in mind that these are general criteria and do not necessarily hold true for all applications that use primers. Your laboratory exercise for PCR primer design today will contain criteria appropriate for the CRISPR-mediated gene editing assay.

- Primer length is generally 18–30 nucleotides. This varies based on the purpose of the primers.
- Primer GC content is 45%–60%.
- Keep the melting temperatures (T_ms) of primer pairs within 5°C of each other.
- Avoid homology within each primer. Regions more than three bases long that have a complementary sequence within the same primer can result in primer dimers rather than primer annealing to the target sequence.
- Avoid stretches of four or more of the same base and dinucleotide repeats (ATATATAT, for example).

Originally designed only by hand, primers are now typically designed with the help of online tools. One of the most important properties of a PCR primer is its specificity. You want to have primers anneal to a designated spot on a particular template. Computer algorithms typically perform a specificity check on all primers. This utilizes an alignment tool like BLAST (see Lab Session 8) to look for complementary sequences between a given primer and a particular genome or other deposited sequence. Today's lab exercise will walk you through primer design using the Primer3 tool that is built in to Benchling, although other online primer design tools could be used, as well.

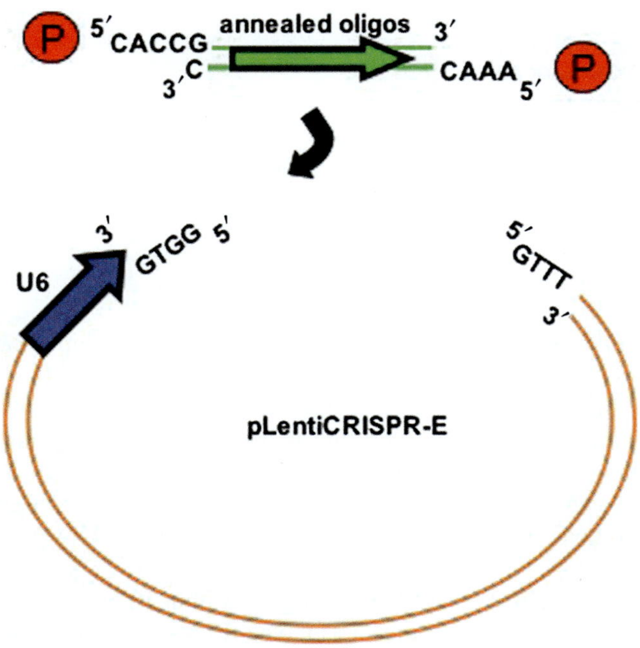

FIGURE 25.2 Schematic of phosphorylated, annealed oligos to be ligated into a BsmBI-digested, dephosphorylated pLentiCRISPR-E. The blue arrow represents the U6 promoter. The green arrow represents the annealed oligos that will be transcribed into an sgRNA. The red "P's" represent phosphoryl groups. *No Permission Required.*

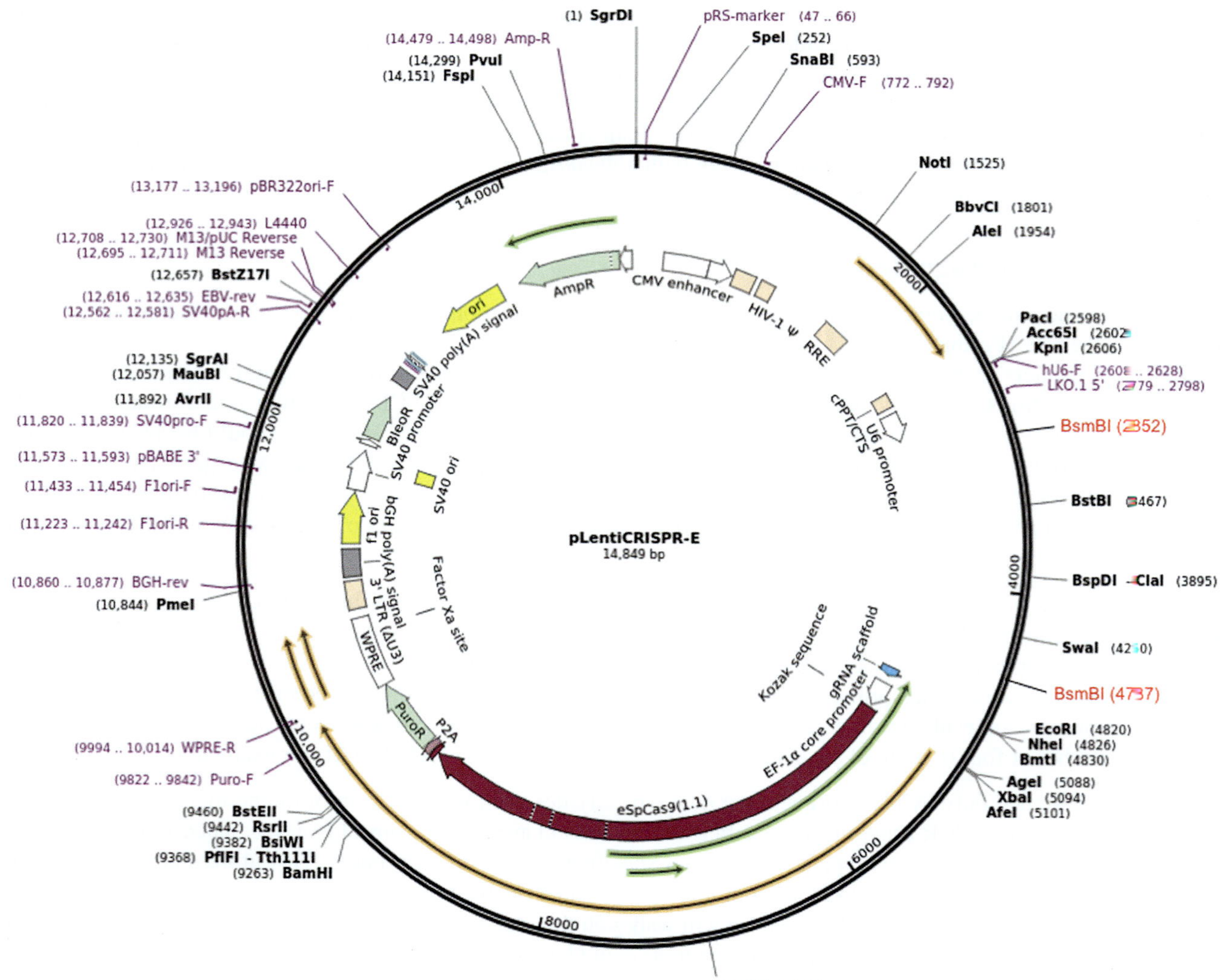

FIGURE 25.3 Vector map of pLentiCRISPR-E. Unique sites as well as the two BsmBI sites utilized for cloning are shown. *Used with permission from Dr. Philip Abbosh.*

II. LABORATORY EXERCISES

A. sgRNA Design and Analysis

Note: The following lab sessions are written for human gene experiments only in the HEK293 cell line. Use of a different cell line, particularly a non-human mammalian cell line, will require modifications from these protocols.

1. Visit the Benchling website (https://www.benchling.com) to create your free account or follow your instructor's directions to access an already established account.
2. Select the plus sign in the Benchling left-hand tool bar, then select CRISPR, CRISPR guides (Figure 25.4A).

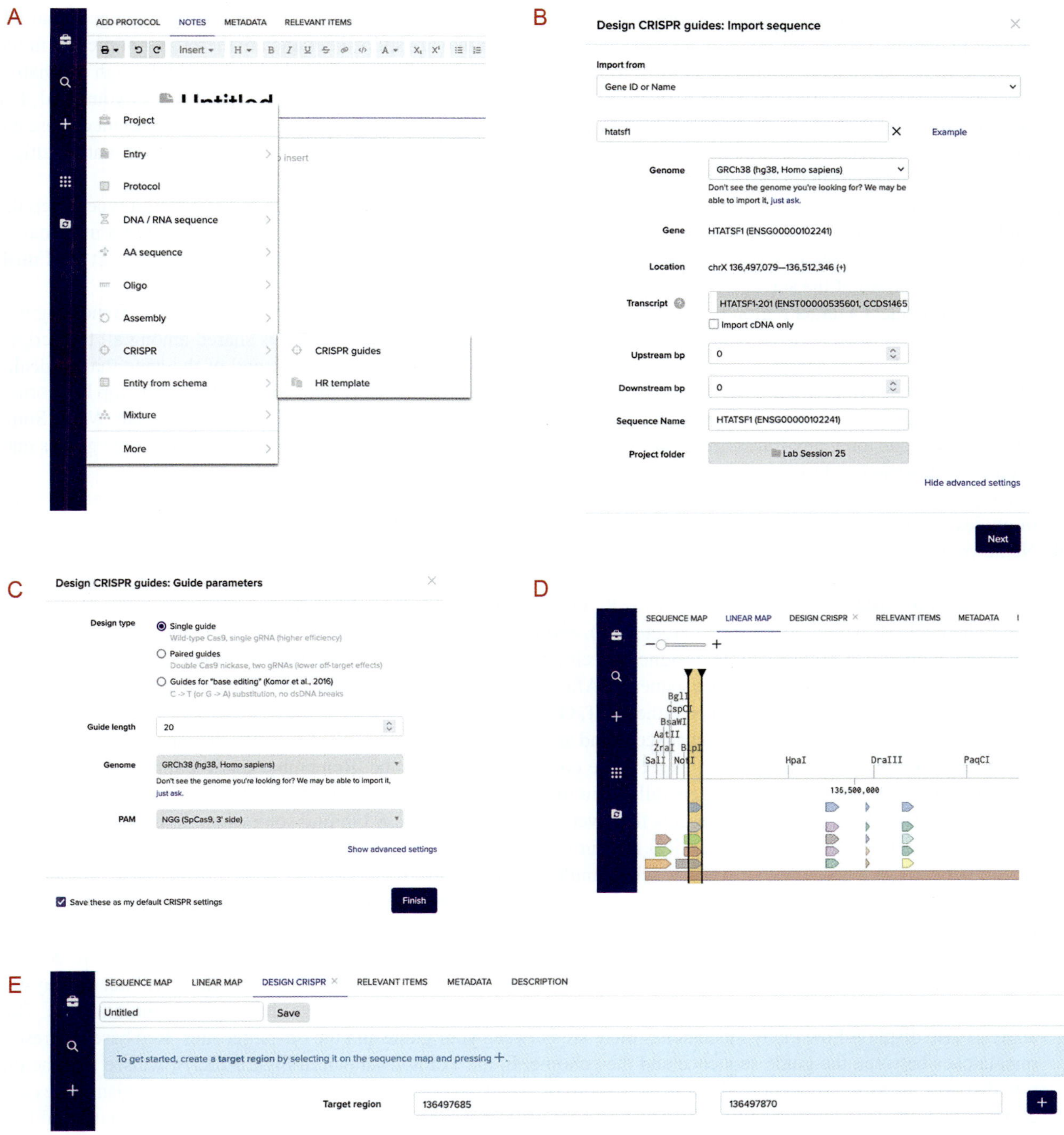

FIGURE 25.4 Benchling screenshots of CRISPR tool. (A) Selecting the CRISPR tool to design guides. (B) Importing sequences. (C) Setting guide parameters. (D) Visualizing transcripts. (E) Determining the target region. *Used with permission from Benchling.*

3. There are multiple ways to start working with a gene's sequence. In the Design CRISPR guides: Import sequence window (Figure 25.4B), try entering a gene ID or gene name to automatically import a gene of interest's sequence. Alternatively, choose a different import method from the drop-down menu. For example, you can choose a Sequence File, and Benchling will have you upload that Genbank or FASTA file. You could also choose raw bases from the drop-down menu and paste in your sequence from another source.
4. If you chose to enter a gene ID or gene name, genome options will now appear below in the window. Choose the appropriate genome from the drop-down list. *Note: In order to use the pLentiCRISPR-E vector backbone, you must perform your CRISPR experiment in human cells. Please click on "GRCh38 (hg38, Homo sapiens)* (Figure 25.4B). Select "Show Advanced Settings." There, you can use the shift button on your keyboard and click on all transcripts of your gene of interest (Figure 25.4B). This is important for CRISPR design in eukaryotes. Since this is a eukaryotic gene, Benchling will display known exons and introns. Recall that exons contain protein coding sequences that are present in mature mRNA. Introns contain noncoding sequences that are spliced out and are therefore not present in the mature mRNA. Since there are likely multiple transcript variants of your gene of interest, produced through alternative processing, it is best to identify sgRNAs that will target a sequence common to *all* variants if that is your goal. For example, if your gene can produce variants that skip an exon and your sgRNA targets that exon, that variant can escape targeting and proteins will still be produced. Some genes will have so many variants that it will necessitate multiple sgRNAs to target all of them. Select the next button at the bottom right of the window.
5. A guide parameters window will appear (Figure 25.4C). You should select single guide as your design type. Keep the guide length as 20 and the PAM site as NGG (default settings). Since you will clone the sgRNAs into a mammalian expression vector that also encodes spCas9, you must select a PAM site that spCas9 will recognize. Select the Finish button at the bottom of the screen.
6. You will now select a target region in your gene's sequence. Since you selected all transcripts in step 5, Benchling will display annotations for all exons in all transcripts. To look for an exon sequence that is shared among all transcripts, select the "LINEAR MAP" tab. Your goal is to find a region to target that is on the 5' end of the gene that is ideally shared by all transcripts (Figure 25.4D). You may need to zoom in or out using the zoom control in the top left corner. Click and drag to select some sequence. With that region selected, select the "DESIGN CRISPR" tab. *Note:* Some genes will encode too many transcript variants to be targeted by just one guide. If that is the case, multiple guides may need to be designed and used to increase the chances of gene editing in all transcripts.
7. The region you selected is now shown as the Target region. Select the plus sign button to have Benchling create a series of guides specific to that region (Figure 25.4E).
8. Name your CRISPR guide design strategy in Benchling. The top left corner has a text box to title your work. Select the Save button.
9. Guides are now listed on the "DESIGN CRISPR" tab (Figure 25.5A). By default, they are listed in order of cut position (where a DSB would occur in the target gene). Sequences at the top of your list will be the most 5' on your target sequence. Notice that guides can be designed on either strand of DNA (+ or -). That is because Cas9 will cut both DNA strands after binding to the target sequence. After the guide sequence, you will see a column of PAM sites. These should all be "NGG," where N could be either A, T, G, or C.
10. Finally, the guides each have an **on-target score** and an **off-target score**. You should change the ranking of your guides by on-target score by selecting the on-target score column heading in blue. Remember that the higher the score, the better the predicted activity. You will generally want to test guides with scores of 60 or higher that also have a high off-target score. The off-target score represents the inverse probability of Cas9 binding somewhere in the human genome other than your targeted region. It is critical in your design strategy to evaluate the potential off-target effects. Err on the side of caution if faced with a scenario where multiple guides have high on-target scores. From those options, you will want to choose those guides that also have the best (highest) off-target scores. If you select the actual off-target score for any guide, Benchling will show you the off-target sites in the human genome for that guide (Figure 25.5B). The off-target analysis shows if the off-target site is within a gene or not (between genes in the genome). Guides that target a gene will list that gene's name, while guides that target intergenic regions (between genes) will not display any gene name. Try to find guides that have a high off-target score and whose off-target hits are not within a gene. This analysis also displays how many mismatches there are between your guide and the off-target sites. Red bases represent mismatches between the guide sequence and the genome. In the example shown in Figure 25.5B, we see that the top row displays the target sequence with a perfect score of 100.0. Every off-target site has multiple red mismatches and correspondingly low scores. Additionally, most of these off-target sites do not fall within a known gene, so it is unlikely that this particular guide will silence unintended genes, even if similar sequences exist in the genome.

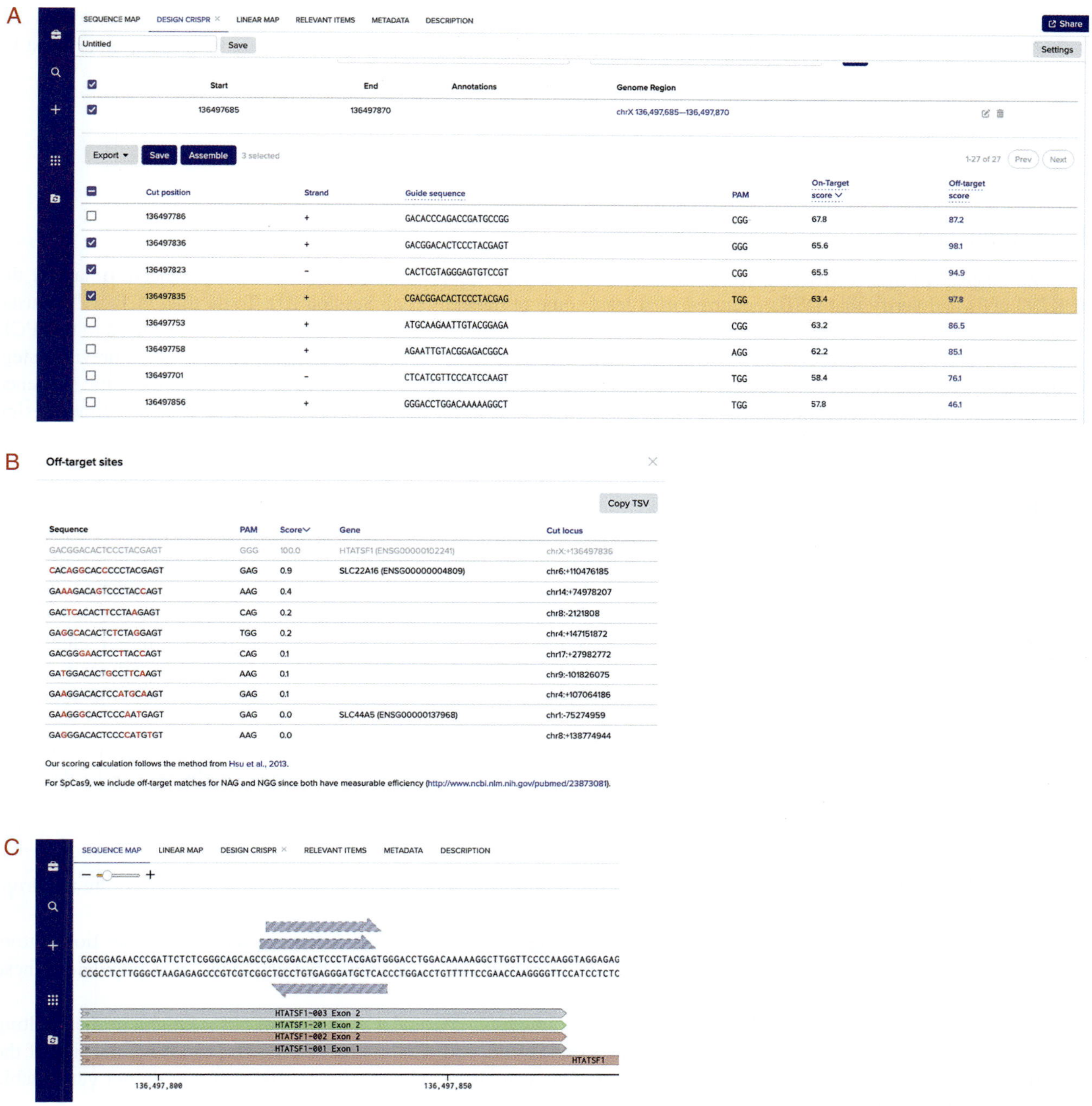

FIGURE 25.5 Benchling screenshots displaying CRISPR guide sequences and information. (A) Viewing your CRISPR guide sequences. (B) Analyzing off-target sites. (C) Viewing guide sequences on the sequence map. *Used with permission from Benchling.*

11. Select the checkbox for a few or your top-scoring guides and then select the "SEQUENCE MAP" tab to visualize where the guides would anneal and target your gene of interest. Zooming in will allow you to see DNA bases. Following each guide sequence from 5' to 3', you will find that a PAM site (NGG) follows each sequence (Figure 25.5C).
12. On the "DESIGN CRISPR" tab, you have two options for sharing the list of potential guide sequences to be reviewed by your instructor. If you click the blue Save button, you can save each of the selected guide sequences as DNA oligos in Benchling. If you click the Export button, it will copy your selected guides in tab or comma-delimited form to paste in Excel. Your instructor will notify you how to share your potential guides. They will also help determine if they are appropriate, unique from others in the class, as well as how many will be ordered for each gene.[4]

13. To create overhangs for cloning purposes, add the following sequences to the 5' end of your oligos, where "N" is one of the 20 nucleotides in your oligo. Remember that two oligos that are reverse complements of each other must be designed to create one guide sequence that will be transcribed into a sgRNA (Figure 25.1).

Example:
Oligo 1: 5′ CACCG NNNNNNNNNNNNNNNNNNNN 3′
Oligo 2: 5′ AAAC NNNNNNNNNNNNNNNNNNNN C 3′

B. Primer Design

After your sgRNA oligos arrive and you proceed with the CRISPR experiment, you will eventually need to harvest the HEK293 cells and verify that DSBs occurred in at least some of the cells (Lab Session 29). To do this, PCR will be used to amplify the genomic sequence surrounding the predicted DSB. Gene-specific primers (oligos) will be used in this PCR assay. Therefore, PCR primers should be designed and ordered concurrently with your sgRNA oligos in the previous step. If you used Benchling in Part A to design your CRISPR guides, you can easily design PCR primers on the same sequence file of your target gene. Benchling has the Primer 3 program built in for this purpose. Alternatively, many primer design tools exist online, including NCBI Primer-BLAST (https://www.ncbi.nlm.nih.gov/tools/primer-blast/).

1. Use Benchling to access the sequence file of your target gene. To view your saved CRISPR guides, you can select the CRISPR icon on the right-hand toolbar. Select your saved guides (Figure 25.6A).
2. Rather than have the program choose primers at any location within the gene, you will want to specify the range of bases where primers should anneal. This would be the region that includes the PAM site immediately downstream of where CRISPR/Cas9 will potentially create a DSB. Choose a range of sequence so that the potential cleavage site near the PAM site is not in the center of the PCR product. Cleavage would then result in two fragments of DNA that are the same length. It is critical for your assay in Lab Sessions 29 and 30 that cleavage fragments are different lengths. To specify this range, you can simply click and drag within the sequence map to highlight bases. You will want a range that is approximately 500 bp or more. The length of your selected sequence will automatically show at the bottom of the screen (Figure 25.6B).
   Note that if your group designed more than one CRISPR guide, you may need more than one primer pair in this step to span the potential cleavage site. Also, if other groups are designing guides to target nearby regions of the same gene, one primer pair may be able to be used for analyzing multiple cleavage sites. Consult with your instructor.
3. With your 500 bp or more region selected, choose the double-arrow primer icon on the right-hand toolbar. Select the Create Primers button and choose the Wizard function.[5]
4. The "PRIMER WIZARD" tab will contain all of the settings for your primer design (Figure 25.6C). At the top drop-down menu called Task, choose the first option, PCR.
5. In the Region section, select "Show advanced." To the right of the "Included" boxes, select the Use selection button. This will specify that your primer pair should include the region you selected and that region will now populate in these boxes (Figure 25.6C).
6. Below that, in the Primer section, set the GC% to a minimum of 45% and a maximum of 60%. Enter a primer melting temperature (T_m) with a minimum of 58, optimum (opt) of 60, and maximum of 62°C. You can leave the size of the primers and the 3'GC clamp as the default settings. You can always change these settings if you do not get suitable primer options for your particular gene (Figure 25.6C).
7. In the Amplicon section, enter the size range of 400 for the minimum and 500 as the maximum. All other sections in this window can be left as default (Figure 25.6C).
8. Select the Generate Primers button at the top right corner (Figure 25.6C). *Note: In some instances the program may not be able to identify suitable primers. If this is the case, please consult with your instructor as you will need to adjust your parameters before moving forward.*
9. A "PRIMER3 RESULTS" tab will now display primer pairs that meet the above specifications (Figure 25.6D). You can easily visualize where these primers will anneal to your target region by selecting each check box in the first column and then clicking on the "SEQUENCE MAP" tab.
10. In the "PRIMER3 RESULTS" tab, you should select "Save Selected Primers" to save them in Benchling. Saving them allows you not only to visualize them and search for them in the future, but also to analyze them further in the next step. You can also export them as a CSV file, which is very helpful for ordering purposes.
11. To further analyze candidate primers, right-click on the primer arrow and choose "Check primer structure." Scroll down to the Verify section to see new properties of your candidate primer: Min ΔG Homodimer and Min ΔG Monomer

FIGURE 25.6 Benchling screenshots displaying primer design steps. (A) Finding your saved CRISPR guides (B) Selecting a region of sequence to design primers to (C) Setting primer specifications (D) Primer design results (E) Analyzing potential primer dimers. *Used with permission from Benchling.*

(Figure 25.6E). This shows the free energy values for the primer, which report on how easily any secondary structure could be melted. Single-stranded primers can hydrogen bond with themselves (monomer) or two molecules of themselves (homodimers aka primer dimers). You want primers with ΔG values close to 0. As a guideline, those with values of −9 kcal or more negative can be problematic because if they anneal into secondary structures, they are doing that instead of annealing to your intended template DNA.

12. If you analyze the specificity of a single ~20 nt candidate primer, it is practically guaranteed to be a perfect match to other human genes due to its short length. Rather than using BLAST for one primer at a time, you can do the following to assess how likely both primers in a pair would anneal to and amplify human genes other than the one you intend to amplify. Right-click on a candidate forward primer arrow and select "Submit to NCBI BLAST." Your forward primer's sequence will automatically be in the Query box (Figure 25.7A). Type in 20 "N"s in a row, followed by a candidate reverse primer (also written 5' to 3'). To add the reverse primer, right click on the reverse primer arrow, click copy, and then choose the DNA sequence. Paste that into the BLAST Query box. Now you will be performing a BLAST search using both primers simultaneously, separated by some bases of any sequence.[6]
13. Under the BLAST Choose Search Set section, choose the database "genomic + transcript." Then in the drop-down menu underneath the database, select "human genomic plus transcript (human G+T)." Scroll to the bottom for Program Selection. Here optimize the search for "somewhat similar sequences blastn." Select the BLAST button to run this analysis. BLASTn will analyze the possible off-target amplifications that could occur with that particular primer pair (Figure 25.7A).
14. Evaluate the primer pairs that are generated in consultation with your instructor. It may be beneficial to order two primer pairs in case one does not work well.

C. Restriction Enzyme Digestion and Dephosphorylation of the CRISPR Vector

These steps will allow the BsmBI digestion of the CRISPR vector to occur first, followed by the dephosphorylation of the CRISPR vector. This protocol is written for using "time-saver" restriction enzymes like those purchased through New England Biolabs. Note that BsmBI-v2 is most active with a 55°C incubation (not 37°C), while Quick CIP will require 37°C. If you are using reagents from a different manufacturer, your instructor may make modifications to the protocol.

1. You will receive an aliquot of pLentiCRISPR-E. Record the concentration. Calculate the volume that will give you 5 μg.
2. In a microcentrifuge tube, combine the reagents in the order listed in Table 25.1.
3. Quickly vortex the tube to mix, then spin down briefly in a microcentrifuge to collect the liquid at the bottom of the tube.
4. Incubate the tube in a 55°C water bath for 15 minutes.
5. Spin down briefly in a microcentrifuge to collect the liquid at the bottom of the tube. Carefully add 1 μL Quick CIP to the tube.
6. Quickly vortex the tube to mix, then spin down briefly in a microcentrifuge to collect the liquid at the bottom of the tube.
7. Incubate the tube in a 37°C water bath for 10 minutes.

D. DNA Agarose Gel Electrophoresis

1. While the CIP reaction in Part C is proceeding, prepare a 0.6% agarose gel. This is similar to what is described in Lab Session 3, except that only 0.6 g of agarose will be dissolved per 100 mL of 1× TBE.
2. Remove your digested, dephosphorylated pLentiCRISPR-E plasmid from the water bath and perform a quick spin in a microcentrifuge to bring the liquid to the bottom of the tube.
3. Add 12 μL of 6× DNA loading dye (or other appropriate DNA loading dye) to the reaction and mix by pipetting up and down.
4. Prepare an undigested plasmid control by adding 200 ng of undigested pLentiCRISPR-E plasmid to a new microcentrifuge tube. Add 6× DNA loading dye and sterile water to bring the sample up to a volume between 10 and 20 μL.
5. Place your agarose gel in an electrophoresis tank. Cover the gel with 1× TBE buffer and slowly remove the comb.
6. Load 0.5 μg NEB Quick-Load 1 kb Extend DNA Ladder. This particular DNA ladder includes very large DNA fragments since your vector is larger than those used in previous lab sessions.
7. As the volume of your sample plus loading dye (72 μL) is greater than what one well of the gel will hold, load your entire sample into *two adjacent wells* of the gel. It is fine if the entire volume does not fit. Just pipette carefully in each adjacent well until it is full.
8. Place the lid on the electrophoresis tank and plug it into a power supply.

A

B

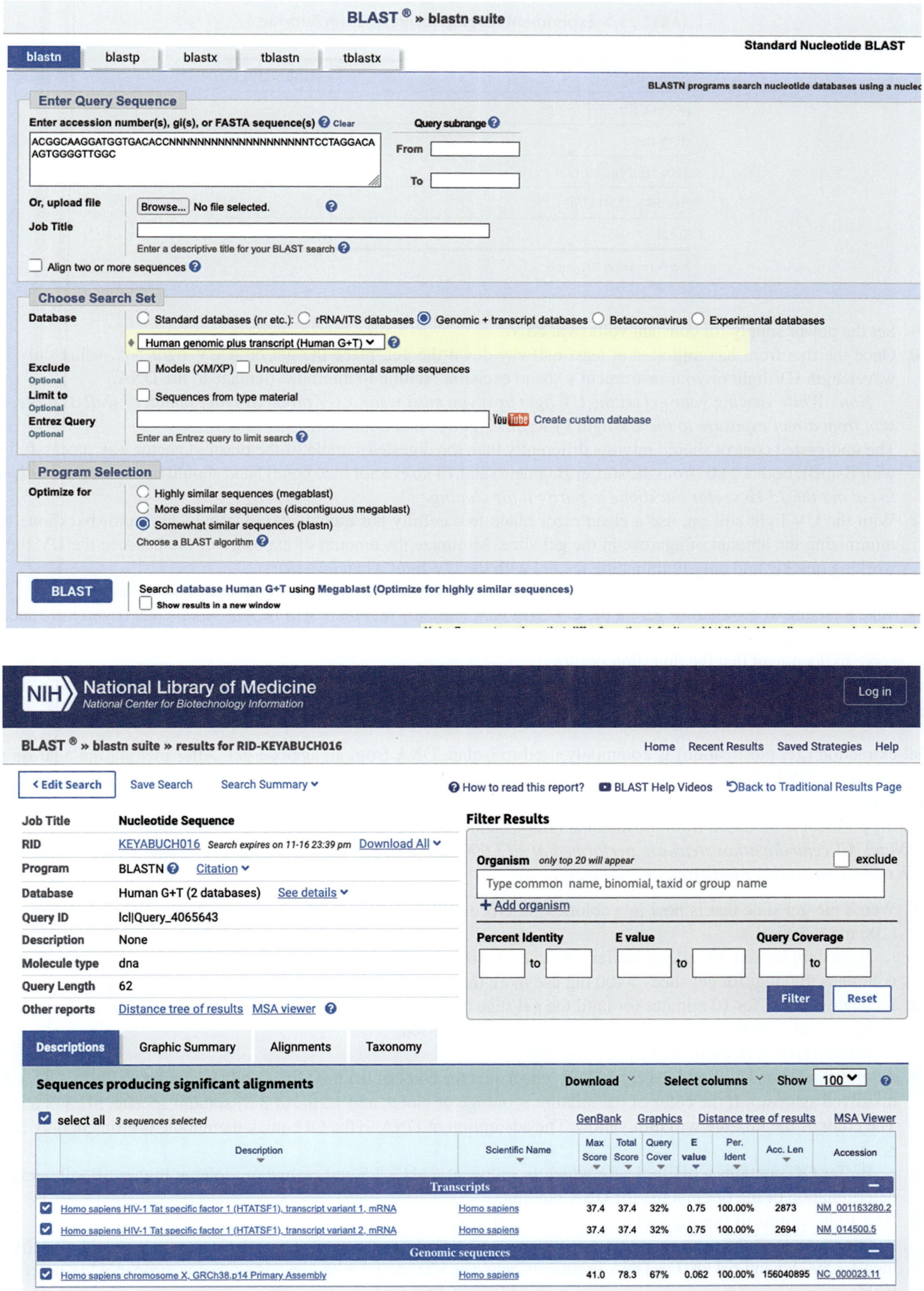

FIGURE 25.7 BLAST screenshots. (A) BLAST settings. (B) BLAST results. *No Permission Required.*

TABLE 25.1 Experimental Setup for Restriction Enzyme Digestion

Reagent	Volume (μL)
pLentiCRISPR-E (5 μg)	
dI Water	
10× NEB buffer r3.1	6
BsmBI-v2 (10 U/μL)	3
Total	60

No Permission Required.

9. Set the power supply for constant voltage at 85 V.
10. Once the dye front has migrated at least halfway down the gel, place the gel on a UV light box, using only long-wavelength UV light or your instrument's "band excision" setting to minimize damage to the DNA.
    Note: While viewing your gel on the UV light box, you must wear a UV protective faceshield, as well as protect any skin from direct exposure to the UV light by wearing gloves and a lab coat.
11. The undigested control should migrate differently than the digested sample. If the plasmid vector was successfully cut with BsmBI, both a 2 kb (from the stuffer fragment) and 13 kb (vector backbone) band should be visible. *You only want to cut out the 13 kb vector backbone to purify it for cloning.*
12. With the UV light still on, use a clean razor blade to carefully but quickly excise the 13 kb vector backbone band, minimizing the amount of agarose in the gel slice. Minimize the amount of exposure of the DNA to the UV light by working quickly and only illuminating the gel with the UV light when necessary.
13. Place the gel slice in a sterile, colorless, labeled microcentrifuge tube that has been tared on a balance.
14. Take a picture of the agarose gel on the UV light box, realizing that there will be a missing slice. If you have successfully extracted the 13 kb vector backbone, all that should remain is the ~2 kb fragment in those lanes. This image will serve to document that the digestion occurred.

E. Gel Extraction of Digested CRISPR Vector Backbone

Gel extraction (gel purification) is commonly used to isolate DNA from an agarose gel. After melting the agarose slice containing the DNA of interest, the protocol includes steps that are similar to those followed in other silica-membrane spin columns.

This protocol is modified from the manufacturer's handbook (QIAquick Spin Handbook 01/2020).[7]

Note: All centrifugation steps are performed at ~13,000 rpm in a conventional microcentrifuge. Balance your tube(s) with those from another lab group, if necessary.

1. Weigh the gel slice that is now in a colorless microcentrifuge tube. Add 3 volumes of Buffer QG to 1 volume of gel (100 mg = 100 μL).
    For example, add 300 μL of Buffer QG to each 100 mg of gel. The maximum amount of gel slice per QIAquick column is 400 mg; for gel slices >400 mg use more than one QIAquick column.
2. Incubate at 50°C for 10 minutes (or until the gel slice has completely dissolved). To help dissolve the gel, mix by vortexing the tube every 2–3 minutes during the incubation period.
    IMPORTANT: Solubilize agarose completely.
3. After the gel slice has dissolved completely, check that the color of the mixture is yellow (similar to Buffer QG without dissolved agarose). If the color of the mixture is orange or violet, add 10 μL of 3 M sodium acetate, pH 5.0, and mix. The color of the mixture will turn yellow. The adsorption of DNA to the QIAquick membrane is efficient only at a pH ≤7.5.
    Buffer QG contains a pH Indicator which is yellow at a pH ≤7.5 and orange or violet at higher pH, allowing easy determination of the optimal pH for DNA binding.
4. Add 1 gel volume of isopropanol to the sample and mix.
    For example, if the agarose gel slice is 100 mg, add 100 μL isopropanol. This step increases the yield of DNA fragments <500 bp and >4 kb. For DNA fragments between 500 bp and 4 kb, the addition of isopropanol has no effect on yield.

5. Place a QIAquick spin column in a provided 2 mL collection tube.
6. To bind DNA, apply the sample to the QIAquick column and centrifuge for 1 minute.
   The maximum volume of the column reservoir is 800 μL. For sample volumes more than 800 μL, load 800 μL, centrifuge, remove the flow-through from the collection tube, then add the remainder of the volume and spin again.
7. Discard flow-through and place QIAquick column back in the same collection tube.
   Collection tubes are reused to reduce plastic waste.
8. To wash, add 0.75 mL of Buffer PE to QIAquick column and centrifuge for 1 minute.
9. Discard the flow-through and centrifuge the QIAquick column for an additional 1 minute.
   IMPORTANT: Residual ethanol from Buffer PE will not be completely removed unless the flow-through is discarded before this additional centrifugation.
10. Place QIAquick column into a clean 1.5 mL microcentrifuge tube.
11. To elute DNA, add 30 μL Buffer EB to the center of the QIAquick membrane, let the column stand for 1 minute, and then centrifuge for 1 minute.
   IMPORTANT: Ensure that the elution buffer is dispensed directly onto the QIAquick membrane for complete elution of bound DNA.

F. DNA Concentration Determination

1. Use a Nanodrop microvolume spectrophotometer or spectrophotometer to determine the concentration of the purified vector backbone. Be sure to use Buffer EB as a blank.
2. Record the concentration of the gel extracted vector backbone, as well as the A_{260}/A_{280} ratio to assess DNA purity.
3. Save the DNA in your −20°C freezer box.

REFERENCES

1. F. Yanfang, A.F. Jennifer, K. Cyd, L.M. Morgan, R. Deepak, J. J. Keith, D.S. Jeffry, High-frequency off-target mutagenesis induced by CRISPR-Cas nucleases in human cells, Nature Biotechnology 31 (9) (2013) 822–826.
2. P.D. Hsu, D.A. Scott, J.A. Weinstein, F.A. Ran, S. Konermann, V. Agarwala, Y. Li, E.J. Fine, X. Wu, O. Shalem, T.J. Cradick, L.A. Marraffini, G. Bao, F. Zhang, DNA targeting specificity of RNA-guided Cas9 nucleases, Nature Biotechnology 31 (9) (2013) 827–832.
3. K. Michael, T. Kärt, B. Allan, Repair of double-strand breaks induced by CRISPR–Cas9 leads to large deletions and complex rearrangements, Nature Biotechnology 36 (8) (2018) 765–771.
4. Design guide RNAs (gRNAs). https://help.benchling.com/hc/en-us/articles/9684282104333-Design-guide-RNAs-gRNAs
5. Design primers with the primer wizard. https://help.benchling.com/hc/en-us/articles/9684281285389-Design-primers-with-the-primer-wizard
6. Megan Cartwright, Three Tips (and Two Tricks) for Using BLAST (2021). https://bitesizebio.com/9951/three-tips-and-two-tricks-for-using-blast/
7. QIAGEN QIAquick Spin Handbook, https://www.qiagen.com/us/resources/resourcedetail?id=a9de0fd4-e405-4bb7-b3a0-a74b336d613e&lang=en

DISCUSSION QUESTIONS

1. Variants of Cas9 have been engineered for their different features. The Benchling CRISPR tool has a menu for you to select the Cas9 variant you are working with. Why is it important to know which variant of Cas9 you are using when designing CRISPR guides?
2. Instead of gel purifying the vector backbone, suppose you accidentally purified the ~2 kb band for use in your cloning procedure. Would that DNA have BsmBI overhangs? Could your oligos be ligated to it? Could Cas9 be expressed from the resulting recombinant DNA?

Lab Session 26

Advanced CRISPR: Part II

Goal: You will prepare your insert DNA, ligate to your prepared CRISPR vector, and transform into *E. coli*.

I. INTRODUCTION

Please refer to Lab Session 5 for information on ligation and transformation.

Now that you have prepared a dephosphorylated, BsmBI-digested version of the pLentiCRISPR-E vector, you are ready to prepare your insert. This insert DNA is the CRISPR guide sequence you designed to target a particular gene. Two processes will occur: a kinase reaction and annealing of your oligos.

A kinase is an enzyme that catalyzes the transfer of a phosphoryl group. In our case, T4 polynucleotide kinase and an appropriate buffer will be incubated with both of the oligos you ordered. At 37°C (the optimal temperature for this enzyme), phosphoryl groups will be transferred from molecules of ATP to the 5′ ends of the oligos. That way, the 5′ ends on these insert molecules can later form phosphodiester bonds with the 3′ hydroxyl groups on the linearized vector molecules.

After the kinase reaction occurs, annealing of your oligos will occur in the same tube. Annealing refers to complementary bases on opposite DNA strands forming hydrogen bonds with each other. This is possible because the oligos you designed and ordered were reverse complements of each other. Unlike the kinase reaction, this step is not enzyme-catalyzed. By starting at 95°C and gradually lowering the temperature of the reaction, the phosphorylated oligos will anneal to each other, forming double-stranded DNA.

As in Lab Session 5, you will need to ligate vector with insert. You will set up one experimental ligation (phosphorylated, annealed oligos as insert and dephosphorylated, BsmBI-digested pLentiCRISPR-E as vector). You will also have three control ligations to control for bacterial cell competency, presence of uncut vector, and recircularization of vector. This last control is included to confirm that transformed bacterial colonies did not arise from the cut pLentiCRISPR-E vector ligating back on itself.

You will then transform your ligation reactions into chemically competent *E. coli*. However, this time you will transform into an *E. coli* strain that does not recombine as frequently, as lentiviral vectors have long terminal repeats that result in homologous recombination events. The Stbl3 strain is deficient in a recombinase enzyme. Therefore, it has reduced recombination frequency and maintains these unstable repeated sequences in the plasmids. Another distinction from the BL21(DE3) strain is that the Stbl3 strain used here is not designed to express recombinant protein. Instead, our goal is to use a strain that is appropriate for propagating vector DNA.

II. LABORATORY EXERCISES

A. Oligo Annealing and Phosphorylation

Note: This experimental procedure is for one reaction (one pair of oligos). If multiple pairs of oligos are to be annealed and phosphorylated, you will be required to prepare multiple reactions.

1. Obtain the two single-stranded DNA oligos (CRISPR guides) that you designed from your instructor. They should each be 100 μM.
2. On ice, in a clean PCR tube, combine in order the reagents listed in Table 26.1.

Molecular Biology Techniques. DOI: https://doi.org/10.1016/B978-0-443-26714-7.00043-0

TABLE 26.1 Experimental Setup for Kinase/Oligo Annealing Reaction

Reagent	Volume (μL)
Oligo 1 (100 μM)	1
Oligo 2 (100 μM)	1
10× T4 DNA ligase buffer[a]	1
Sterile water	6.5
0.5 μL T4 PNK (polynucleotide kinase) (10 U/μL)	0.5
Total	10 μL

[a]Be sure to use T4 DNA ligase buffer, as ATP is required for this reaction. If you use the buffer supplied with T4 PNK, you must supplement to 1 mM ATP. Keep your reaction(s) on ice until the start of the reaction, since ATP is heat labile.
No Permission Required.

TABLE 26.2 Kinase/Oligo Annealing Thermal Cycler Program

Temperature (°C)	Time	Cycles
37	30 min	1
95	5 min	1
25	1 min; Decrease 5°C per cycle	14

No Permission Required.

3. Program a thermal cycler so that following a 30-minute incubation at 37°C for the kinase reaction, the temperature will increase to 95°C and then slowly ramp down to room temperature. This allows the now phosphorylated oligos to anneal to each other. This program can be found in Table 26.2.
4. Annealed oligos can be left at room temperature until you are ready to ligate and transform bacteria. If you will perform those steps on a different day, oligos can be saved at 4°C.

B. Ligation of sgRNAs to Digested CRISPR Vector

Note: This experimental procedure is for one sgRNA per group. If multiple sgRNAs were annealed and phosphorylated, you will be required to prepare multiple reactions.

1. Obtain your BsmBI digested, calf intestinal alkaline phosphatase (CIP)-treated, gel-purified vector from Lab Session 25. Allow the tube to thaw at room temperature.
2. Label a new microcentrifuge tube and dilute each of the annealed oligo reactions from Part A 200-fold in sterile water (1 μL reaction plus 199 μL water). Mix well.
3. In new, labeled microcentrifuge tubes, set up the following four ligation reactions, adding enzyme last and pipetting up and down to mix (Tables 26.3–26.6).
4. Spin each of the four tubes briefly in a microcentrifuge and incubate at room temperature for 10 minutes to allow ligation to occur. *Note: follow manufacturer instructions for ligation times and temperatures.*

C. Transformation Into Stbl3 *E. coli*

Competent cells are extremely fragile, and you must keep them on ice. Never vortex or pipette competent cells vigorously because they will lyse. Never warm competent cells to room temperature because they will start repairing their membranes and will lose their competency. Do not centrifuge the competent cells before use, because you will be unable to get them back into suspension. Set up four transformations (one for each ligation) as described below:

TABLE 26.3 Control Reaction for Viability and Compentency

50 ng uncut pLentiCRISPR-E	______ μL
Sterile dH_2O	______ μL
	20 μL total

No Permission Required.

TABLE 26.4 Control Reaction for the Presence of Uncut Vector

50 ng pLentiCRISPR-E cut with BsmBI	______ μL
Sterile dH_2O	______ μL
	20 μL total

No Permission Required.

TABLE 26.5 Control Reaction for Vector Recircularization

50 ng pLentiCRISPR-E cut with BsmBI	______ μL
Sterile dH_2O	______ μL
10× T4 DNA ligase buffer	2 μL
T4 DNA ligase (400 U/μL)	1 μL
	20 μL total

No Permission Required.

TABLE 26.6 Experimental Ligation Reaction

50 ng pLentiCRISPR-E cut with BsmBI	______ μL
Diluted oligo duplex from this Lab session	1 μL
Sterile dH_2O	______ μL
10× T4 DNA ligase buffer	2 μL
T4 DNA ligase (400 U/μL)	1 μL
	20 μL total

No Permission Required.

1. Set the heat block or water bath to 42°C. Verify the temperature using a thermometer, as digital displays are not always accurate.
2. Obtain four tubes of Stbl3 *E. coli* from your instructor containing 50 μL of competent cells. Label them 1–4 (designations from ligation). Chill and keep on ice at *all* times.

   $ Money-Saving Tip: Stbl3 cells shipped from the manufacturer are in 50 μL aliquots. The instructor may choose to make 25 μL aliquots for students to increase the number of transformations possible.
3. Carefully add 2 μL of the ligation mixtures from Part B to the appropriately labeled competent *E. coli* tube, being sure to add the DNA directly into the bacterial suspension. Mix the tube gently by finger flick.
4. Incubate these vials on ice for 5 minutes.
5. Heat-shock the cells by incubating them for 45 seconds in a 42°C water bath or heat block.
6. Place the vials back on ice for about 2 minutes.
7. Add 250 μL of room temperature SOC or LB medium to each vial. Close each of the vials tightly.
8. Shake horizontally at 37°C for ~1 hour at 225 rpm.

9. Obtain four pre-warmed LB/amp plates from your instructor and label each accordingly. Spread 50 μL (or the entire volume of cells used in each transformation) on the corresponding plate.
10. Incubate overnight at 37°C.
    Time-Saving Tip: Alternatively, you can incubate plates at room temperature over ~3 nights.
11. You may choose to save the remaining transformation mix at 4°C in order to plate a different volume the next day, if needed. Alternatively, you can plate two different volumes of your experimental transformation mix the same day to ensure at least one plate will have isolated colonies. In most cases, 50 μL will produce isolated colonies by the next morning.

DISCUSSION QUESTIONS

1. Phosphatases and kinases were discussed in the last two lab sessions. Differentiate between these two classes of enzymes.
2. Suppose you forgot to keep the T4 DNA ligase buffer on ice during all of the steps performed today. How could that affect your ligation and transformation steps?
3. Two different types of bonds are formed during the kinase reaction and annealing of oligos. What are they? Name two ways they are different.

Lab Session 27

Interim Laboratory Session

Molecular Biology Techniques. DOI: https://doi.org/10.1016/B978-0-443-26714-7.00033-8

Lab Session 27a

Interim Laboratory Session

COUNTING TRANSFORMANTS AND INOCULATING CULTURES FOR MINIPREP DNA

Goal: Today you will count and record the number of transformants on each plate. You will also make a replica plate of putative positive clones and inoculate liquid cultures. Plasmid DNA will be purified from these liquid bacterial cultures that can later be sequenced to confirm your insert is correctly cloned.

I. INTRODUCTION

Review Lab Session 6a, Interim Laboratory Session for information on replica plating.

II. LABORATORY EXERCISE

1. Count and record the number of transformants on each of your ligation/transformation plates. You should have many colonies on the plate from the uncut vector control ligation/transformation (Reaction 1) and several colonies from your experimental ligation (Reaction 4). You should have extremely few, if any, colonies on the plates where the bacteria were transformed with the other controls.
2. Assuming your controls gave you the expected results, obtain an LB/amp plate. Adhere a grid sticker to the back and label with your initials and date. *Always label the backs of Petri dishes rather than the lids to avoid mix-ups*

   Money-Saving Tip: In lieu of grid stickers, a Sharpie marker can be used to draw grids on the back of the Petri dish.
3. Label three culture tubes (containing 2 mL LB/amp) with your initials, date, and transformant number.
4. You will inoculate three isolated colonies from your transformation plate 4 (experimental ligation) onto the new agar plate, as well as into a liquid culture. Do this as you did in Lab Session 6a, Interim Laboratory Session, using a sterile pipette tip for each transformant. First, pick a small amount of bacterial growth from each isolated colony you wish to analyze; streak it gently on one of the squares of the grid plate. Then, simply drop that same tip into the correctly labeled tube. Be sure to change pipette tips between each transformant you pick. If you had unexpected results in your controls, you may need to select colonies from another group's plate.

   Note: This inoculation method does not utilize the principle of aseptic technique. It typically works well for inoculating cultures for minipreps because the antibiotics in the medium keep most contaminants from growing. Furthermore, the inoculum is so large that it would likely outcompete any contaminant. In cases where a pure culture is critical, such as creating a freezer stock, ***do not*** *utilize this method.*
5. Place your tubes and plates in the appropriately labeled area at the front of the lab. Your instructor will place these in the refrigerator at 4°C until the afternoon prior to your lab session and will then incubate them at 37°C overnight so cultures will be fresh for your lab session.

   Time-Saving Tip: Alternatively, transformation replica plates can be incubated at room temperature over 2–3 nights.

Lab Session 27b

Advanced CRISPR: Part III

Goal: You will isolate the plasmid DNA from the transformants you inoculated in the interim lab session and prepare these plasmids for sequencing reactions to confirm that you have at least one positive clone.

I. INTRODUCTION

After completing the annealed oligo cloning, your next challenge is to confirm that the transformants have your CRISPR guide insert properly cloned into the BsmBI sites of pLentiCRISPR-E. Although there are multiple screening techniques that can be employed (see Lab Sessions 6–8), you will use Sanger sequencing to identify positive clones in the next lab sessions. One reason why Sanger sequencing is well suited for screening is that the insert you are trying to detect is extremely small. Remember that the oligos you designed were just over 20 nt each. If you were to use BsmBI to digest the purified vector and liberate your insert, it would be difficult to visualize a fragment that small on an agarose gel. This is in contrast to the ~700 base pair *egfp* insert you used in PCR cloning. Please review the process of Sanger sequencing in Lab Session 7.

Miniprep DNA will be required for both sequencing analysis AND transfection of cells. If necessary, please review background information on plasmid purification from Lab Session 2.

II. LABORATORY EXERCISES

A. Isolation of Miniprep DNA

This protocol is modified from the QIAGEN QIAprep Miniprep Kit handbook (June 2020).

Note: All protocol steps should be carried out at room temperature.

Note to instructor: Numerous other miniprep kits are available for DNA purification. If you choose to use a different kit, follow the protocol for a high-copy plasmid. Be sure to follow the instructions with the kit to prepare the reagents prior to use.

1. Label three sterile microcentrifuge tubes to match the labels on your overnight culture tubes. Be sure to include your initials.
2. Using a micropipette, transfer ~1.5 mL of each of your three cultures into the correctly labeled microcentrifuge tube.
3. Harvest the bacterial cells by centrifugation at >8000 rpm (6800 × *g*) for 3 minutes.
   Remove all traces of supernatant by decanting or micropipetting the media into a liquid waste container with 10% bleach. (*Note: If you wish to stop the protocol and continue later, freeze the cell pellets at −20°C.*)
4. Resuspend the bacterial pellet in 250 μL Buffer P1 (resuspension buffer). The bacteria should be resuspended completely until no cell clumps remain.
5. Add 250 μL Buffer P2, and mix gently but thoroughly by inverting 4–6 times.
   Do not vortex, as this will result in shearing of genomic DNA. If necessary, continue inverting the tube until the solution becomes viscous and slightly clear. Do not allow lysis reaction to proceed for more than 5 minutes. *Buffer P2 contains a detergent (SDS) for cell lysis and sodium hydroxide for denaturation of DNA.*
6. Add 350 μL Buffer N3 to the lysate and mix immediately and thoroughly (but gently) by inverting 4–6 times. After addition of Buffer N3, a fluffy white precipitate containing genomic DNA, proteins, cell debris, and SDS becomes visible. The buffers must be mixed completely. If the mixture still appears viscous and brownish, more mixing is required to completely neutralize the solution. *Buffer N3 neutralizes the solution, causing plasmid DNA to reanneal, and acts to precipitate the chromosomal DNA and insoluble proteins.*
7. Centrifuge for 10 minutes at 13,000 rpm (17,900 × *g*) in a microcentrifuge.
8. Apply the supernatant (liquid above pellet) from Step 7 on the QIAprep spin column (containing the silica membrane) by pipetting, taking care not to transfer the fluffy white precipitate. Place column in accompanying tube.

9. Centrifuge for 30 seconds. Discard flow-through.
10. Wash the QIAprep spin column by adding 0.5 mL Buffer PB and centrifuge for 30 seconds. Discard the flow-through.
11. Wash the QIAprep spin column by adding 0.75 mL Buffer PE and centrifuge for 30 seconds.
12. Discard flow-through, and centrifuge for an additional 1 minute to remove residual wash buffer.
13. Place each column in a clean 1.5 mL microcentrifuge tube. Label the tubes with pLentiCRISPR-sgRNA clone number_____, the date, and your initials.
14. To elute DNA, add 50 µL Buffer EB to the center of each column, let stand for 1 minute, and centrifuge 1 minute.
15. Use a Nanodrop microvolume spectrophotometer or spectrophotometer to determine the concentration and purity of your DNA samples. Be sure to write the DNA concentration (including units) directly on the labeled microcentrifuge tube.

Note to instructor: The following instructions are for preparing sequencing samples for submission to the GENEWIZ sequencing facility. There are many different sequencing facilities available throughout the world, with many academic institutions having their own facilities. If you use a different facility, follow their specific sample preparation guidelines and instructions.

B. Preparation for DNA Sequencing

1. Obtain three thin-walled PCR tubes in which you will set up your sequencing reactions. Your TA or instructor will give you details on how to label the tubes.
2. Using your miniprepped DNA, determine the volume of each sample that will give you 1000 ng. The sequencing facility will want that approximate mass amount in 10 µL. (*Note: If your DNA has a low concentration, use as much as possible.)*
3. For sequencing reactions, we will utilize a free sequencing primer that most facilities keep on hand. This primer is specific to the U6 promoter. In the pLentiCRISPR-E vector, the human U6 promoter is immediately upstream of where you putatively cloned your sgRNA (Fig. 23.3 from Lab Session 23: CRISPR-Mediated Knockout of EGFP: Part I and Fig. 23.4 from Lab Session 23: CRISPR-Mediated Knockout of EGFP: Part I). The short size of your sgRNA insert increases the chances that the sequencing reaction will yield enough high-quality sequence with just one sequencing primer. The sequencing facility will add this primer to your submitted DNA at their facility when preparing the reactions.

   Note: If your facility does not provide a sequencing primer that anneals to the U6 promoter, you can have them synthesize the following primer for use in the sequencing reactions:

   5′GACTATCATATGCTTACCGT3′
4. Give your tubes to your instructor for submission to GENEWIZ or another sequencing facility.
5. Save the miniprepped plasmid DNA in your freezer box at −20°C.

DISCUSSION QUESTIONS

1. Examine the vector map of pLentiCRISPR-E (Fig. 25.3 from Lab Session 25: Advanced CRISPR: Part I). Where else could a sequencing primer be designed to anneal that could confirm the presence of your sgRNA?
2. Compare and contrast the selectable markers of pLentiCRISPR-E and pEGFP-N1. Predict what would happen if you inoculated your CRISPR transformants in LB/kan.

Lab Session 28

Advanced CRISPR: Part IV

Molecular Biology Techniques. DOI: https://doi.org/10.1016/B978-0-443-26714-7.00030-2

Lab Session 28a

Advanced CRISPR: Part IV

I. INTRODUCTION

Refer to Lab Session 7 d and Lab Session 8 for Sanger DNA sequencing information.

II. LABORATORY EXERCISES

A. Computational Analysis of DNA Sequences

1. To visualize your sequencing results, you will need the files containing them. They are typically saved as .ab1 files. There are many different programs that can view these files. The instructions in this lab session are specific for Chromas, which is available as freeware via Technelysium Pty Ltd (http://technelysium.com.au/wp/chromas/), but the general use of most programs is similar and they can perform the same functions. For Mac users, Finch TV is popular freeware that works similarly.
2. Visually scan the chromatogram to determine the length of sequence containing high-quality data with no (or only a few) Ns. The sequencing facility may also provide you with quality and length values for each sample.
3. The exact region containing your insert sequence cannot be found by searching for the BsmBI site because it was destroyed during cloning. Instead, since the insert is very short, search for the insert sequence itself. You can use the find button on the toolbar to search for a nucleotide sequence. Verify that the sequence obtained matches the guide sequence (20 mer) you designed and that there are no ambiguous nucleotides in this region.

   To identify the 5′ end of your guide sequence within the sequencing read, search for the CACCG sequence. Recall that immediately upstream of your 20 mer, you engineered the CACCG overhang into the forward oligo to create complementary ends that could be ligated to the BsmBI-digested vector. To find the 3′ end of your guide sequence, search for the gRNA scaffold. It is part of the pLentiCRISPR-E vector, as shown in Figs. 23.3 and 25.3. This sequence begins with GTTTT on the top strand. Verify that the 20 mer sequence that is found between the 5′ CACCG and the 3′ GTTTT matches your designed CRISPR guide. Record your observations in your notebook. Repeat this analysis for each sequencing file your group received.
4. Perform a nucleotide BLAST: https://blast.ncbi.nlm.nih.gov/Blast.cgi?PROGRAM=blastn&BLAST_SPEC=GeoBlast&PAGE_TYPE=BlastSearch

   This is different than the earlier BLAST you may have performed with sequencing results from *egfp* cloning. You are not interested in performing a BLAST search of your sequencing result with *all* genes, since you already designed the insert to be specific to one *particular* gene. Instead, this nucleotide BLAST allows you to align two designated sequences. Make sure to check the box that says "Align two or more sequences" (Fig. 28.1). That will display two boxes: one will be a "Query sequence," and the other one will be a "Subject sequence." Paste your sequencing results into the query sequence box and paste or enter the accession number of your gene of interest in the subject sequence box. Leave the program selection as "Highly Similar Sequences (Megablast)." Your results should display a short stretch of sequence that is 100% aligned to your gene of interest. Repeat this analysis for each sequencing file your group received.
5. Show your nucleotide BLAST results to your instructor to verify that your sequencing results showed the correct insert (top strand of CRISPR guide sequence you designed). You may find that all of the transformants you screened by sequencing are positive. Choose the one that had the best A_{260}/A_{280} ratio in Lab Session 27b, because you only need one to continue with the experiment. If your group did not have high-quality sequencing reads or any transformants that contained your insert, you will want to go back and screen additional transformants or use plasmid DNA from another group after consulting with your instructor.

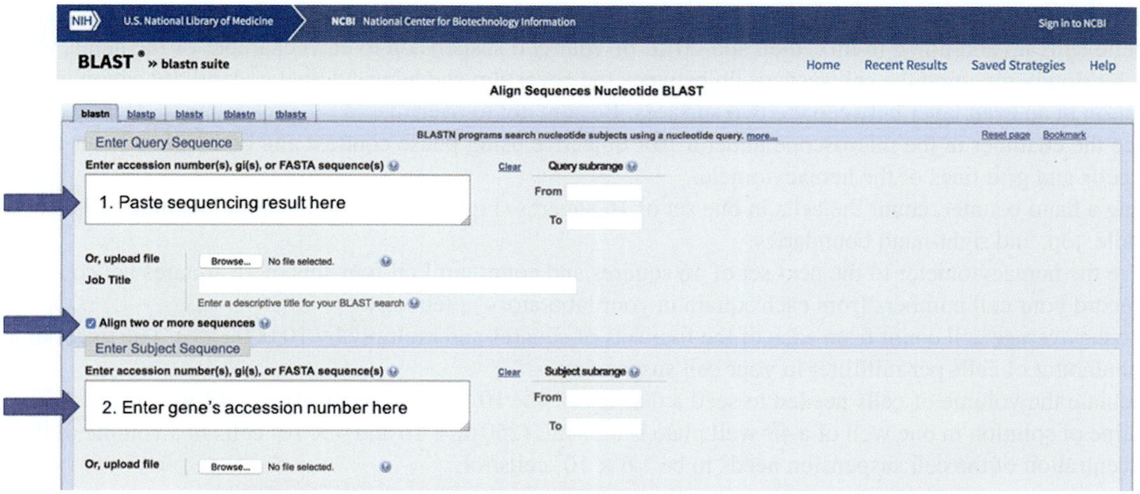

FIGURE 28.1 A screenshot of NCBI's Nucleotide BLAST. *No Permission Required.*

B. Counting and Plating EGFP Expressing Mammalian Cells

Please use aseptic technique when working with cells in culture (for a review, see Lab Session 18).

Note: If your group is analyzing more than one experimental CRISPR guide, you will need to plate more wells accordingly. The following instructions are written to include appropriate controls and three experimental wells for optimizing the transfection of one CRISPR guide that you designed.

1. As you did in Lab Session 18, you will turn on, clean, and prepare the biological safety cabinet before you begin working with cells. Collect all of the consumables you will need for these steps: sterile pipette tips, sterile Pasteur pipettes, sterile serological pipettes, and a sterile 48-well plate. Micropipettes and pipette aids should remain in the biological safety cabinet.
2. Obtain a 10 cm^2 dish or flask of HEK293 cells provided by your instructor. Observe their confluency and morphology and write the details in your lab notebook. If you observe any signs of contamination, notify your instructor immediately. Also observe the color of your cell culture *complete growth medium*. The complete medium for HEK293 cells is high glucose (4500 mg/L) Dulbecco's Modified Eagle's Medium (DMEM) supplemented with 10% fetal bovine serum, 0.1 mM nonessential amino acids, 2 mM L-glutamine, 1 mM sodium pyruvate, 1500 mg/L sodium bicarbonate, and 1% penicillin-streptomycin. It contains phenol red, a pH indicator. If the medium drastically changes color to yellow (acidic) or purple (basic), notify your instructor.
3. Using a sterile Pasteur pipette, aspirate the medium from the dish. Do this by gently tipping the dish toward you and placing the tip of the Pasteur pipette to the bottom of the vessel until all the medium is aspirated. Remember the spot where you touched the dish. You will always aspirate from this spot to avoid aspirating cells from your dish.
4. Gently add ~2 mL of sterile 1× PBS to the dish and swirl carefully. Whenever you remove or add liquid to your cells, be sure to do so from the same location to avoid dislodging cells.
5. Aspirate the PBS from the dish.
6. Add 2.5 mL of pre-warmed 0.25% Trypsin-EDTA solution to the dish and swirl to ensure adequate coverage on the bottom. Place dish back in the incubator and set your timer for 2 to 3 minutes.
7. Observe your cells under the microscope; detached cells will float in the medium and appear rounded. You may want to gently but firmly tap your dish on a sterile surface to help further detach your cells. If you still have 50% or greater cells attached to the dish, return it to the 37°C incubator for another 2 minutes.
8. Once cells appear detached, add 5 mL of pre-warmed complete medium (DMEM) *without any antibiotics* to inactivate the trypsin. Gently pipette the medium over the cell layer surface several times to ensure detachment. Vigorous pipetting can cause cell damage, so be careful.
9. You are now ready to count your cells using a hemacytometer.

   Time-Saving Tip: Alternatively, an automated cell counter such as a Cellometer from Nexcelom can be used.

   Hemacytometers are devices used to count cells. The following procedure is appropriate for glass hemacytometers, although plastic disposable ones can be used as well.

10. Clean the chamber and cover slip with 70% EtOH. Dry and put the cover slip in position.
11. Pipette cells up and down to mix, then add 20 µL of your cell suspension to each chamber of the hemacytometer. Do this by slowly pipetting the cell suspension between the cover slip and hemacytometer; capillary action will spread the solution in an even layer between the two surfaces. Be sure not to overfill.
12. Place the chamber in the microscope under a 10× objective using phase contrast and course/fine focus to distinguish the cells and grid lines of the hemacytometer.
13. Using a hand counter, count the cells in one set of 16 squares (Fig. 18.3). When counting, include cells that fall in the middle, top, and right-hand boundaries.
14. Move the hemacytometer to the next set of 16 squares and count until all four sets of 16 squares are counted. Be sure to record your cell numbers from each square in your laboratory notebook.
15. Take the average cell count from each of the four sets of squares and multiply by 10,000 (10^4). The final number is the total amount of cells per milliliter in your cell suspension.
16. Calculate the volume of cells needed to seed a density of 4×10^4 cells in one well of a 48-well plate. An appropriate volume of solution in one well of a 48-well plate is 0.25 mL (250 µL). To add 4×10^4 cells in a volume of 0.25 mL, the concentration of the cell suspension needs to be 1.6×10^5 cells/mL.

$$\text{Using } C_1\ V_1 = C_2\ V_2$$

where

C_1 is the concentration of cells counted (cells/mL):______________

V_1 is the amount of cell suspension (mL) required (unknown):______________

C_2 is the desired concentration of cells: 1.6×10^5 cells/mL

V_2 is the desired total volume: 2.5 mL

17. Add the appropriate volume of cell suspension to a sterile 15 mL conical. Bring the total volume up to 2.5 mL with complete medium (DMEM) *without any antibiotics*. Mix well by carefully pipetting up and down (avoid forming bubbles) and immediately add 250 µL of cell suspension per well of your 48-well plate for a total of six wells. You may be asked to share this plate with other student groups.
18. In addition, plate 125 µL of cell suspension to six different wells.
19. Add 125 µL of complete medium (DMEM) *without any antibiotics* to those wells for a total of 250 µL/well. You should now have six wells with 4×10^4 cells each and six wells with half that amount (2×10^4 cells each).
20. Write your initials as well as the date on the lid of your plate. Return your plate to the incubator. Your instructor will tell you how to dispose of your extra cells

DISCUSSION QUESTIONS

1. When performing the BLASTn, the majority of your DNA sequencing read does not align to your gene of interest. Why?
2. What is a major limitation of annealed oligo cloning? Could you have used it in place of cloning by PCR in Lab Session 3? Why?

Lab Session 28b

Interim Laboratory Session

TRANSFECTION OF CRISPR VECTORS

I. INTRODUCTION

Now that you have analyzed your group's sequencing data and confirmed that a CRISPR guide of interest was cloned into pLentiCRISPR-E, you can transfect those plasmids into HEK293 cells stably expressing EGFP. This transfection will allow the cells to:

1. Transcribe the sgRNA(s) specific to your gene of interest in the human genome.
2. Transcribe and translate the Cas9 nuclease that will complex with the sgRNA and create double-strand breaks in your gene of interest to activate the NHEJ pathway.

Please review Lab Session 19 for more information about transfecting mammalian cells.

II. LABORATORY EXERCISE

Note: If your group is analyzing more than one experimental sgRNA, you will need to perform more transfections accordingly. These instructions are written to include appropriate controls and three experimental wells for optimizing the transfection of one sgRNA that you designed.

Please refer to Fig. 19.3 from Lab Session 19, Transient Transfection of Mammalian Cells to familiarize yourself with the overall workflow of the transfection experiment. This procedure was modified with permission from Promega's FuGENE HD transfection reagent protocol:

1. Remove the FuGENE HD reagent from the 4°C refrigerator and allow to reach room temperature. Mix gently by inversion before use.
2. Construct a table in your notebook similar to Table 28.1.
   a. Your negative controls will be Opti-MEM only (no plasmid or FuGENE HD) and pLentiCRISPR-E (which produces no sgRNA, only Cas9).
   b. Your positive control for CRISPR will be a validated sgRNA against *egfp*. This control will not target your gene of interest and will be provided by your instructor (or you may have saved it from a previous lab session).
   c. Look up the concentration of each mini-prepped pLentiCRISPR-sgRNA plasmid recorded in Lab Session 27. You will need to decide which experimental plasmid DNA to transfect, based on your sequencing results. Record its concentration in the last three rows of Table 28.1 (pLentiCRISPR-sgRNA).
3. Obtain your group's 48-well plate of HEK293 cells that were plated the day prior. Remove the medium from each well and replace with 250 µL of fresh serum-containing complete medium (DMEM) *without any antibiotics*.
4. Using the microscope and 10× objective lens, observe and record the morphology and confluency of your cells under white light. Your cells should be 60%–80% confluent with few floating cells in the medium. You plated cells at two different densities. You will need to choose the density which resulted in 60%–80% confluency, as you will only transfect those six wells. If you notice any signs of contamination or if none of your cells are in the 60%–80% confluent range, notify your instructor. Return your plate to the cell culture incubator.
5. In the biological safety cabinet, label six sterile microcentrifuge tubes to match the labels in Table 28.1. It is best to transfect different amounts of plasmid to optimize transfection efficiency, and therefore overall gene editing. Therefore, you will prepare three different transfection complexes with increasing amounts of your recombinant CRISPR vector (pLentiCRISPR-sgRNA): 0.28, 0.56, and 0.84 µg. Label the tubes accordingly. You will keep the µL to µg ratio of FuGENE HD:DNA at 5:1.
6. Add the appropriate amount of each plasmid DNA to the correctly labeled tube. For all tubes that receive 0.28 µg of plasmid DNA, you will add OptiMEM to a total volume of 12 µL.

   $ Money-Saving Tip: Serum-free media can be used in place of OptiMEM.

TABLE 28.1 Experimental Setup for CRISPR-mediated Knockout of a Gene of Interest

Tube	Plasmid	Conc. (µg/µL)	Vol. to Add for 0.28 µg Plasmid (µL)	Vol. to Add for 0.56 µg Plasmid (µL)	Vol. to Add for 0.84 µg Plasmid (µL)	Opti-MEM (µL)	FuGENE HD (µL)
Opti-MEM only (negative control)	No plasmid	0	0	0	0	12	0
Empty (negative control)	pLentiCRISPR-E			N/A	N/A	Up to 12 total	1.4
egfp (positive control for CRISPR, negative control for targeting your gene of interest)	pLentiCRISPR-EGFP			N/A	N/A	Up to 12 total	1.4
Experimental #1	pLentiCRISPR-sgRNA			N/A	N/A	Up to 12 total	1.4
Experimental #2	pLentiCRISPR-sgRNA		N/A		N/A	Up to 24 total	2.8
Experimental #3	pLentiCRISPR-sgRNA		N/A	N/A		Up to 36 total	4.2

7. For the experimental tubes that receive 0.56 µg of plasmid DNA, you will add OptiMEM to a total volume of 24 µL.
8. For the experimental tubes that receive 0.84 µg of plasmid DNA, you will add OptiMEM to a total volume of 36 µL (see Table 28.1).
9. To maintain a 5:1 FuGENE HD:DNA ratio while increasing the mass amount of DNA in some samples, the volume of FuGENE HD must also increase. Carefully add the indicated volume of FuGENE HD reagent (Table 28.1) directly into the Opti-MEM diluted DNA in each tube. *DO NOT allow the FuGENE HD reagent to directly contact the plastic walls of the tube*, as this will drastically reduce transfection efficiency. Change tips for each new aliquot.
10. Carefully mix the contents in each tube by pipetting up and down or by vortexing briefly. Leave in the biological safety cabinet for 15 minutes. This time will allow the formation of DNA:FuGENE HD complexes, thus enabling the DNA to enter the cells.
11. Remove your group's 48-well plate from the incubator. Label (on the lid) what each well will contain using the names given in Table 28.1.
12. Add the entire volume of the transfection complexes drop-wise to the corresponding well of the 48-well plate. Rock the plate back and forth gently to ensure complete mixing of the FuGENE HD:DNA complexes throughout the medium and return the plate to the 37°C incubator. During your next lab session, you will harvest gDNA from these transfected cells to begin assaying CRISPR-mediated gene editing.

Lab Session 29

Advanced CRISPR: Part V

Goal: Today you will lyse the cells transfected with CRISPR vectors and PCR amplify the genomic DNA (gDNA) region you intended to target with your sgRNAs.

I. INTRODUCTION

Not all sgRNAs are equally effective at targeting a gene of interest. Recall that a computer algorithm was used to help design your sgRNAs in order maximize cleavage events within your gene of interest, while minimizing off-target effects. Due to this fact, it is recommended that researchers design and test multiple sgRNAs. This experimental design strategy is similar to that involving RNAi. Even though an sgRNA is intended to create a double-strand break in the gene and lead to an insertion or deletion, one must consider other possible outcomes. Transfection of the CRISPR vector that encodes the sgRNA and Cas9 is never 100% effective. Furthermore, since you transiently transfected cells, the CRISPR vector was only produced in a subpopulation of the cells. In order to maximize transfection efficiency, you were instructed to test multiple amounts of CRISPR vector in your transfections. Even in cells that did receive the vector, the sgRNAs may only target one of the two alleles for your gene of interest. This would lead to DSBs in a heterozygous, not homozygous manner. Thus, CRISPR experiments hinge not only on transfection efficiency, but also cleavage efficiency.

A researcher must confirm that an sgRNA created an indel where it was intended to. The CRISPR experiment aimed at knocking out *egfp* (Lab Sessions 23 and 24) was unique in that a fluorescent, easily detectable gene product was targeted. In most cases, your gene of interest will not be detectable by eye. One must use biochemical techniques to analyze if the targeting was accurate. A quantitative measure of this gDNA cleavage will allow you to determine the most effective sgRNA to move forward with in future CRISPR-mediated gene editing experiments. This lab exercise and the next will serve as primary screens, which are outlined in Fig. 29.1. Today, you will isolate gDNA from your CRISPR vector-transfected cells since it is the gDNA you aimed to disrupt. Then, you will use PCR to amplify a fragment of that gDNA containing the locus where you expect a double-strand break. Remember, this is immediately upstream of a PAM site. In the next lab session, you will denature and reanneal the strands of the PCR product and add an enzyme that detects mismatches in the DNA. A mismatch signals that a DNA strand with an indel has re-annealed with a strand that does not have an indel or has a different indel (see Fig. 29.1 Steps 3–5). By using this assay, you will be able to determine to what extent each of the tested sgRNA conditions resulted in cleavage of a genomic locus.

II. LABORATORY EXERCISES

The following steps are modified from the GeneArt Genomic Cleavage Detection Kit Manual MAN0009849, 2014. If you are using a different commercially available indel detection assay, please refer to the manufacturer's instructions.

Note to instructors: Cells should be harvested ~48 hours post-transfection (Lab Session 28: Advanced CRISPR: Part IV). Longer incubations will result in overly confluent cells. If harvesting as a class cannot occur 48 hours post-transfection, instructors should follow the steps under "A: Harvesting Cells" and freeze pellets at −80°C for use on the next lab day.

If students are going to continue to maintain CRISPR vector-transfected cells with the goal of creating a clonal population (see Introduction in Lab Session 30), they should harvest and analyze just half of the cells in every transfected well. The other half of the cells in each well can continue to be cultured while this indel detection assay is performed. The instructions in Lab Sessions 29 and 30 are written so that students will stop with the indel detection assay. For that reason, all of the cells are harvested for lysates. In addition, lysates can be made from just the pLentiCRISPR-EGFP

Molecular Biology Techniques. DOI: https://doi.org/10.1016/B978-0-443-26714-7.00025-9

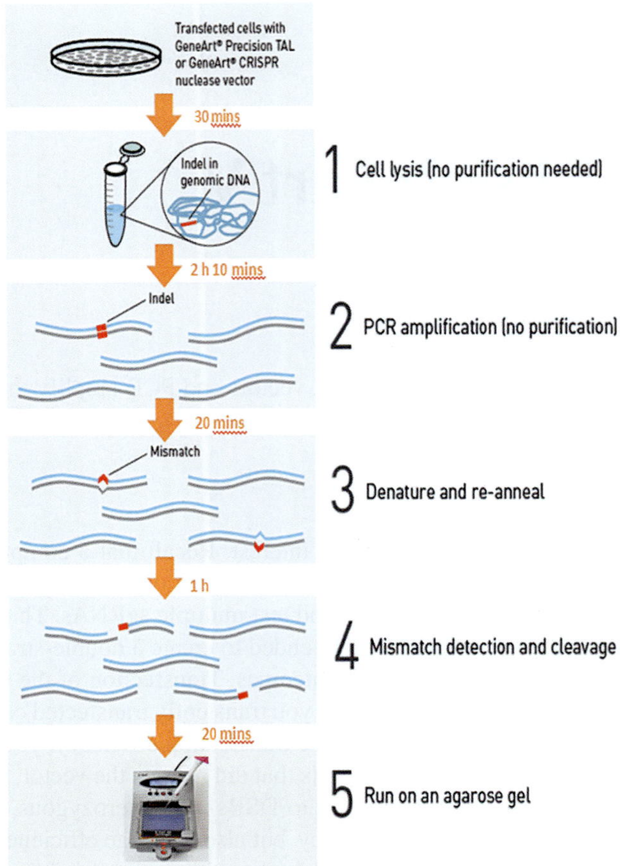

FIGURE 29.1 An overview of the GeneArt Genomic Detection Kit. Lab Session 29, Advanced CRISPR: Part V, will involve Steps 1 and 2 Lab Session 30, Advanced CRISPR: Part VI, will involve Steps 3–5.

(negative control for targeting your gene of interest) and each well transfected with pLentiCRISPR-sgRNA (experimentals) for four lysates per group.

A. Harvesting Cells

1. Obtain your group's 48-well plate of HEK293 cells that were transfected in Lab Session 28.
2. Using the microscope and 10× objective lens, observe and record the morphology and confluency of your cells under white light. Your cells should be extremely confluent at this point.
3. Take your plate of transfected cells to the bench. You will not work with them in a sterile environment, as the cells will be lysed today. *Note: If you will also be maintaining these cells in culture to create a clonal population, cells must only be manipulated in the biological safety cabinet.*
4. Aspirate all of the medium using an in-house vacuum at the bench or a self-contained aspirating pump. Whenever you remove or add liquid to your cells, be sure to do so from the same location to avoid dislodging cells.
5. If using a 48-well plate, gently add ~200 µL of 1× PBS to the wells: pLentiCRISPR-EGFP and each transfected with pLentiCRISPR-sgRNA. Rock the plate gently. Aspirate.
6. Add ~200 µL of pre-warmed 0.25% Trypsin-EDTA to each well and incubate on the bench for ~2 to 3 minutes.
7. Add ~200 µL of complete media to each well.
8. Pipette up and down to mix and create a cell suspension.
9. Transfer each cell suspension to a labeled microcentrifuge tube.
10. Spin down the cells at 200 × *g* for 5 minutes at 4°C.
11. Carefully remove the supernatant with a pipette tip and dispose of all liquid in a waste container containing 10% bleach. Do not attempt to aspirate at this step, because you may lose the pellet of cells!

B. Lysis and Extraction

1. Thaw Cell Lysis Buffer and Protein Degrader at room temperature. You will need to mix 50 μL Cell Lysis Buffer with 2 μL Protein Degrader for each cell lysate you make. Since you have four lysates, calculate the volume of each you will need, being sure to prepare approximately 20% extra to account for pipetting variability. Pipette these into a sterile microcentrifuge tube and pipette up and down to mix.
2. Add 50 μL of Cell Lysis Buffer/Protein Degrader mix to each cell pellet and resuspend the pellet by pipetting up and down.
3. Transfer all of the resuspended cell pellet to a labeled PCR tube.
4. Run this program in a thermal cycler:
   68°C 15 minutes
   95°C 10 minutes
   4°C Hold*

**Following completion, immediately proceed to PCR amplification or store at −20°C.*

Even though this step uses a thermal cycler, PCR is not occurring. The increased temperature helps lyse the cells. Using the instrument is a convenient way to incubate for specific times and temperatures, just like it was used for reverse transcription reactions in Part IV of this book.

While the cells are lysing, proceed to pouring agarose gels, as you will need three per group in the next session to confirm: (1) PCR resulted in one specific product from your gene of interest at the correct molecular weight and (2) an indel was detected in a proportion of your cells.

C. Pouring Agarose Gels

1. Each group will need an agarose gel next week that has enough lanes for:
   a. one DNA ladder.
   b. one positive control reaction.
   c. one lane for each lysate (pLentiCRISPR-EGFP, pLentiCRISPR-sgRNA × 3).
2. Each group will additionally need enough lanes for:
   a. one DNA ladder.
   b. two lanes for the control reaction.
   c. two lanes for each lysate (eight total).
3. Prepare three separate agarose gels today to accommodate the number of lanes needed next week. You should prepare a 2% agarose gel solution (2 g agarose in every 100 mL TBE). The higher percentage of agarose will create smaller pore sizes to better resolve DNA fragments of similar molecular weight. You will still add 10,000× GelRed stock to 1× final concentration.
4. Once solidified completely, each group can wrap the entire tray, gel, and comb in plastic wrap and store at 4°C until the next lab session. Your instructor may provide you with different directions for storing gels so that trays and combs can be used by other groups.

D. PCR

1. Briefly vortex cell lysates once the thermal cycler program is complete.
2. In addition to your lysates, you should also run a positive control that is provided with the kit (five PCR reactions total). This control includes both DNA template and forward and reverse primer mix in one reagent. The gene-specific primers you need to add were designed and ordered in Lab Session 25. Label five new PCR tubes for a control PCR and lysates 1–4. Use labels consistent with what was transfected in each well.
3. The following components will be added to a PCR tube. Create a table in your notebook like Table 29.1 for a master mix containing calculated volumes of everything except lysate (forward and reverse primers, AmpliTaq Gold 360 Master Mix, and water). Vortex the master mix. Distribute 48 μL of that master mix per tube, then add 2 μL of the appropriate cell lysate. Set up the control reaction separately with only control template and primers, AmpliTaq Gold 360 Master Mix, and water. Vortex to mix.
4. Keep your PCR tubes on ice until all groups are ready.
5. Run the thermal cycler using the program shown in Table 29.2.

TABLE 29.1 Experimental Setup for PCR

Reagent	Amount Per Reaction (μL)	Amount for 4 reactions (μL)	Amount Per Control Reaction (μL)
Cell lysate (serves as DNA template)	2	–	–
Forward gene-specific primer (10 μM)	0.5		–
Reverse gene-specific primer (10 μM)	0.5		–
Control template and primers (provided in kit)	–	–	1
AmpliTaq Gold 360 Master Mix	25		25
Water	22		24
Total	50		50

No Permission Required.

TABLE 29.2 Thermal Cycler Program

Stage	Temperature (°C)	Time	Cycles
Enzyme activation	95	10 min	[illegible]×
Denature	95	30 s	[illegible]0×
Anneal	55–60	30 s	
Extend	72	30 s	
Final extension	72	7 min	[illegible]×
Hold	4	Indefinitely	[illegible]×

No Permission Required.
Note: The above conditions assume that previous primers designed in Lab Session 25, have similar melting temperatures (T_ms) around 60°C and result in products of ~400–500 bp in length. The annealing temperature should be adjusted based on the primers used.

After the program is complete, your instructor or TA will store your PCR tubes in the −20°C freezer until your next lab session.

DISCUSSION QUESTIONS

1. How does the cell lysis buffer used in the genomic cleavage detection experiment differ from that used in protein purification experiments? What is the purpose of these two buffers?
2. If your sgRNAs failed to target a particular genomic locus, will you get a PCR product? Why?

Lab Session 30

Advanced CRISPR: Part VI

Goal: Today you will detect indels in your gene of interest and quantify the effectiveness of each sgRNA.

I. INTRODUCTION

As you recall from the last lab session, you PCR amplified a genomic region (locus) that contained the site where your sgRNA should have created a DSB. These specific products will be confirmed using agarose gel electrophoresis.

One way a cell can repair a DSB is by making an indel. In order to detect indels and how frequently they occurred, you will denature your double-stranded PCR products into single strands. After allowing strands to randomly reanneal by decreasing the temperature to 25°C, mismatches will occur when the sequence on one strand is not complementary to the sequence on the opposing strand (see Fig. 29.1 from Lab Session 29: Advanced CRISPR: Part V). Reannealed PCR products will be treated with an enzyme capable of recognizing mismatches and cleaving the mismatched DNA, then analyzed on an agarose gel. Finally, the efficiency of the gene modification will be quantified using NIH ImageJ.

Please be aware that an entire CRISPR gene knockout workflow would take several months to achieve in a research lab. For this reason, the laboratory exercise that follows gives you a solid foundation for a successful CRISPR experiment. Should you wish to create a knockout cell line, you would already have the tools available: an sgRNA to target your gene with high efficiency (as determined in today's lab session), Cas9 expressed on the same plasmid, and a human cell line to transfect. The screening involved to complete the workflow would take additional time if you wanted to clonally expand CRISPR-edited cells (Fig. 30.1); however, the fundamental methods involved in screening have already been conducted in the DNA sequencing (Lab Session 7d) and western blotting sessions (Lab Sessions 9–10) of this manual.

Today's indel detection assay (a mismatch cleavage assay) is a first step in confirming CRISPR-mediated gene knockout. Fig. 30.1 shows a typical CRISPR workflow. We are aiming to functionally knockout a gene through the cell's NHEJ pathway, so we will focus on the left side of Fig. 30.1. The right side gives an overview of the workflow using homology-directed repair (HDR) and donor DNA. At the top, the two main components of the CRISPR-Cas9 system are introduced into a cell: Cas9 and one or more guide RNAs. Although these can be transfected separately, the pLentiCRISPR vector you used encoded both. If NHEJ resulted in indels in your gene of interest, you will detect those today with a mismatch cleavage assay. This serves as a primary screen and is depicted by the gel image in Fig. 30.1. You are analyzing multiple sgRNA transfection conditions (and possibly multiple sgRNA sequences) because CRISPR success hinges on two types of efficiencies: transfection efficiency and cleavage efficiency. You are already familiar with transfection efficiency if you completed the lab exercise where EGFP was expressed in HEK293 cells (Lab Session 19). This type of efficiency is a quantification (in percent) of how many eukaryotic cells took up a nucleic acid. Cleavage efficiency, as it is used here, is a quantification of how many molecules of target genomic DNA have been genetically modified. Cleavage efficiency can have a wide range of values, but it is important to realize that testing a population (a pool) of transfected cells will not show 100% cleavage efficiency because transfection efficiency is not 100%. Even if your sgRNA works well in the cells that it enters (high cleavage efficiency), it may not be present in a large percentage of the cells analyzed (low transfection efficiency). Therefore, optimizing transfections and testing multiple sgRNAs is performed. This strategy is similar to designing multiple siRNAs to knockdown a gene product, in that some molecules work better than others.

Achieving high transfection as well as cleavage efficiency is important because it ultimately means that fewer clones will need to be screened to find cells with both copies of your gene knocked out. In other words, the more cells that

Molecular Biology Techniques. DOI: https://doi.org/10.1016/B978-0-443-26714-7.00036-3

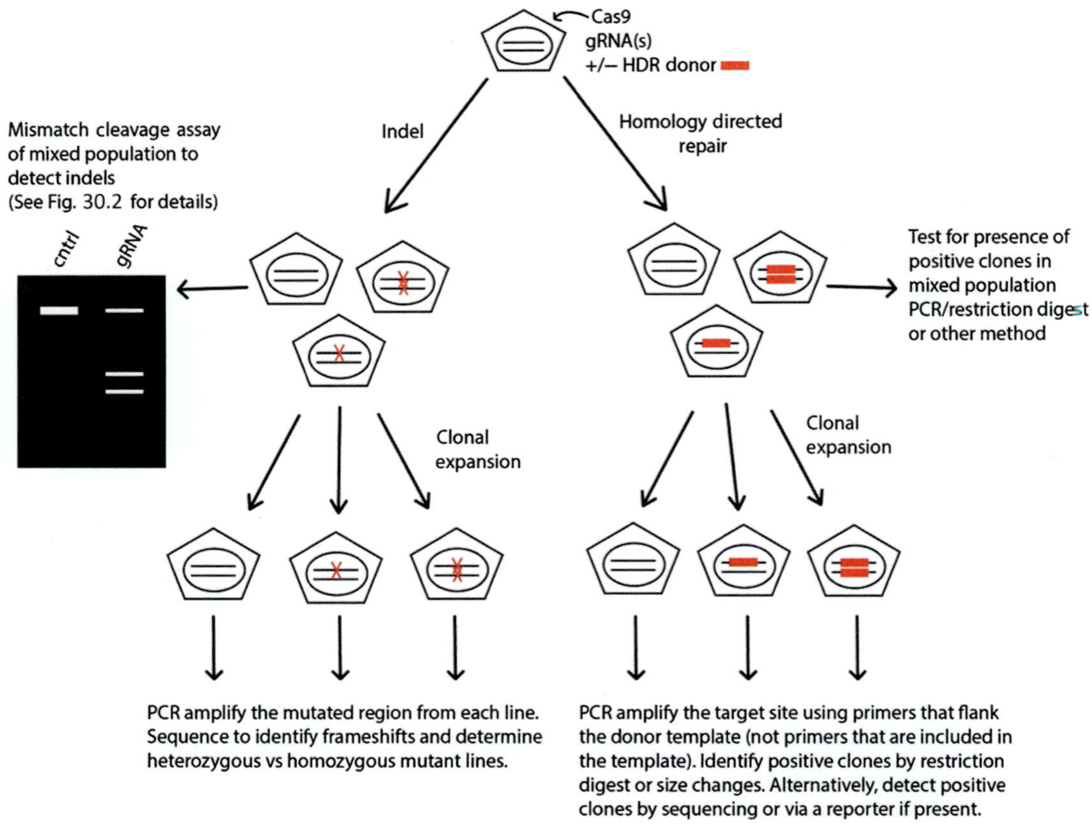

FIGURE 30.1 CRISPR workflow in mammalian cells. *Used with permission from Addgene.*

received the transfected CRISPR vector and also had indels created, the higher the likelihood of finding your desired knockout. Rather than working with a mixed population of cells like you are here, you can later isolate single cells that were transfected with an sgRNA and allow them to expand. This is called clonal expansion, as illustrated in Fig. 30.1. Depending on how well the sgRNA targeted DNA and how efficient the transfection was (obtained from the primary screen), you would screen a number of individual mammalian cell clones to identify a homozygous genome edit in your clonal population (a secondary screen). This is often preferable to a heterozygous genome edit. In Fig. 30.1, one cell is illustrated as having no edit to either copy of a gene. A second cell shows a heterozygous edit (one red X), and a third shows a homozygous edit (two red X's). This secondary screen could be conducted using Sanger DNA sequencing. In this way, you would obtain nucleotide-level resolution of the indel created. This data would reveal the exact nature of the mutation. To do this, the same PCR products of your gene of interest created in Lab Session 29 could be TA-cloned into a vector specially designed for sequencing PCR products (see Introduction to Lab Session 3) and sequenced. Returned sequences would be compared to the wild-type sequence of the gene of interest. A successful gene edit in a knockout experiment will show one or more base pairs inserted or deleted where the DSB occurred. Heterozygous vs homozygous edits can be determined by analyzing the resulting chromatograms.

In addition to sequencing, western blotting can confirm ablation of the protein product in transfected cell lines. Finally, to help rule out off-target effects, whole genome sequencing can be applied. This is quite expensive, though. At the front end of CRISPR experiments, care was taken to synthesize sgRNAs predicted in silico to not target additional targets. This remains the best approach to minimize off-target edits. In addition, if you perform experiments with more than one sgRNA targeting the same gene and get the same biochemical effect, it adds confidence to the specificity of the observed effect.

Online Resources

https://www.thermofisher.com/us/en/home/life-science/sequencing/sanger-sequencing/applications/crispr-talen-genome-editing-sanger-sequencing.html
http://www.genecopoeia.com/wp-content/uploads/2015/07/Downstream-work-07.pdf

II. LABORATORY EXERCISES

A. Verifying the PCR Product

1. Obtain a 2% agarose gel that you made in Lab Session 29, as well as your frozen PCR products. Allow them to thaw at room temperature.
2. Label microcentrifuge tubes to match the labels used for the PCR products.
3. Mix 3 μL of each PCR product with water and the correct volume of DNA loading dye. A minimum volume of 10 μL should be used.
4. Load 150 ng of NEB 50 bp ladder (catalog number N3236). This DNA ladder is more appropriate for viewing fragments less than 500 bp than the previously used 1 kb ladder.
5. Load the entire prepared volume of each loading dye mixed with PCR product on the 2% agarose gel. You should keep the remaining PCR reactions at −20°C if you intend to sequence them.
6. Run the gel at a constant 100 V until the dye front is three quarters of the way down the gel.
7. Image your gel on a UV light box. If a single band of the correct size is present, proceed to the denaturing and reannealing step (Part B below).

Note: A single band of expected size is crucial for accurate cleavage detection. Optimize the PCR conditions including primers, annealing temperature, and amount of lysate volume until you obtain good-quality PCR products. Alternatively, you may have success gel-purifying your expected PCR band away from other bands (see Lab Session 25). This gel-purified product could be used moving forward with denaturing and reannealing.

B. Denaturing and Reannealing Reactions

It is good practice to include a negative enzyme control for each sample, i.e., make duplicate reannealing reactions for each sample to be analyzed in order to distinguish the background bands from the expected cleavage product.

1. Obtain sterile PCR tubes and label them appropriately. You will need two tubes per PCR sample (10 total).
2. Combine 3 μL of experimental PCR products with 1 μL of 10× Detection Reaction Buffer in a PCR tube. For the positive control reaction, use 1 μL of the control PCR product with 1 μL of 10× Detection Reaction Buffer. Make duplicate reactions that are identical at this point in the procedure.
3. Bring each to a volume of 9 μL with water.
4. Briefly centrifuge to ensure that no bubbles are present.
5. Place the PCR tubes in a thermal cycler with a heated lid and run this program (Table 30.1):

   Note that this is **not a PCR reaction**. You are simply using a thermal cycler to control the time and temperature of each incubation step required for denaturing and reannealing PCR products.

C. Enzyme Digestion

1. Add 1 μL detection enzyme to all test samples, including the control (half of your total reactions). Mix well by pipetting up and down.

   The DNA strands were denatured and allowed to randomly reanneal by decreasing the temperature to 25°C. Mismatches will occur when the sequence on one strand is not complementary to the sequence on the opposing

TABLE 30.1 Denaturing and Reannealing Program

Stage	Temperature (°C)	Time	Temperature/Time
1	95	5 min	–
2	95–85	–	−2°C/s
3	85–25	–	−0.1°C/s
4	4	∞	Hold[a]

[a]Following completion of reannealing reaction, immediately proceed to enzyme digestion or store reaction at −20°C.
No Permission Required.

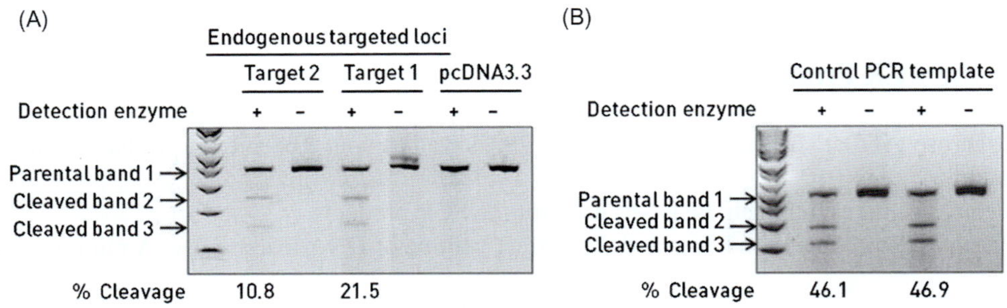

FIGURE 30.2 (A) Gel image of Genomic Cleavage Detection Assay using transfected cells. 293 FT cells were transfected with GeneArt CRISPR All-In-One vectors targeting different regions of the human AAVS1 locus (Target 1 and Target 2) using Lipofectamine 2000. A negative control sample for gene modification was also prepared by transfecting with pcDNA3.3. These samples were PCR amplified using the same set of primers flanking the region of interest. After reannealing, samples were treated with and without Detection Enzyme and run on a 2% E-Gel EX Gel. (B) Gel image of Genomic Cleavage Detection Assay using Control Template and Primers. After reannealing samples were treated with and without Detection Enzyme and run on a 2% E-Gel EX Gel. *Figure used with permission from Thermo Fisher Scientific.*

strand (see Fig. 29.1*). Reannealed PCR products are now treated with an enzyme capable of recognizing mismatches and cleaving the mismatched DNA, producing two fragments.*

2. Add 1 μL water to all negative control samples (no detection enzyme controls). Mix well by pipetting up and down.
3. Incubate in a 37°C water bath or heat block for 1 hour. Alternatively, you can place your tubes in a thermal cycler programmed for this time and temperature.
4. Vortex briefly and spin down all tubes in a microcentrifuge (with PCR tube adaptors) for a few seconds to collect liquid at the bottom. If your microcentrifuge does not fit PCR tubes, just tap the tube on the bench to collect most of the liquid at the bottom.
5. Proceed directly to gel analysis. Only if you are going to perform your gel analysis on a different day, add 1 μL of 100 μM EDTA to each reaction to stop the digestion and store samples at −20°C.

D. Gel Analysis

1. Add the appropriate volume of DNA loading dye to each 10 μL sample.
2. Obtain 2% agarose gel(s) saved from Lab Session 29. The number of gels needed to accommodate today's 10 samples plus a ladder will depend on the size of the gel and comb used previously.
3. Load 150 ng of NEB 50 bp ladder (catalog number N3236).
4. Load each sample on the 2% agarose gel, keeping the negative control (no detection enzyme) sample next to the corresponding detection enzyme-treated sample.

   Note: Avoid any bubbles in the well as this generates an uneven band and might skew the cleavage measurements.
5. Run the gel at a constant 100 V until the dye front migrates approximately three quarters of the way down the gel. Be sure to turn off the power before unplugging your gel apparatus.
6. View the gel using a UV transilluminator and save a picture of your gel as a .JPG or .TIFF for analysis using ImageJ. Refer to Fig. 30.2 for an example of a gel image.

   Note: In some instances, there is a band visible above the parent amplicon after denaturing and reannealing. For example, in Fig. 30.2A, when no detection enzyme is added to Target 1, there is a band migrating slower than the parent band. This occurs when a heteroduplex forms, but it will still get cleaved, resulting in its disappearance once treated with detection enzyme. Therefore, it will not skew quantification.
7. Use NIH ImageJ to determine the relative proportion of DNA contained in each band. If cleavage occurred, you will see three bands: the parental band, plus the two cleaved bands. Refer to Lab Session 17 for details on quantifying band areas. You will paste your quantities in Microsoft Excel for all calculations.
8. Use this equation to calculate the cleavage efficiency:

$$\text{Cleavage Efficiency} = 1 - [(1 - \text{fraction cleaved})^{1/2}]$$

Fraction Cleaved = sum of two cleaved band intensities/(sum of the cleaved and parental band intensities)

Note: Digestion of hybrid control PCR from control template and primers produces two cleavage products of 225 and 291 bp in size (see Fig. 30.2B). Also note that cleavage efficiencies may underestimate actual events, as the detection enzyme used does not recognize indels of only one nucleotide.[1]

9. Consult with your instructor about which sgRNA to use in future CRISPR experiments. You may want to compare your efficiencies with those of other groups targeting the same gene.

REFERENCE

1. L. Vouillot, A. Thélie, et al., Comparison of T7EI and surveyor mismatch cleavage assays to detect mutations triggered by engineered nucleases, G3 5 (3) (2015) 407–415.

DISCUSSION QUESTIONS

1. It is likely that cleavage of your gene of interest will not be 100% efficient. Give at least two reasons why that is the case.
2. Describe what you would observe in the kit's provided control reaction if the detection enzyme failed to work properly.
3. If you tested multiple sgRNAs or others in the class tested sgRNAs to the same gene, compare efficiencies. Also compare on-target scores for these sgRNAs from Lab Session 25. Do the efficiencies correspond with those scores (the higher the score from Benchling, the higher the cleavage efficiency by this assay)? If not, speculate as to why.

Part VI

In Silico Tools

As high-throughput sequencing technologies have become more popular, a massive amount of biological data has been generated. To effectively analyze and interpret this continually-expanding dataset, scientists have turned to computational tools, giving rise to the concept of in silico studies. In silico techniques allow researchers to perform virtual experiments on computers, complementing and accelerating traditional laboratory-based research.

In silico tools encompass a wide range of applications for researchers to explore the intricacies of biological systems. These applications include sequence analysis, gene prediction, protein structure prediction, and pathway analysis. By leveraging algorithms and databases, researchers can identify genes, predict protein functions, design primers for polymerase chain reaction (PCR), and even simulate molecular interactions. In this section, lab sessions involve gene cloning (Lab Session 31), differential gene expression of publicly available RNA-seq data (Lab Session 32), and structure-activity modeling between a drug and protein (Lab Session 33).

By performing in silico virtual experiments, scientists can quickly optimize experimental conditions and test basic hypotheses before attempting "wet" experiments in the laboratory. This can help save both valuable time and resources, while also allowing scientists to try a wider range of conditions than would be feasible to do in the actual laboratory. One of the significant advantages of in silico tools is their ability to accelerate research timelines.

Lab Session 31

Gibson Assembly In Silico Cloning

I. INTRODUCTION

You learned about expression vectors in Part I, DNA Manipulation (Lab Session 2). Recall that expression vectors are a specialized type of cloning vector that are synthetically designed in the laboratory to allow the transcription of the cloned gene and its translation into protein. Currently, there are a variety of methods scientists may use to clone a gene into an expression vector. We will discuss a few of the more popular methods to create recombinant DNA below. However, for the purposes of this lab, we will simulate cloning using Gibson assembly.

A. Gibson Assembly

Gibson assembly was created by Dr. Daniel Gibson and his colleagues at the J. Craig Venter Institute as a relatively easy and efficient way to clone multiple DNA fragments in a single isothermal reaction[1–3] (Fig. 31.1). DNA fragments containing 20–40 base pair homology at their ends (added through PCR) can be ligated together in one reaction. This technology takes advantage of homologous recombination to assemble these DNA fragments using three enzymatic activities: a 5′-3′ exonuclease, a DNA polymerase, and a DNA ligase. In the first step of the reaction, the 5′-3′ exonuclease generates long single-stranded 3′ overhangs at the ends of the DNA fragments. The overhangs that are formed contain complementary sequences that will then anneal to one another, thus creating the double-stranded recombinant DNA of interest. In step two, the DNA polymerase will extend the 3′ ends and DNA ligase will seal the nicks, creating one piece of dsDNA (Fig. 31.1).

The optimal length of homology can vary depending on the number and length of DNA fragments in the assembly reaction but are typically between 20 and 40 base pairs in length. Gibson assembly can be used to simultaneously clone multiple DNA fragments and does not depend on fragment length or end compatibility, thus making it useful for multiple types of DNA manipulations.

B. Restriction Enzyme Cloning

Restriction enzyme (RE) cloning is discussed in Lab Session 2. It is considered one of the classic methods of DNA cloning, which employs the use of restriction enzymes naturally found in bacteria to predictably cleave double-stranded DNA (dsDNA) at specific sequences. In this method, selected REs are used to cut dsDNA of interest into fragments that form precise 5′ or 3′ overhangs (also known as sticky ends) or no overhang (blunt ends). Fragments of DNA that contain complementary nucleotide sequence overhangs (or blunt ends) can be ligated together using a DNA ligase.[4]

RE cloning remains popular today due the plethora of restriction enzymes available for cloning, many of which are relatively cheap. Moreover, almost all commercially available plasmids used for DNA cloning contain a number of RE recognition sites in a region known as the multiple cloning site (MCS). This enables scientists to relatively easily clone any piece of DNA that is flanked with RE sites into any plasmid that has the same or complementary sites in its MCS. Please refer to Part I, Manipulation of DNA for more information on RE cloning.

C. Gateway Cloning

Gateway© cloning involves moving a piece of DNA from one plasmid to another using recombination. To perform this type of cloning procedure, the fragment of DNA you wish to clone must be flanked by special recombination sites specific for Gateway cloning by attaching the ~25 bp attB1 and attB2 sites on the 5′ and 3′ ends of the DNA sequence via PCR.[5] Now the DNA sequence can be cloned into any Gateway donor plasmid that contains compatible attP sites using a Gateway-specific BP clonase enzyme mix. This proprietary enzyme mix contains both Integrase and Integration host factor proteins.

Molecular Biology Techniques. DOI: https://doi.org/10.1016/B978-0-443-26714-7.00038-7

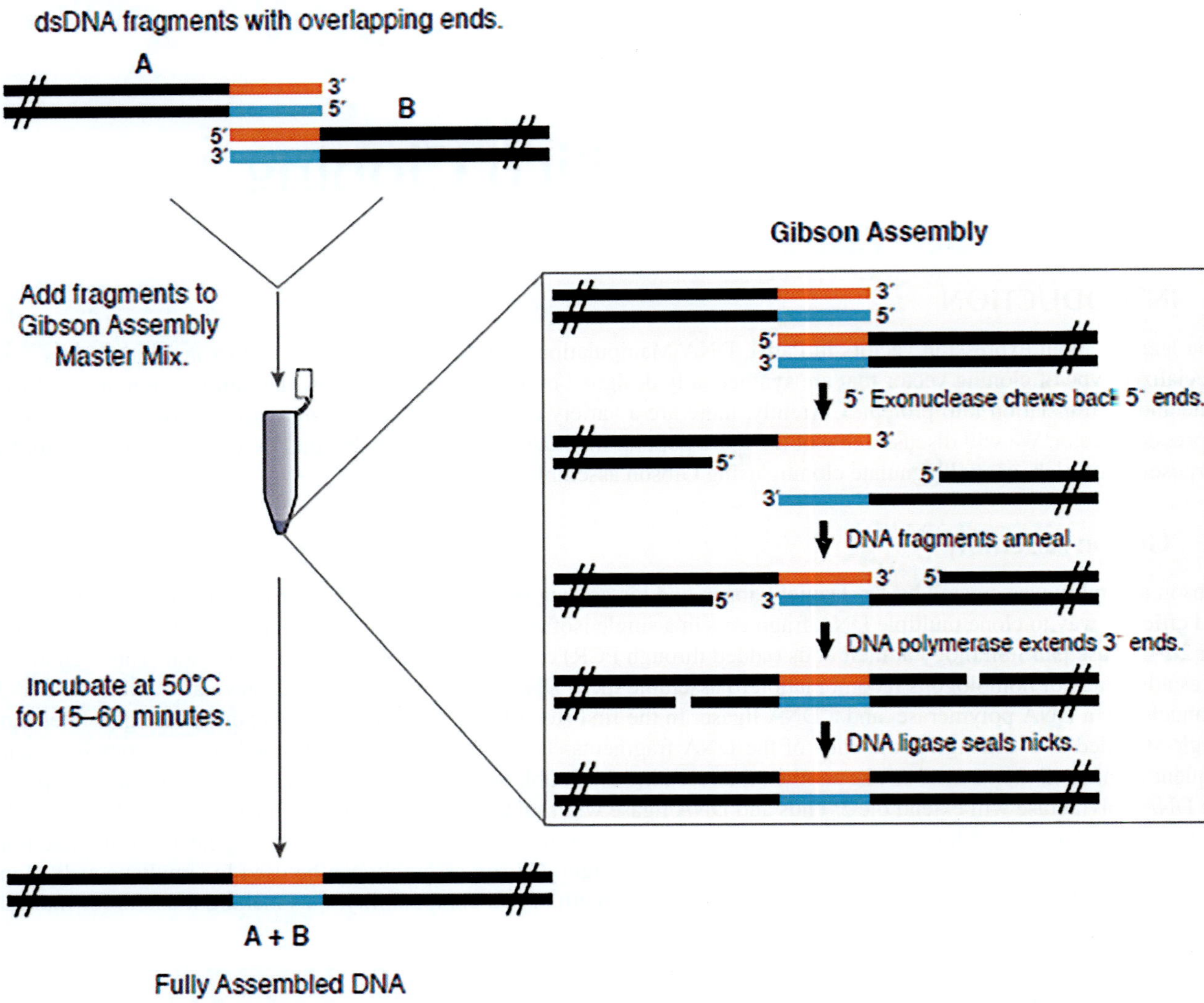

FIGURE 31.1 Gibson assembly workflow. *Reprinted from* www.neb.com *(2024) with permission from New England Biolabs, Inc.*

Together, they catalyze the in vitro recombination of PCR products/DNA fragments that contain attB and attP sites, resulting in an entry clone with attL sites flanking the DNA fragment of interest (Fig. 31.2, top panel).

The benefit of this cloning method is that once your DNA fragment has been cloned into the donor vector, it can quickly and easily be inserted into any compatible Gateway Destination vector that contains attR sites via LR clonase enzymes (Fig. 31.2, bottom panel). This proprietary enzyme mix contains a blend of Integrase, Integration host factor, and Excisionase enzymes that creates the expression clone. Therefore, combining your entry clone with your selected Gateway Destination vector will only require a recombination reaction. Since thousands of Gateway Destination vectors exist that contain different promoters, fluorescent proteins, and tags, it can reduce the time needed for cloning by other methods.[6]

D. TOPO Cloning

Topoisomerase based cloning (TOPO cloning) is a one-step method of DNA cloning that uses the enzyme DNA topoisomerase I, which functions both as a restriction enzyme and as a ligase. *Taq* polymerase has a non-template-dependent terminal transferase activity. When combined with a linearized PCR product (i.e., gene of interest), *Taq* will add a single deoxyadenosine to the 3′ ends. In the same reaction, a linearized vector is included that has single overhanging 3′ deoxythymidine residues. Collectively, this affords the PCR inserts to ligate efficiently with the vector. On a molecular level, *Vaccinia* virus topoisomerase I recognizes the pentameric sequence 5′-(C/T)CCTT-3′.[7] Topo I will bind dsDNA and digest it at this sequence in one DNA strand, thereby causing the DNA supercoil to unwind. After this cleavage event, Topo I will

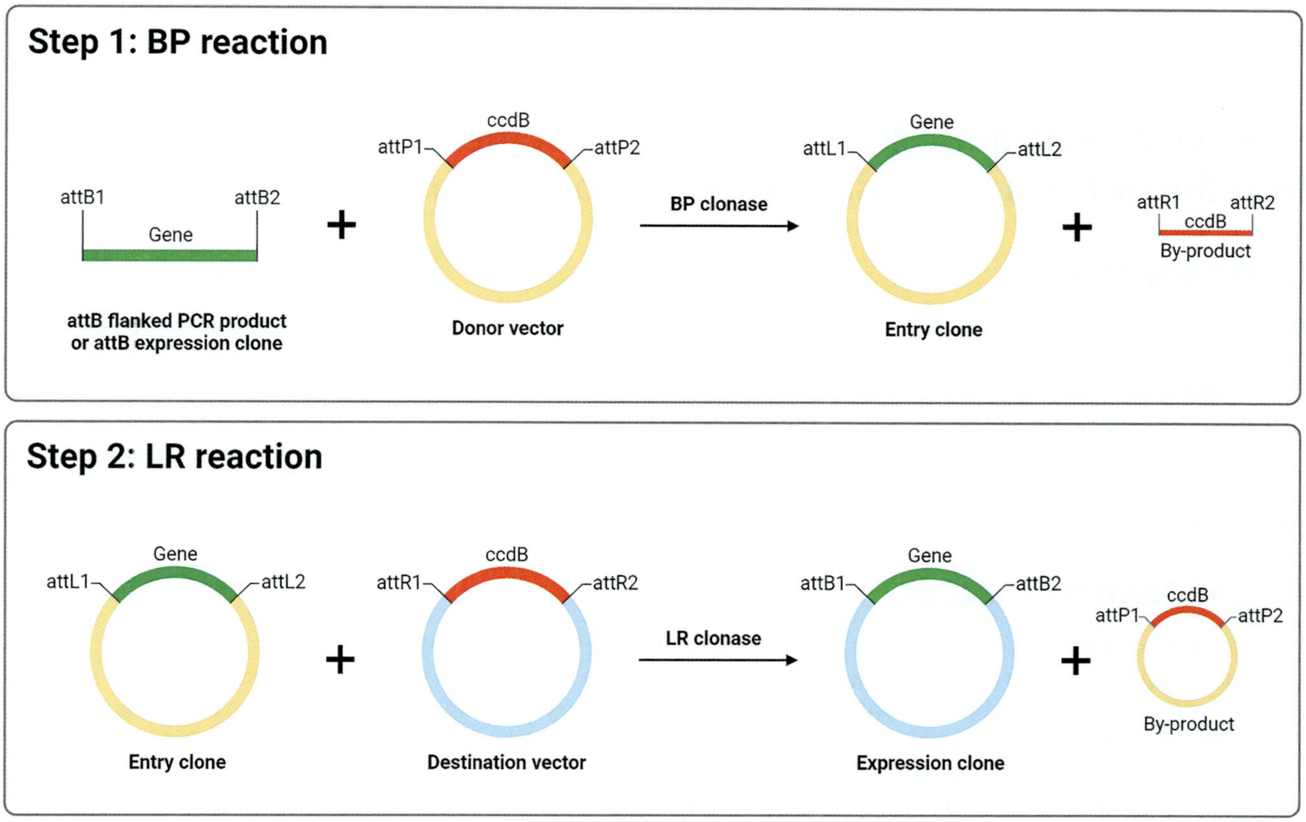

FIGURE 31.2 Gateway cloning workflow. BP reaction, top panel; LR reaction, bottom panel. *Created with BioRender.com.*

form a protein-DNA cleavage complex where a covalent bond is formed between its catalytic tyrosyl residue (Tyr-274) and the free 3′ phosphate from the cleaved DNA strand. Next the phospho-tyrosyl protein-DNA bond is free to attack by the 5′ hydroxyl of the original cleaved DNA strand, thus reversing the reaction and releasing TOPO I.[8] Topo cloning uses this natural reaction to incorporate DNA fragments via the cloning process.

Lab Session 31a

In Silico Design and Cloning Procedure of the *S. aureus pbp2* Gene Into the pET-41a Expression Vector Using Gibson Assembly

I. INTRODUCTION

For this lab session, you will be designing Gibson assembly primers to clone a gene of interest, the *S. aureus* penicillin binding protein 2 (*pbp2*). As its name implies, PBP2 plays a role in antibiotic susceptibility in *S. aureus*. It is a bacterial protein that binds penicillin and other similar antibiotics. It is critical for bacterial cell wall synthesis.[9] PBP2 and *S. aureus* will also be examined throughout the next two laboratory sessions.

This gene of interest will be cloned into the pET-41a expression vector. As discussed earlier, there are many different types of methods that can be employed to clone your gene of interest into a cloning vector. The methodology of choice will depend on your specific experimental situation.

II. LABORATORY EXERCISE

1. Create a free online account with Benchling[10] by selecting this URL: https://www.benchling.com
2. On the home screen, create a new project by selecting the toolbox icon on the left side of the panel.
3. A new panel will appear showing projects. Select the "+" sign on the top right of this panel to create a new project.
4. A pop-up window will appear. Name the project "Gibson Assembly Cloning of pbp2" for penicillin binding protein 2, which is the gene that will be cloned in this activity.
5. Select "create project" button in the bottom right of the screen.
6. Open a new browser tab, and navigate to the NCBI database to find the nucleotide coding sequence for *pbp2*: https://www.ncbi.nlm.nih.gov/
7. To find the coding sequence from NCBI, select the "nucleotide" database from the drop-down menu at the top left of the screen.
8. In the search window, type the accession number for *S. aureus pbp2*: Y17795
9. The *S. aureus* sequence information for *prfA* and *pbp2* genes will now appear.
10. Scroll down until you see information specific for the *pbp2* gene. It will appear as "gene" on the left-hand side of the screen next to "gene="pbp2" on the right side of the screen.
11. Below this, select "CDS" on the left-hand side for the *pbp2* gene. This will now highlight the nucleotide sequence for the *pbp2* gene only from all the gene information listed in this accession file.
12. On the bottom right corner of the screen, select the "FASTA" display format. This will open up a new window that only includes the nucleotide sequence for *pbp2*.
13. Highlight this entire sequence starting with the "ATG" and ending with the "TAA" and copy it.
14. Return to the Benchling tab.
15. Under your project panel on the left, select the "+" sign. Scroll down and select "DNA/RNA sequence" and then New DNA/RNA sequence.
16. A window will appear. Name your sequence "pbp2."
17. Set the nucleotide type to DNA and the topology to linear.
18. Under "Set project folder" select the project name "Gibson Assembly Cloning of pbp2," which you specified earlier.
19. Under the base entry form, you will paste the selected *pbp2* nucleotide coding sequence you retrieved from the NCBI database. You may leave the select schema option at the blank default.
20. Next, select the "Create" button in the bottom right of the pop-up window.
21. The screen will change to show the sequence map of *pbp2* on the left and the linear map featuring all restriction enzyme sites in the sequence on the right. *Note*: *If you do not see the linear map on the right, select "split workspace" on the bottom right of the screen.*
22. Next, you will add an annotation to this sequence so you can track its location during the cloning procedure.
23. To add an annotation, highlight the entire sequence in the sequence map view. Right-click on your mouse and select the "Create annotation" option in the micro window.
24. In the annotation window you will name the sequence "pbp2 CDS." It will show you the nucleotide range in the position column. Verify that the nucleotide range is accurate based on the sequence. You can also select the color of your choosing. Under "strand" select "forward," indicating that this sequence is on the top strand of the dsDNA.

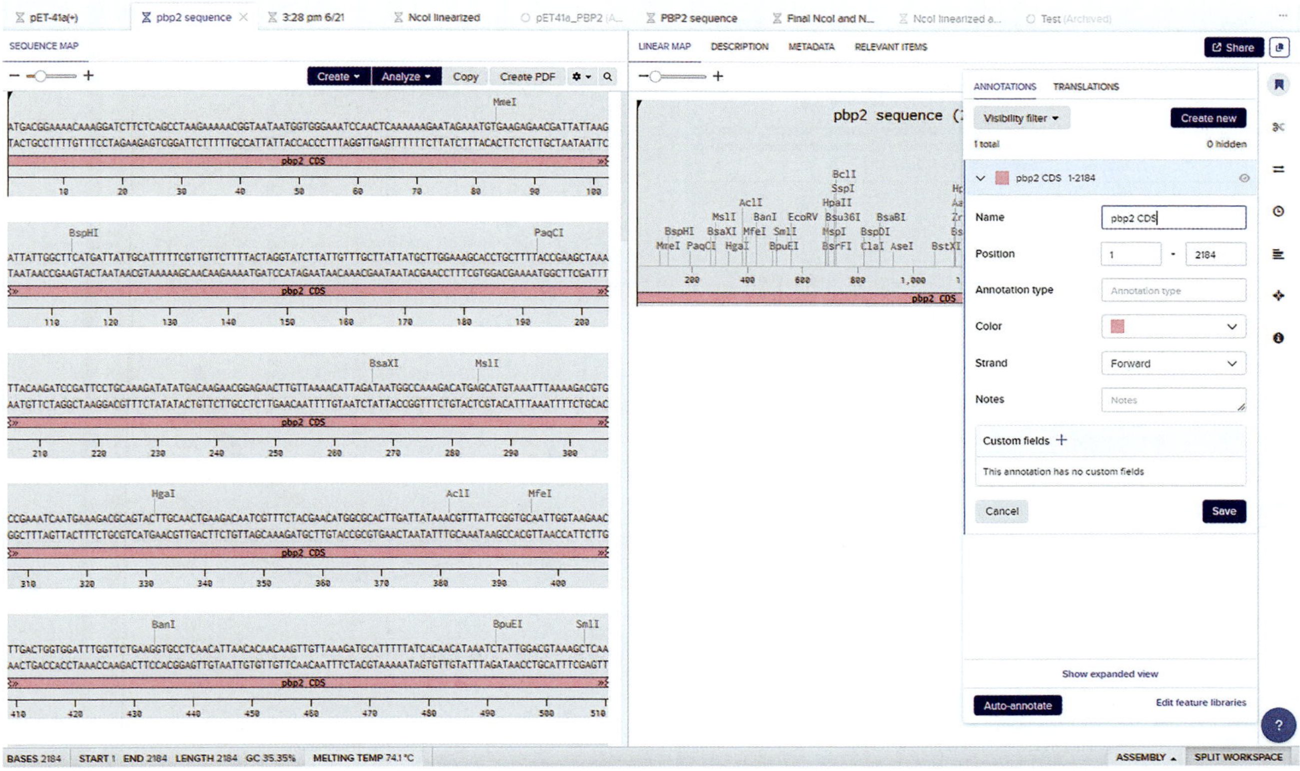

FIGURE 31.3 Benchling screenshot of the annotation window. *Used with permission from Benchling.*

25. Select "Save" in the bottom right of the annotation pane. A colored label should now appear below your *pbp2* sequence and include its directionality (i.e., 5′-3′) (Fig. 31.3).
26. Next, we will add in the sequence of the expression plasmid pET-41a.
27. Under the left toolbar, select the "+" sign. Scroll down and select "DNA/RNA sequence," and then "Import DNA/RNA sequences."
28. In the top of the new window, select the upload files tab.
29. Open a new browser window tab, and navigate to this URL: https://www.snapgene.com/plasmids/pet_and_duet_vectors_(novagen)/pET-41a(%2B)
30. This will open the plasmid map for pET-41a from the SnapGene website. Select "download plasmid."
31. Return to the Benchling tab. Click the "UPLOAD FILES" tab. Drag and drop the downloaded pET-41a.dna file from your downloads folder into the import DNA sequences window.
32. Select the nucleotide type as "DNA" and verify that the project folder is correct. Next, select "Close."
33. A new screen will appear that shows the pET-41a expression vector sequence map on the left and the circularized plasmid map on the right (click "Split Workspace" if necessary). *Note: If you do not see the circularized plasmid map on the right workspace, select the plasmid tab at the top.* Notice that a number of plasmid features have been annotated in this file, including the MCS, GST tag, *lacI* gene, and others. Take a moment to explore the different unique features of this plasmid map.
34. Now you have both the *pbp2* and pET-41a sequence, and you can run the in silico Gibson assembly reaction.
35. Navigate to the two tabs in the lower right-hand side of the screen.
36. Select the "ASSEMBLY" tab, which will open the pop-up Assembly Wizard window. Here, you will select the assembly strategy of choice. In this lab, you will select "Gibson" and next select "Start."
37. Select the word "Backbone" in the bottom left of the screen (Fig. 31.4). We will next need to determine the appropriate restriction enzymes to use to linearize the plasmid. Ideally you will choose two enzymes that only cut once in the MCS and that do not leave compatible cohesive ends. This is important to prevent the vector from re-circularizing.
38. To determine the appropriate enzymes, we will first select the MCS region where we wish to clone the gene of interest. Using the mouse, scroll over the MCS feature in the plasmid map and select it.

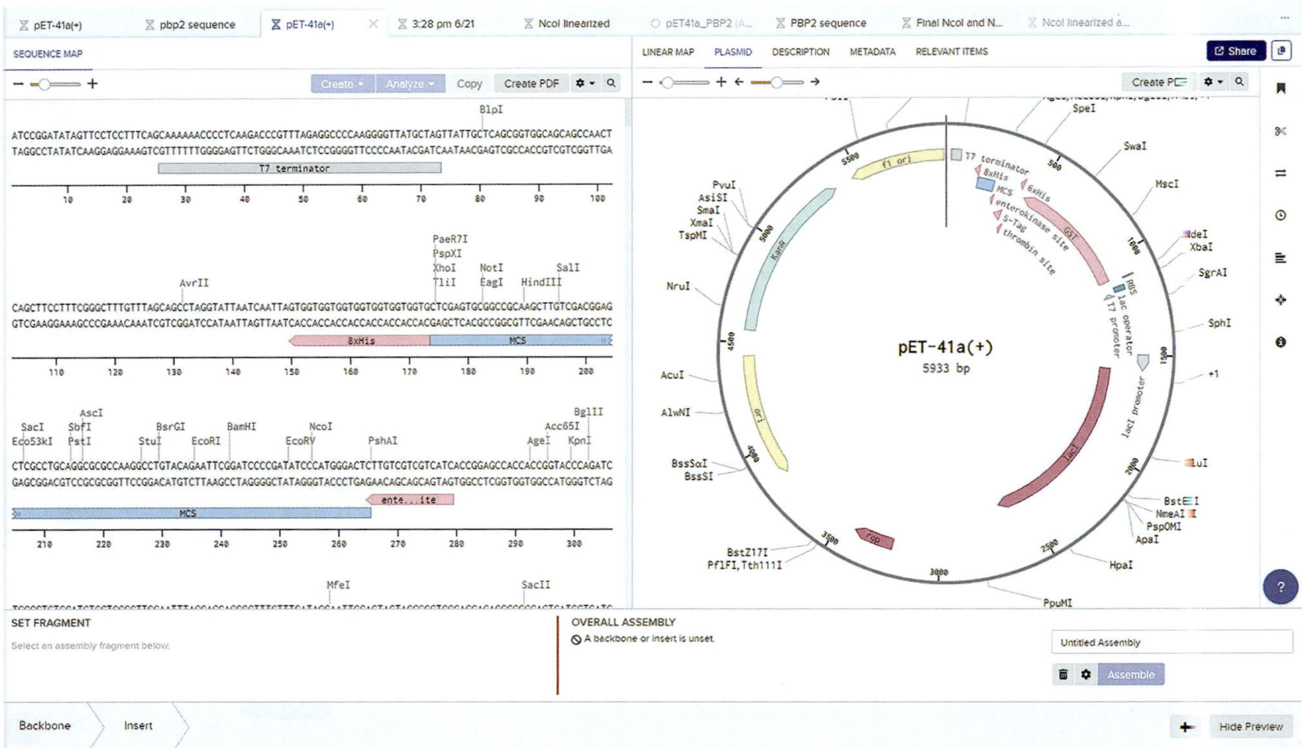

FIGURE 31.4 Benchling screenshot indicating the Assembly Wizard window where you will select the DNA fragments you wish to use in Gibson assembly cloning. *Used with permission from Benchling.*

39. Next, select the scissors icon on the far right of the screen. A small panel will appear. Ensure that the tab on the top is selected for a new digest. Under enzyme lists, from the drop-down menu select "NEB." Under cut sites visible on the maps, use the drop-down to select "singles cutters." This will bring up a list of enzymes that only cut once within our MCS (Fig. 31.5).
40. Based on these results, we will use the restriction enzymes NotI and NcoI as these are two enzymes that do not leave compatible cohesive ends to one another and only cut the plasmid once within the MCS.
41. We will next linearize our pET41a plasmid by digesting it with NotI and NcoI. Using the sequence map on the left-hand side of the screen, select the word "Backbone" at the bottom left of the screen (Fig. 31.5 and Fig. 31.6).
42. Next, navigate to the MCS feature on the plasmid. To linearize the backbone sequence with restriction enzyme cut sites, click the word "NotI" first on the plasmid sequence map. Hold shift and click the "NotI" site again. Keeping the shift key engaged, next select the word "NcoI." Now you have set the plasmid to be linearized by the double digest with both NotI and NcoI. The complete sequence on the plasmid map will now be highlighted except for the region that will be removed by the restriction enzymes.
43. At the bottom left of the screen, select the "Set Fragment" button. A statement will appear that will show you the current fragment after the digest, where you can verify your restriction enzyme choices. Press "Set Fragment" to set the backbone.
44. Next, select "Insert" at the bottom left of the screen.
45. Select the sequence of our *pbp2* gene of interest file from our project folder. In some cases, the project window will be open at the left panel of the screen. This will now bring up the *pbp2* sequence in the main screen.
46. Press shift and hold to select the entire sequence. Then select the "Set fragment" button using the blue button at the bottom of the screen. It will now allow you to change the insert orientation if required by selecting the double-headed arrow icon. *Note: Gibson assembly can be used to clone multiple DNA fragments at once. If you wished to clone more than one fragment, at this step you could select the "+" on the bottom right of the screen and add the additional fragment(s) you wish to clone or use the combinatorial assembly tool in Benchling* (Fig. 31.7).
47. In the bottom right of the screen, Benchling software will check your cloning for any issues with the overall assembly. If everything checks out, you can name your new file pET-41a_pbp2 and select the assemble button.

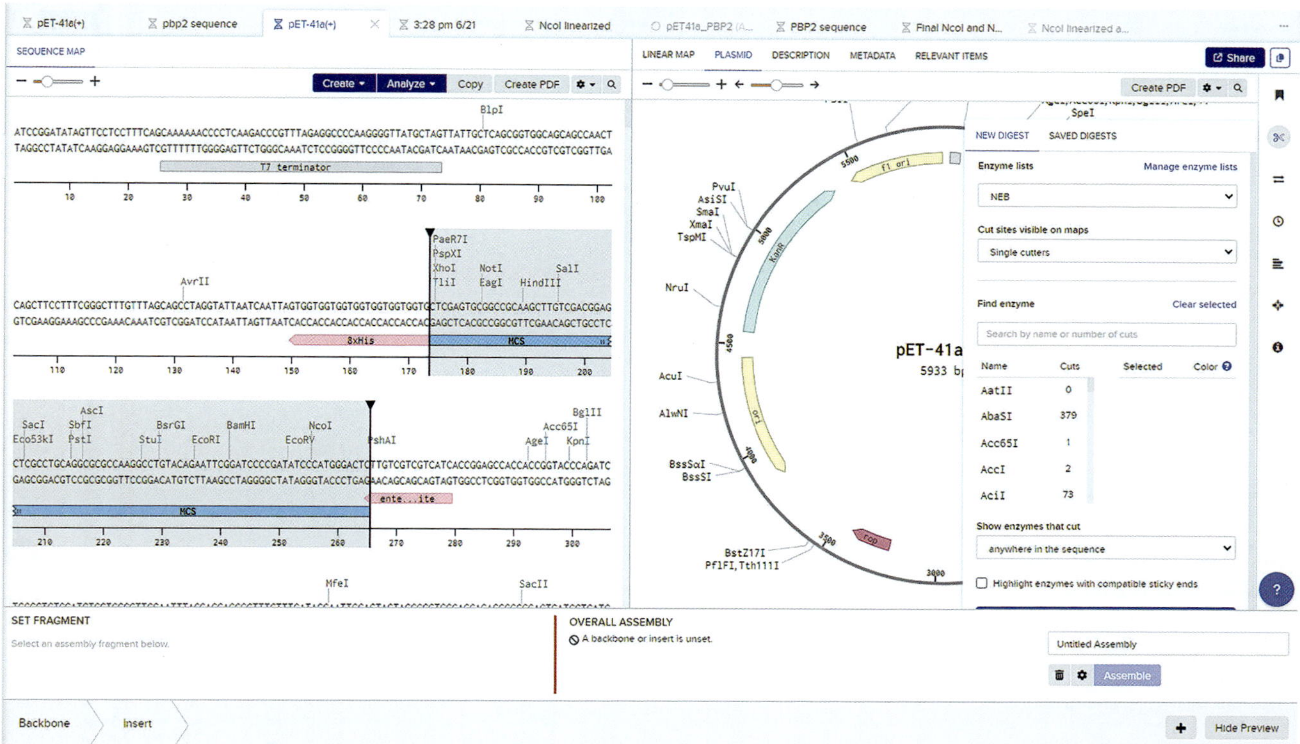

FIGURE 31.5 Benchling screenshot indicating the location of the restriction enzyme sites used to linearize the pET41a plasmid. *Used with permission from Benchling.*

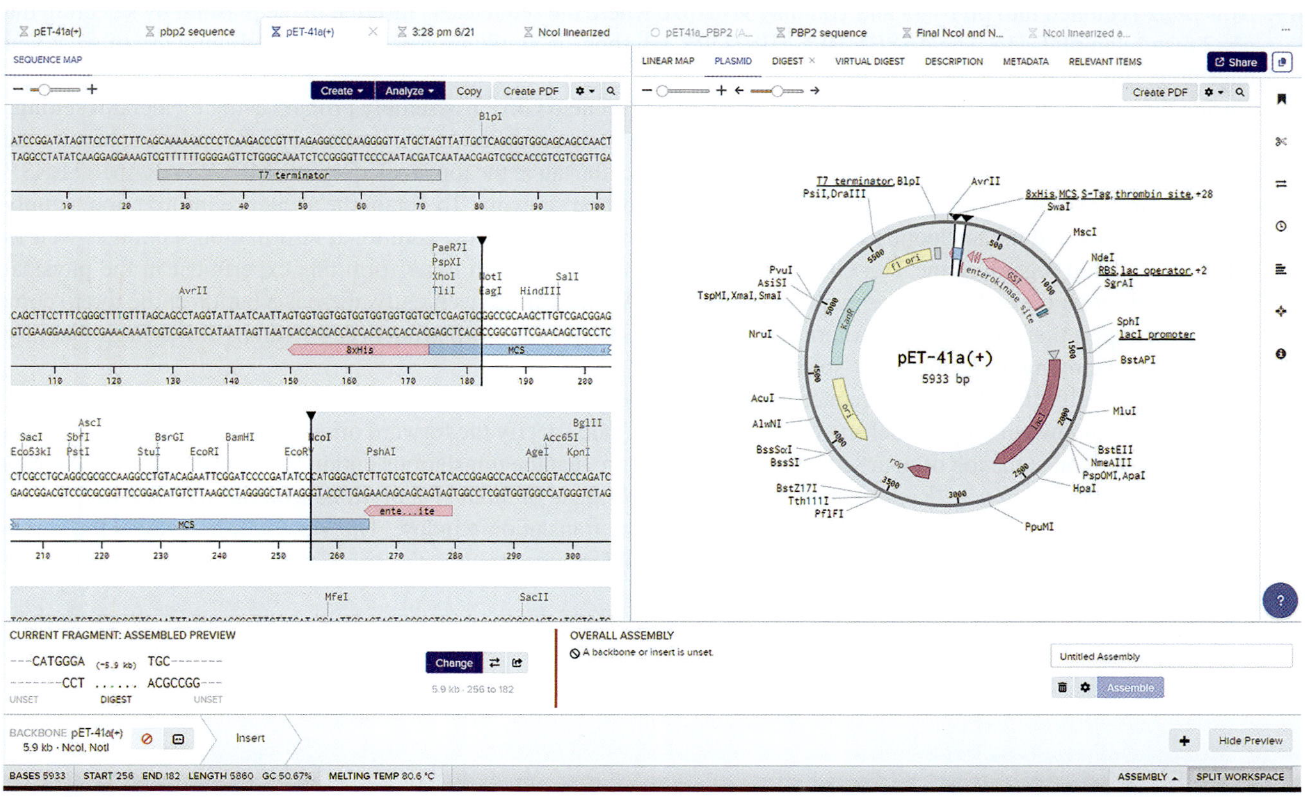

FIGURE 31.6 Benchling screenshot of linearized pET41a vector after cutting with restriction enzymes NotI and NcoI. *Used with permission from Benchling.*

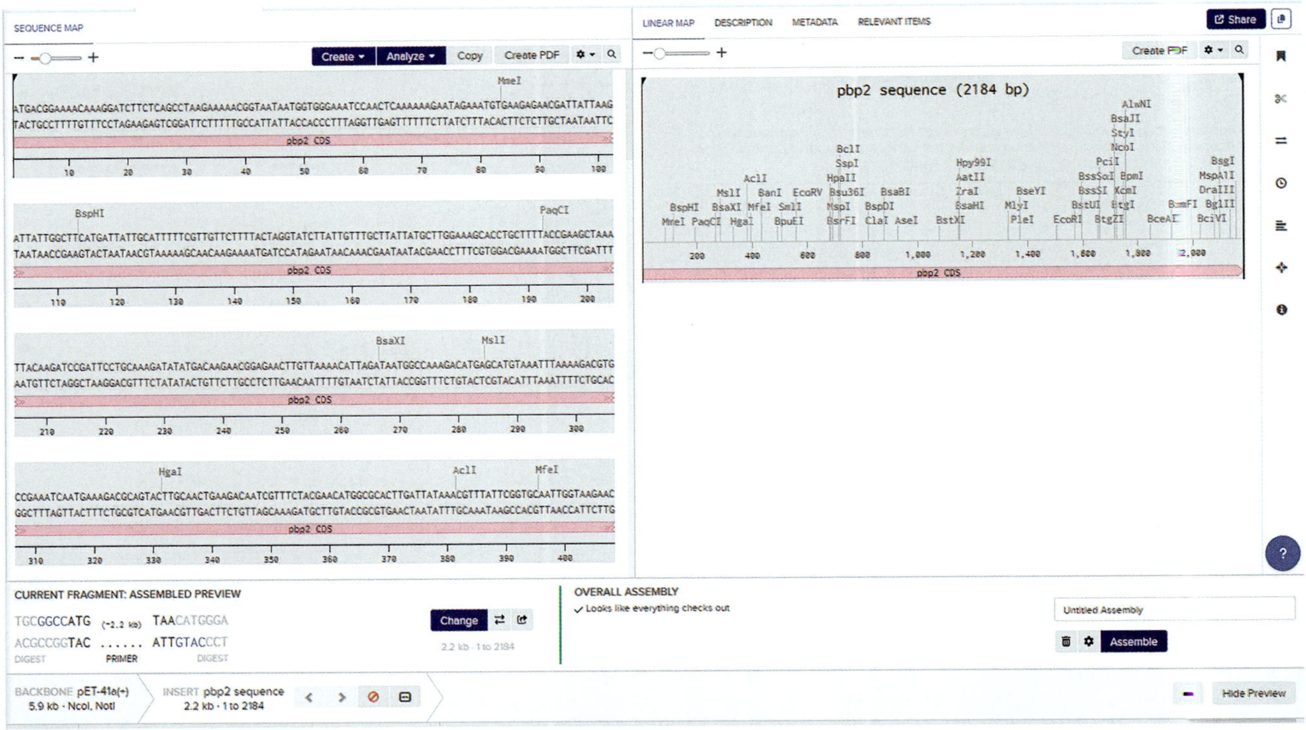

FIGURE 31.7 Benchling screenshot with the *pbp2* DNA sequence selected as the insert fragment in the cloning workflow. *Used with permission from Benchling*.

48. Choose the folder for the assembly (pbp2) and select "Create."
49. Now *pbp2* is cloned into pET-41a and you may visualize where the sequence is inserted in the plasmid by selecting the *pbp2* feature on pET-41a. The ASSEMBLY HISTORY tab shows a graphical representation and summary of what you have done (Fig. 31.8).
50. During the virtual cloning process, Benchling software has created Gibson assembly primers using the default settings (i.e., ~20 bp overhang, annealing Tm ~50 C). To retrieve the sequence information for the *pbp2* GA primers, be sure the plasmid view is showing in the right part of the screen. Select the tab at the top labeled "ASSEMBLY PARAMETERS."
51. The GA forward (FWD) and reverse (REV) primers will now be showing. To obtain the sequence information, simply select the primer and the nucleotide sequence information will appear in the additional information section as well as other information regarding the Tm, GC content and length. If you wish to perform this experiment in the physical lab, you will order these designed primers for use in the GA reaction. Conveniently, there is a button at the top to copy primers or export them to a .csv file. Alternatively, you may design your own primers and virtually test their probability for successful cloning using this same software.
52. We will next verify that the coding sequence is in the correct frame. To do this, right-click on the insert sequence. A new pop-up window will appear. Select "Create translation" and specify the forward orientation. Alternatively, you can select the gear icon at the top of your sequence map and ensure that the translations option is selected. *Note: If you wish to create a tagged version of the protein, you must ensure the gene of interest is in frame with the selected tag.*
53. In the new window, select "Save" in the bottom right of the translation window. The view will now show the single AA sequence for the predicted translated region. *Note: When interpreting the translated sequence, be mindful of directionality.*

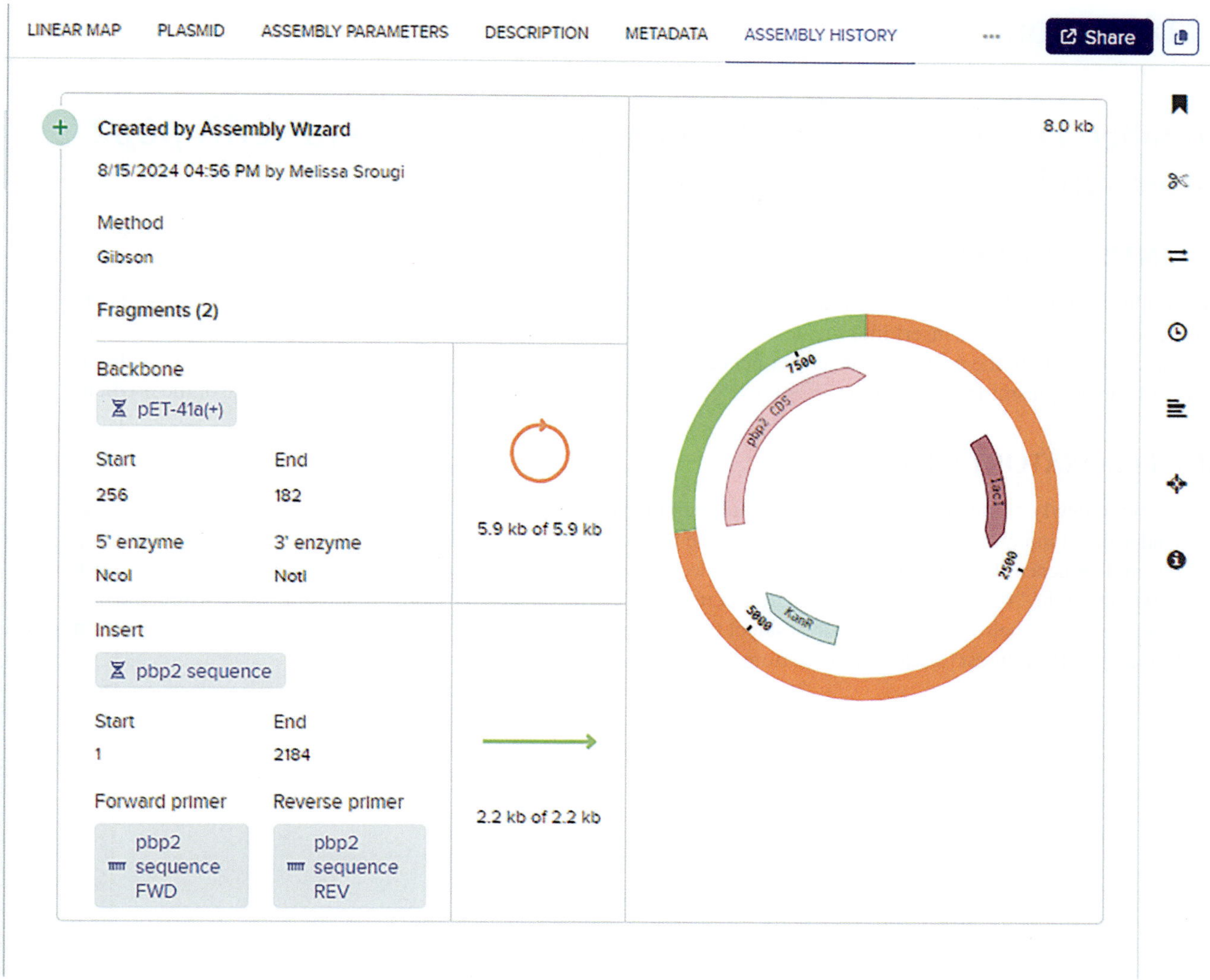

FIGURE 31.8 Benchling screenshot of the assembly history verifying the final in silico designed recombinant DNA. *Used with permission from Benchling*.

Lab Session 31b

In Silico Diagnostic Restriction Enzyme Digestion to Verify *pbp2* Insertion Into the pET-41a Vector

I. INTRODUCTION

As you learned in Lab Session 7a, restriction enzyme digestion is an easy screening technique that can determine the presence and direction of your insert into the desired plasmid after a cloning experiment. To do this, we will select two restriction enzymes that cut the vector and insert sequence only once, respectively. Once the in silico digest is complete, the uncut plasmid and cut plasmid will be run on an agarose gel to verify that the vector backbone and insert are of the predicted sizes.

II. LABORATORY EXERCISE

1. With the newly cloned pET-41a_pbp2 plasmid map opened, select the scissors icon on the far right of the screen.
2. Under the "Find enzyme" entry, write in PvuI—the restriction enzyme PvuI, which will cut only once within the pET-41a plasmid sequence. Then below, use the cursor to select this enzyme. Make sure that NcoI and NotI are no longer selected here.
3. Next, under the "Find enzyme" entry, write in ZraI. The restriction enzyme ZraI will cut only once within the *pbp2* DNA sequence. Then below, use the cursor to select this enzyme. Both REs will now be added to the list of selected enzymes. Select the "Run digest" button at the bottom (Fig. 31.9).

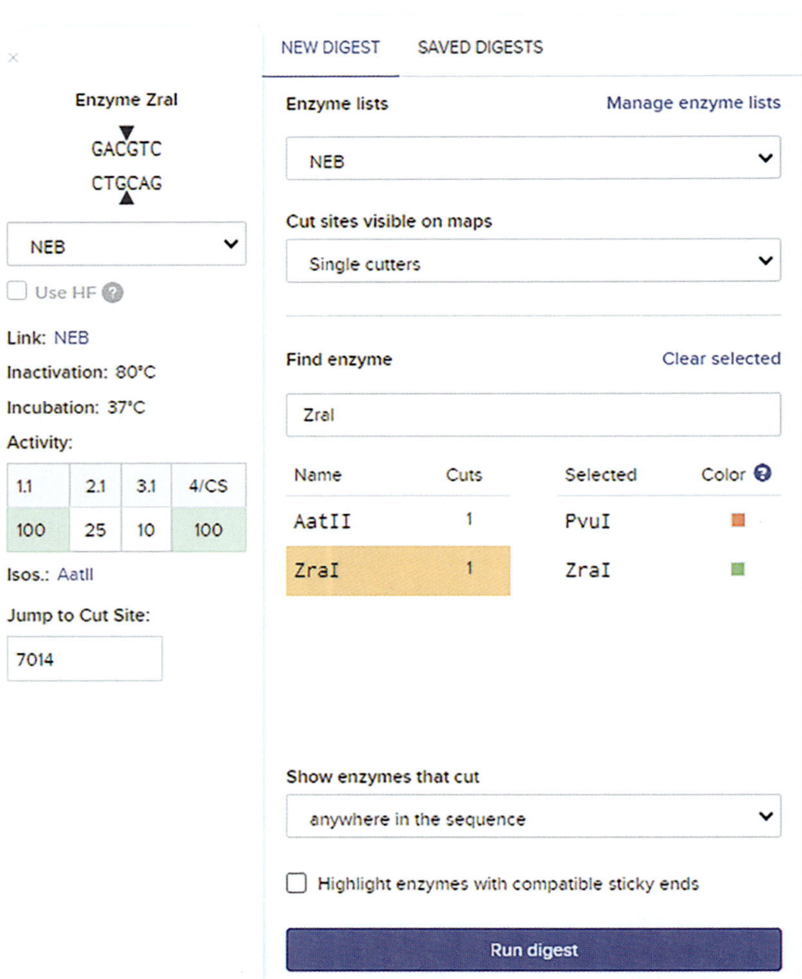

FIGURE 31.9 Benchling screenshot of the virtual DNA digest window. *Used with permission from Benchling.*

4. In the top panel, select the "VIRTUAL DIGEST" tab. This will visualize the predicted DNA band sizes of 5.8 kb and 2.3 kb if the *pbp2* insert was cloned successfully into the pET-41a vector alongside a molecular weight ladder. You can determine the band sizes by placing your cursor on top of the selected band.
5. Next, we will compare this to the digested empty vector. Select the pET-41a vector map from the list of files at the right hand of the screen.
6. Select the scissors icon on the far right.
7. Under the new digest select "PvuI" and "ZraI" enzymes. *Note: It will show that ZraI does not cut the pET-41a vector because this restriction enzyme site is only found in the pbp2 insert sequence.*
8. Select "Run digest" and choose the "VIRTUAL DIGEST" tab at the top of the screen.
9. The digested empty pET-41a vector is now run alongside the digested pET-41a_pbp2 vector in the virtual agarose gel. Notice that the digested pET-41a vector only contains one band at 5.9 kb. *Note: While restriction enzyme diagnostic digests are helpful preliminary screening tools, you should always sequence your clones to verify that no changes in nucleotide sequence occurred during the cloning process.*

DISCUSSION QUESTIONS

1. Compare and contrast Gibson assembly cloning to forced cloning. What are the advantages and disadvantages to each of these cloning methods?
2. As with any cloning method you choose, it is important to include the appropriate controls. If you were to run the in silico cloning in the physical lab, list and describe three controls for the Gibson assembly reaction. Be sure to explain the purpose of each control.
3. Once you have cloned your insert into your plasmid vector in the physical lab, what screening methods in addition to the diagnostic restriction enzyme digestion can you perform to determine if your cloning procedure was successful? *Hint:* Refer to the lab sessions on screening transformants!

REFERENCES

1. W.M. Barnes, PCR amplification of up to 35-kb DNA with high fidelity and high yield from λ bacteriophage templates. National Academy of Sciences, United States Proceedings of the National Academy of Sciences of the United States of America. 00278424. 91 (6) (1994), 2216–2220.
2. D.G. Gibson, H.O. Smith, C.A. Hutchison, J.C. Venter, C. Merryman, Chemical synthesis of the mouse mitochondrial genome, Nature Methods. 1548-7091. 7 (11) (2010) 901–903.
3. D.G. Gibson, L. Young, R.Y. Chuang, J.C. Venter, C.A. Hutchison, H.O. Smith, Enzymatic assembly of DNA molecules up to several hundred kilobases, Nature Methods. 15491676. 6 (5) (2009) 343–345.
4. C. Brownlee, Danna and Nathans: Restriction enzymes and the boon to modern molecular biology, Proceedings of the National Academy of Sciences of the United States of America. 00278424. 102 (17) (2005) 5909.
5. A. Landy, Dynamic, structural, and regulatory aspects of λ site-specific recombination, Annual Review of Biochemistry. 0066-4154. 58 (1) (1989) 913–941.
6. J.L. Hartley, G.F. Temple, MA Brasch, DNA cloning using in vitro site-specific recombination, Genome Research 10 (11) (2000) 1788–1795.
7. S. Shuman, Recombination mediated by vaccinia virus DNA topoisomerase I in Escherichia coli is sequence specific, Proceedings of the National Academy of Sciences. 0027-8424. 88 (22) (1991) 10104–10108.
8. S. Shuman, Novel approach to molecular cloning and polynucleotide synthesis using vaccinia DNA topoisomerase, American Society for Biochemistry and Molecular Biology Inc., United States Journal of Biological Chemistry. 00219258. 269 (51) (1994) 32678–32684.
9. T.A. Łęski, A. Tomasz, Role of penicillin-binding protein 2 (PBP2) in the antibiotic susceptibility and cell wall cross-linking of Staphylococcus aureus: Evidence for the cooperative functioning of PBP2, PBP4, and PBP2A, Journal of Bacteriology. 187 (5) (2005) 1815–1824.
10. Benchling Software (2024).

Lab Session 32

Transcriptomic Analysis

Goal: Today you will use publicly available RNA-seq data to investigate differential gene expression in *S. aureus*.

I. INTRODUCTION

Gene expression is a critical series of events that occurs in every cell in every organism. You learned in Part IV (Analysis of mRNA Levels) that the central dogma of molecular biology describes how information flows from DNA to RNA to protein. While DNA (what most organisms' genomes are made of) is considered a relatively static molecule that faithfully stores genetic instructions, RNA is much more dynamic. The expression levels of different types of RNA change under different conditions, during development of an organism, and are cell and tissue type dependent. You can think of RNA levels as responding to cellular cues by either staying at the same level, increasing, or decreasing through sophisticated regulatory methods.

While scientists have been able to investigate and quantify RNA levels (i.e., transcript levels) for several decades, the scale and the ways in which they can answer gene expression questions have changed in recent years. You learned in Lab Sessions 16–19 that reverse transcriptase polymerase chain reaction (RT-PCR) is commonly used to measure RNA levels. This method became even more powerful with the advent of quantitative PCR (qPCR). However, these methods are limited by the number of samples that can feasibly be analyzed at one time. For example, if a scientist wants to measure a small subset of genes to examine if expression levels change in response to a drug, they can use RT-qPCR. This necessitates designing either primer pairs or probes specific to each gene of interest (a test gene), as well as one or more reference genes. This process is not feasible at a larger, high-throughput scale where one wants to measure thousands of genes. Scientists can now measure nearly every transcript in a cell or collection of cells using transcriptomic techniques. The transcriptome consists of all the RNAs (i.e., transcripts) in a cell. The most frequently used transcriptomic technique currently is a type of high-throughput sequencing effort called RNA-sequencing (RNA-seq).[1]

RNA-seq simultaneously measures all the RNAs in a sample by sequencing those RNAs and quantifying the resulting sequencing reads. The workflow for a typical RNA-seq experiment is shown in Fig. 32.1. First, cells or tissues of interest are obtained. These may be cultures of cells untreated or treated with a drug, different tissue types, normal and cancerous cells, etc. (Fig. 32.1A). Total RNA is purified from the cells or tissues as explained in Lab Session 13 (Fig. 32.1B). Next, RNA is fragmented, and a complementary DNA (cDNA) library is constructed from each sample using reverse transcription (Lab Session 14) (Fig. 32.1C). This process of converting the RNA into cDNA is necessary to be recognized by DNA-dependent DNA polymerases that amplify the DNA. You can think of the cDNA as representing the RNA content in the sample, but in a more stable chemical form. Various techniques can be used to create the resulting high-throughput sequencing reads. Most platforms involve chemical reactions that ligate adapter sequences (purple, Fig. 32.1D) to the cDNA so that every piece has some sequence in common that can later be recognized by a sequencing primer. Resulting cDNAs with adapters are attached to flow cells (glass slides with lanes). They are then amplified by PCR to have enough genetic material for sequencing and form what are called clusters. In a type of sequencing technology known as sequencing by synthesis, millions of these clusters get sequenced simultaneously by DNA polymerase adding fluorescently tagged dNTPs. The dNTPs are added to the flow cell sequentially and the dNTP complementary to the next one in the template cDNA gets incorporated. A light source is used to excite the fluorescently tagged dNTPs after each cycle. Fluorescence gets detected and recorded (Fig. 32.1D). These steps that involve incorporating and detecting fluorescent dNTPS are similar in principle to traditional Sanger sequencing (Lab Session 7 d).

Unlike Sanger sequencing,[2] millions of short sequencing reads are created in each sample. Among the millions of reads, some will still have adaptor sequences present. Software will trim these sequences to remove bases that came from the previously ligated adaptors. Some reads will be low-quality. For example, base calls will be ambiguous. These reads will be filtered out as another quality control step. These short sequences are like puzzle pieces. Individually, they do not

Molecular Biology Techniques. DOI: https://doi.org/10.1016/B978-0-443-26714-7.00029-6

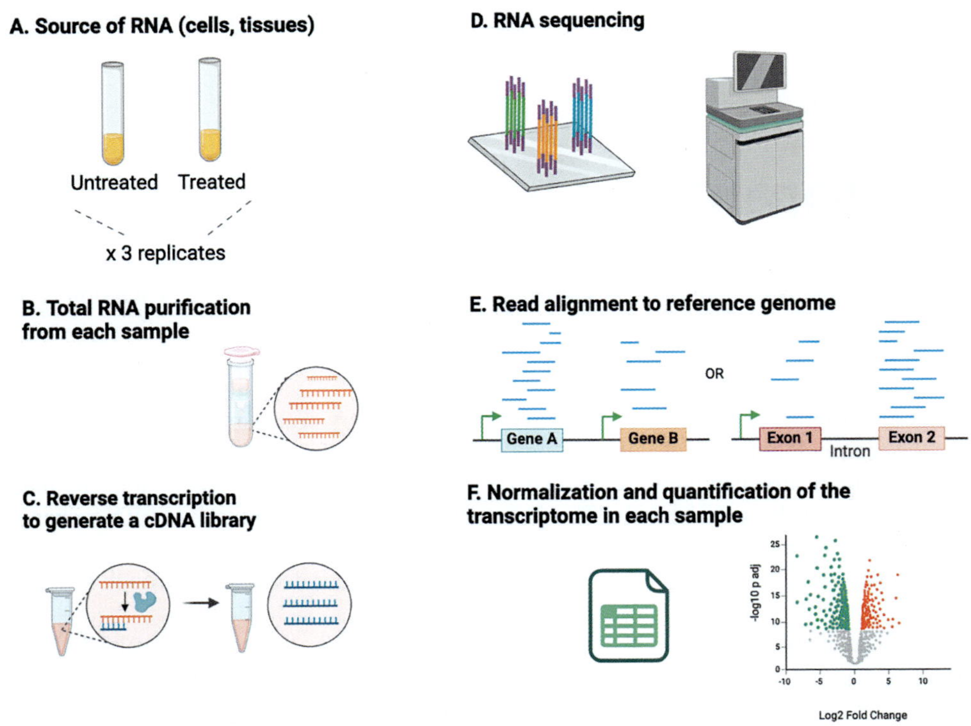

FIGURE 32.1 RNA-seq workflow. *Figure created with Biorender.com.*

give you a sense of the whole picture (the transcriptome); rather, they must be pieced together in a very particular way to make sense. To do this, the puzzle pieces (sequencing reads) are mapped to a reference genome for that organism. The reference genome is a digital map of all of the nucleic acid sequences that make up a particular organism. The nucleic acid sequences are assembled into genes and then annotated, or labeled, with known gene names and locations. The counts of the sequencing reads serve as input for computer programs capable of many steps in the RNA-seq workflow, including normalizing the read counts for gene size. In other words, you would expect a large gene to naturally have more reads created than a small gene without the difference necessarily indicating increased gene expression. Normalizing for gene size is one way to have an equal playing field. If after normalization, many sequencing reads from your experiment are mapped to a particular gene, it indicates there are more RNA transcripts present in the sample compared to a gene with fewer reads. Finally, relative gene expression can be calculated between samples for each transcript in the transcriptome. As you learned in Part IV, relative gene expression compares a transcript's level in one sample to another. These calculations also include statistical analyses to determine if the differentially expressed genes (DEGs) are likely due to chance alone. These steps ultimately create what is referred to as processed data, in contrast to raw data.

RNA-seq can be used not just for measuring differential gene expression, but also to uncover novel genes. Unlike RT-qPCR and DNA microarrays, which rely on sequence-specific primers or probes, RNA-seq sequences all transcripts in a sample regardless of identity, Therefore, RNA-seq can detect novel genes that do not map to any known genes in the reference genome. In eukaryotic organisms, this high- throughput technique can also uncover novel transcript isoforms that undergo alternative processing.[1] In fact, RNA-seq has revealed that approximately 95% of human genes with multiple exons result in alternative isoforms.[3,4] Finally, RNA-seq is useful for not only measuring mRNA levels, but any type of RNA that was present in the sample. For example, if you want to study all of the small RNAs (sRNAs) or long, noncoding RNAs (lncRNAs) in your organism of choice, there is an RNA-seq workflow that will provide you with a rich data set.

RNA-seq data is often stored in a database called Gene Expression Omnibus (GEO) (https://www.ncbi.nlm.nih.gov/geo/). This site is hosted/maintained by the National Center for Biotechnology Information, just like other databases and tools you may be familiar with (BLAST was used in Lab Session 8 when screening your clones). Each entry on GEO is given an accession number, similar to how DNA or protein sequence records are organized. These entries are submitted by scientists who have gathered the data and are sharing it publicly online. Because these entries are publicly available, scientists can also explore the data and "mine" it as they desire.

In Lab Session 31, you designed a cloning experiment with the goal of creating an expression vector containing the *S. aureuspbp2* gene as insert. Today, you are also studying *S. aureus* as a model organism. But, instead of focusing on just one gene of interest, you will use publicly available RNA-seq data to investigate differential gene expression (DGE) across all genes in *S. aureus*. Specifically, you will determine what changes occur when *S. aureus* cultures are treated with the β-lactam antibiotic, oxacillin, by comparing those samples to untreated controls.

II. LABORATORY EXERCISE

A. Understanding the Data Set

Locate the GEO entry called GSE267020 on Gene Expression Omnibus (GEO) (https://www.ncbi.nlm.nih.gov/geo/). There is an Excel spreadsheet of processed data associated with the entry found under the supplementary file at the end of the page. Download this file to your computer. Scroll both left to right and up and down to gain a sense for how much data is present. You will need to read the summary and overall design of this record before you answer the bolded questions below. Record these in your laboratory notebook.

1. Several columns include the term "OxvsUn." **What is the importance of the "vs" in this column title? What do the other terms in this title mean?**
2. Locate the first column in the sheet of RNA-seq data. **What content is found here? What is contained in each row?**
3. Column A is followed by the many columns with a title containing "counts," which are read counts from the sequencing. There are also many columns with the title containing "fpkm." This stands for fragments per kilobase of transcript per million mapped reads. Those columns contain an average (mean) of the normalized RNA-seq count values divided by size factors, taken over all samples. This basically accounts for the fact that longer transcripts (RNAs) will naturally have more sequencing reads than shorter transcripts.

   Scroll further right and you will see columns containing the title "log2FoldChange." Fold change is often used to report the magnitude and direction of differential gene expression. Scientists will report a particular gene's levels in a treated sample compared to a control.

   As a basic example of fold change, if you and your friends fly out of New York LaGuardia Airport, where it is 15°C, and travel to Cancun for spring break, where the temperature when you land is 30°C, then the temperature has increased 2-fold (30/15 = 2). In other words, it is twice as hot in Cancun compared to New York.

   Explain what log2 fold change means using online resources or previous knowledge. Give a numerical example of log2 fold change between treated and untreated samples. How is log2 fold change different from linear fold change?
4. The researchers used software to perform statistical tests on this differential gene expression data and generate p values. You will find many columns with the heading "pvalue." **Explain what p value means. What does a large or small p value indicate? Explain fully.**
5. When doing many comparisons, you have to consider that you will get some false positives. For example, $p < 0.05$ will mean 1 in 20 gene expression changes will be a **false positive**. Not surprisingly, high-throughput (HT) methods are susceptible to both false positive and **false negative** results. **Describe what a false positive AND a false negative would be in this type of experiment.**

B. Understanding Differential Gene Expression

Researchers using HT techniques must define how many false discoveries are acceptable in their experiments. For example, let's say 10% of the genes are incorrectly identified as differentially expressed. The researcher determines this rate is acceptable; therefore we say the false discovery rate (FDR) is 10%.

A p value cutoff can be changed to account for the FDR the researcher determines is acceptable. In gene expression analyses like RNA-seq, many genes (usually thousands) are simultaneously tested with the null hypothesis that the treatment does **not** change gene expression compared to a control. We would say that there are multiple hypotheses being tested, so p values in these experiments are adjusted for that (creating adjusted p values). Scientists sort RNA-seq data based on **adjusted** p values (padj) not p values, even though both are computed. If a 10% FDR is acceptable, then an adjusted p value of less than 0.10 is considered statistically significant. If a 5% FDR is acceptable, then an adjusted p value of less than 0.05 is considered statistically significant.

The treatment comparison we will investigate in this laboratory exercise is "OxvsUn" which means oxacillin-treated bacterial cultures compared to untreated.

To identify those genes with an adjusted p value of less than 0.05:

1. Locate the column titled "OxvsUn_ padj." You can use the "find" function in Excel to look for particular labels in this large worksheet.

2. Select the "sort & filter" button located on the top right of the Excel spreadsheet and "sort from smallest to largest". The sort warning window will pop up asking what you want to do. Choose to "expand the selection". This means that even though you are sorting based on one column, all of the other data in other columns will move with it accordingly. You now have organized the RNA-seq data with genes that change most significantly with oxacillin compared to untreated at the top of the list (smallest adjusted *p* values).
3. Scroll down the "OxvsUn_padj" column and identify the first row that contains the adjusted *p* value of 0.05. All the genes in the list with adjusted *p* values < 0.05 are the ones you will focus your analysis on. Think of this as a subset of your original gene list.
4. Insert a new worksheet next to the one you are working on.
5. Rename the worksheet tab "significant DGE."
6. Copy all of the cells (all columns and rows) containing data for genes with an adjusted *p* value < 0.05 (**do not include values of 0.05 or greater**) and paste it into your new "significant DGE" worksheet.
7. You can highlight the relevant "OxvsUn" columns to find them more easily.
8. Insert the columns "gene name" found in column DI, and "gene description" found in column DP from the original data worksheet next to the highlighted "OxvsUn" columns you are working with in the significant DGE worksheet you created so that it is close by.

   Without regard to upregulated vs. downregulated genes, how many statistically significant DEGs did you find?
9. While staying in your new worksheet of only significant changes, sort this DEG list based on log2 fold change. To do so, place the largest, positive log2 fold change values at the top:
   a. Highlight the entire "OxvsUnlog2foldchange" column by clicking on the column title.
   b. Select the "sort & filter" button and sort from largest to smallest. Expand the selection like you did previously.
   c. Finally, click the "sort" button.

   What is the value of the largest log2 fold change? What does that equal in terms of linear fold change (reported to the hundredths place)? Which gene demonstrated this?
10. **Explain what that fold change in the answer above means experimentally. What effect did oxacillin have?**
11. **Explain why log2 fold changes in gene expression can be positive or negative, while linear fold changes always come out as positive numbers.**
12. **How many total genes were analyzed in this high-throughput experiment?**

    Hint: Consider which worksheet you should refer to.
13. **Based on how many total genes were analyzed (answer from the above question), what percentage showed DGE with oxacillin compared to untreated bacterial cells?**
14. There is something striking about all of the DEGs that occur with oxacillin treatment. **Are they all downregulated or upregulated? How do you know?**
15. You will notice some genes on the original worksheet of all data have "--" listed for log2 fold change. These genes had negligible read counts in this RNA-seq experiment. **What percentage of all the genes analyzed showed negligible read counts?** *Hint:* Use a formula to quickly count these cells: =COUNTIF(range,"--")
16. Click on the "significant DGE" worksheet again. **Looking at the gene names and descriptions, can you find the two genes that are related to each other? What are their names?**
17. **What can you find online about their functions? Can you hypothesize why these genes might get upregulated in the presence of the β-lactam antibiotic, oxacillin?**
18. Some gene descriptions on the significant DGE worksheet contain the words "hypothetical protein." When sequencing reads map to the reference genome, they sometimes map to sequences that likely encode proteins, but they are not well annotated. **How many hypothetical proteins were significantly upregulated with oxacillin in this experiment?**
19. Novel transcripts can also be discovered using RNA-seq. When sequencing reads are mapped to the reference genome, they do not all map to known, annotated genes. If the reads exist in the sample, but do not map to any gene and are not similar in sequence to any gene, they are designated as novel. **How many novel genes were significantly upregulated with oxacillin in this experiment?**
20. **For gene names that are unknown in this experiment, what information is still available about them after the sequencing reads are mapped to the reference genome?**
21. **What is the DEG in the "OxvsUn log2fold change" column with the lowest log2 fold change?**
22. Look at the very last column with the title "tf_family." This is where the transcription factor family to which the gene product belongs is listed, if applicable. This often gives insight into function and is particularly helpful when you are learning about a gene that encodes a hypothetical, unknown, or novel protein. **For the DEG with the lowest log2 fold change, what tf family does its protein product belong to? How does that relate to oxacillin treatment?**

23. **Is it always the case that fold changes that are high in magnitude will also be statistically significant? Explain your reasoning.**
24. In order to enrich for true positives, HT results are sometimes accompanied by validation of a subset of differentially expressed genes with a complementary technique. **Name one validation technique that could be used to measure differential gene expression.**

Note to instructor: An answer key to all of the questions in Laboratory Session 32 is found in the online instructor resources.

DISCUSSION QUESTIONS

1. Compare and contrast RNA-seq with RT-qPCR.
2. What are some specific types of RNA-seq options that you can find using online resources? These are variations on the RNA-seq steps described here.
3. Which sample in this data set served as the control? Why? What other sample comparisons were available in the data set?

ONLINE RESOURCES

RNA-seq video. https://www.youtube.com/watch?v=fCd6B5HRaZ8.
Illumina Next-Gen Sequencing resources: https://www.illumina.com/science/technology/next-generation-sequencing/beginners.html.

REFERENCES

A. Gondane, H.M. Itkonen, Revealing the history and mystery of RNA-Seq. Current Issues in Molecular Biology. 14673045. 45 (3) (2023), 1860–1874,

F. Sanger, A.R. Coulson, A rapid method for determining sequences in DNA by primed synthesis with DNA polymerase. Journal of Molecular Biology. 00222836. 94 (3) (1975), 441–448.

Q. Pan, O. Shai, L.J. Lee, B.J. Frey, B.J. Blencowe, Deep surveying of alternative splicing complexity in the human transcriptome by high-throughput sequencing. Nature Genetics. 15461718. 40 (12) (2008), 1413–1415.

E.T. Wang, R. Sandberg, S. Luo, I. Khrebtukova, L. Zhang, C. Mayr, S.F. Kingsmore, G.P. Schroth, C.B. Burge, Alternative isoform regulation in human tissue transcriptomes. 14764679. Nature 456 (7221) (2008), 470–476.

Lab Session 33

Structure-Activity Relationship Modeling

Goal: Today you will use a number of open-sourced computational tools to virtually model the docking of the antibiotic oxacillin within the active site of the *S. aureus* penicillin binding protein 2 (PBP2).

I. INTRODUCTION

Structure-activity relationship (SAR) modeling is an approach to find relationships between the chemical structure of a compound and its biological activity. Molecular docking is one type of SAR technique that simulates virtually the interaction between a compound and its potential or known target protein to predict binding affinity and activity. SAR modeling is commonly used in drug discovery research to identify novel lead compounds that bind regions of specific protein target(s), predict potential toxicity of new compounds, and for lead optimization. In addition, SAR can be used to understand how chemical modifications can affect a molecule's activity in a biological system, thus aiding in rational drug design. SAR molecular docking in particular can also be used for the high-throughput screening of thousands of molecules virtually. Therefore, SAR modeling streamlines the drug discovery process, while reducing the time and cost of extensive laboratory testing.

For the purposes of this experiment, you will use SAR molecular docking to analyze the predicted binding affinity and identify important intermolecular interactions between two molecules you worked with in the last two laboratory sessions. Oxacillin is the beta-lactam antibiotic introduced in the transcriptomic analysis (Fig. 33.1). *S. aureus* PBP2 is the protein whose gene sequence was cloned in the in silico Gibson assembly laboratory session. You will use SAR modeling of oxacillin (the ligand) within the active site of the PBP2 protein.

II. LABORATORY EXERCISES

There are four main steps to conduct the virtual modeling: (1) configuring the PBP2 protein file, (2) downloading/creating the ligand file for oxacillin, (3) selecting the ligand docking site on the PBP2 protein, and (4) visualizing and analyzing the results. The workflow described in this laboratory exercise was adapted with permission from "Design and Implementation of an Accessible and Open-Sourced *In Silico* Drug Screening Activity for Cancer Drug Discovery," by Fikes and Srougi *J. Chem. Educ*. 2023, 100, 10, 4125-4130. Copyright 2023 American Chemical Society.[1]

For this activity, you will need to download the following free, open-source software prior to beginning the lab:

1. AutoDockTools (part of the MGLTools software suite)*
2. Vina[2]
3. Strawberry Perl
4. PyMol[3]

Download instructions for each of these software programs can be found in the manual's **Appendix E** section. Please refer to this section before beginning the lab.

A. Configuring the protein file

If the crystal structure of the protein has been published, as in the case of PBP2, you can access the structure from RCSB Protein Data Bank (RCSB PDB). This database serves as an archive of experimentally determined 3D protein structures as well as other tools and resources.

Once you have the PBP2 PBD file, you will configure the file for the virtual docking. This step is necessary to prepare both the protein as well as the ligand (in Part B below) files for docking simulations to produce more reliable and accurate results. This process ensures the structures are in the appropriate format for the docking algorithm to function appropriately. Key steps in the configuration process include assigning the atom types and charges to the atoms, which is used

Molecular Biology Techniques. DOI: https://doi.org/10.1016/B978-0-443-26714-7.00044-2

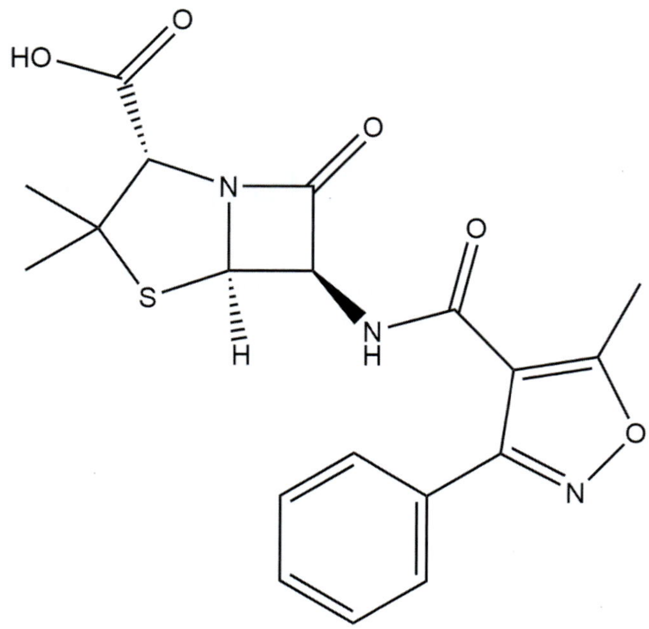

FIGURE 33.1 Structure of oxacillin. *Image made with ChemDraw.*

by the algorithm to determine electrostatic interactions between the two molecules. Another step is removing water molecules as they can interfere with the simulated docking process. Since many PDB files are from crystallographic data, they often lack hydrogen atoms, especially on polar groups. Adding these hydrogen atoms in the configuration process ensures that the protein and ligand have the correct protonation states and hydrogen bonding capabilities. If the binding site on the protein is known, specifying amino acid residues or regions of interest within the protein facilitates the docking algorithm to focus on the relevant area(s) in the protein. Finally, converting the PDB file to PDBQT file format includes crucial information about atom types and charges and torsional degrees of freedom, thus making them ready for virtual docking.

1. Create a file folder on your desktop labeled "**Docking**."
2. Using your web browser, navigate to the **Protein Data Bank** (https://www.rcsb.org/) (accessed 1 Dec 2024)
3. Search for the *S. aureus* PBP2 transferase protein using the 4-digit entry ID: **2OLV** (*S. aureus* PBP2) (accessed 1 Dec 2024).
4. Select the "**Download Files**" button on the top right of the screen and select **PDB Format**.
5. Move the pdb file to your docking folder and rename it **PBP2**.
6. Open **AutoDock Tools** and select "**File**," then "**Read Molecule**."
   a. Open the PDB file for PBP2.
7. Select "**Edit**" and then choose "**Delete Water**."
8. Select "**Edit**" choose "**Hydrogens**" and select "**Add**."
   a. Select "**Polar Only**" and select "**OK**."
9. Select "**Edit**" then "**Charges**" then "**Add KollmanCharges**."
   a. Select "**OK**."
10. To save the protein, go to "**Grid**" select "**Macromolecule**". and then select "**Choose.**" The Grid option is located below the main menu and highlighted in brown.
    a. Highlight "**PBP2**" and select "**Select Molecule**."
    b. A window may appear that states "contains no non-bonded atoms." Select "**OK**".
    c. Choose your Docking folder for the file location.
    d. Select the "**PDBQT files (*.pdbqt)**" as the Save file type.
    e. Name the file "**PBP2**" and select "**Save**".
    i. A warning window will appear that states "these atoms have zero charge." Select "**OK**".
11. Close **AutoDock Tools**.

B. Configuring the ligand file

In this step, you will prepare the oxacillin ligand file for the virtual docking algorithm. Since oxacillin is a well-known compound, its structure is available from PubChem, where you will download the file.

1. Using your web browser, navigate to **PubChem** (https://pubchem.ncbi.nlm.nih.gov/) (accessed 1 Dec 2024).
2. In the search window, type in "**oxacillin**."
3. In the search results, select the oxacillin compound with the compound **CID: 6196**.
4. In the top right of the screen select the "**download**" option.
5. In the 3D conformer section of the dialogue box, select "**Save**" next to "**SDF**."
6. Access the .sdf file in the downloads folder. Rename the file "**oxacillin**" and leave the file in the downloads folder.
7. Next, open the **PyMOL** software.
8. Under "**File**" select "**Open**" and choose the oxacillin .sdf file in the downloads folder. Oxacillin should now appear in the visualization box.
9. Go back to "**File**" and select "**Export Structure**" and then select "**Export Molecule**."
10. A new dialogue box will appear. Using the "**selection**" dropdown menu choose "**all**."
11. Select "**Multi-File**."
12. Select "**one file per object.**"
13. Make sure that the box next to "**Prompt for every file**" is checked and select "**Save**" using the **PDB (*.pdb *.pdb.gz)** file format from the bottom drop-down.
14. Name the file "**oxacillin**" and "**Save**" to the Docking folder. Verify that the file is saved to this folder before moving on to the next step.
15. Close the **PyMOL** program.
16. Next, open the **AutoDock Tools** program again.
17. Go to "**File**" and select "**Read Molecule**."
18. Choose the oxacillin PDB file you just created in step 13 above and select "**Open**."
19. In AutoDock Tools, select "**Ligand**" then "**Input**" and finally "**Choose**."

The ligand option is located below the main menu and highlighted in brown.

1. Select oxacillin in the dialogue box and then choose "**Select Molecule for AutoDock4**."
2. A dialogue box should now appear that summarizes the ligand setup. Select "**OK**."
3. Next, select "**Ligand**" and then "**Output**" and "**Save as PDBQT**." Choose the **Docking** folder as the file location. Select the "**PDBQT files (*.pdbqt)**" as the Save as type.
4. Name the file "**oxacillin**" and select "**Save**."
5. Close **AutoDock Tools**.

C. Select the docking site on PBP2

In this step, you will determine the region on the PBP2 protein that you wish to simulate the docking of the ligand oxacillin. The docking region of oxacillin described below is informed by structurally similar *S. aureus* penicillin binding proteins[4] and PBP2s in other bacterial species.[5]

Note: *In some scenarios, such as novel compound-protein interactions, you will not know the docking region and will run the SAR modeling without this information.*

1. Open **AutoDock Tools**.
2. Go to **File > Read Molecule**.
   a. Select "**PBP2.pdbqt**" file from your docking folder and select "**Open**".
3. Go to **Grid > Macromolecule > Choose** (Fig. 33.2).
   a. Select "**PBP2**" and select "**Select Molecule**"
   b. A dialog box may pop up that asks if you want to preserve charges. Select "**No**".
   c. A warning dialog box might pop up. Select "**OK**".
4. Go to **Grid > Grid Box**.
   a. A **Grid Options** dialog box should pop up along with a red, green, and blue cube in the molecule visualization area. This represents the area where the docking will occur. You can adjust the location and size of the cube by selecting on the red, green, and blue dials in the dialog box and moving your mouse left or right. The top three dials change the box size. The bottom three dials change the box location. Use these dials to center the box over the active site of the protein. You can use the left select option and move your mouse in the visualization window. To move the

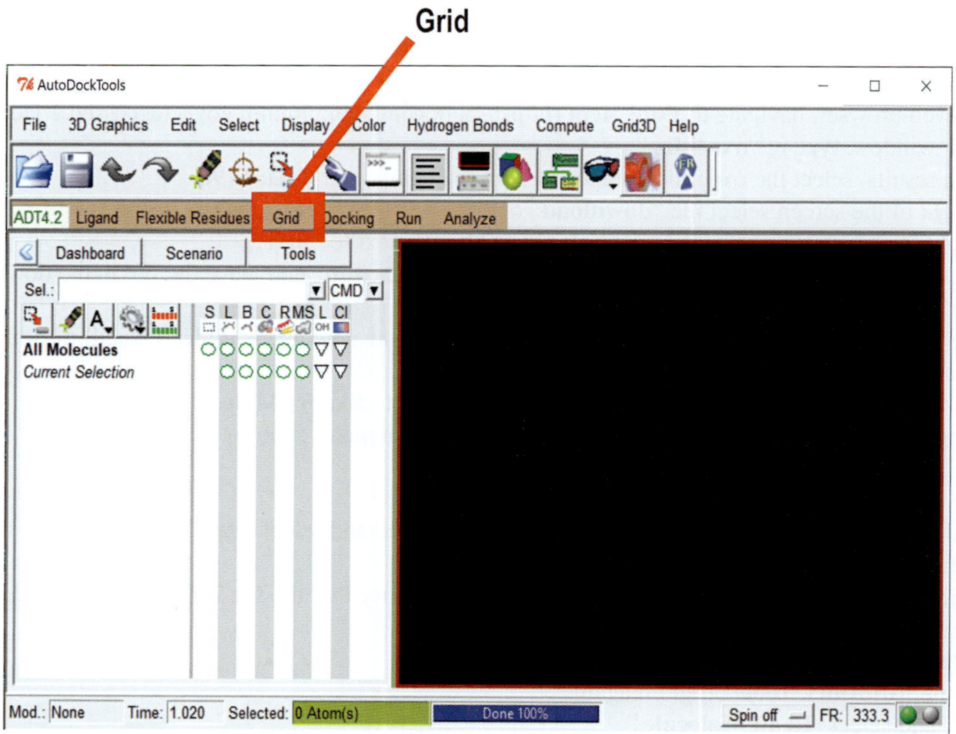

FIGURE 33.2 Screenshot of the AutoDock tools interface. Indicated in the red box is the grid function. *No Permission Required.*

protein, make sure the grid box is in the appropriate location. *Note: for the purposes of this activity, the grid box parameters have been provided for you in step 4c below.*

b. To change the view of the protein, left select to rotate, right select to move, and move two fingers away from each other or towards each other to zoom in and out, respectively.

c. **In order to get consistent results, you should use the same area each time you perform docking on a particular protein. You should use the following values:**
   i. Number of points in the x-dimension: **40**
   ii. Number of points in the y-dimension: **40**
   iii. Number of points in the z-dimension: **40**
   iv. X center: **-17.869**
   v. Y center: **24.64**
   vi. Z center: **-9.923**
   vii. Spacing (angstrom): **0.375**

5. Once you have your box in the correct location, in the **Grid Options** dialog box go to **File > Output grid dimension file.**
   a. Select your **docking folder**.
   b. Select "**.txt (*.txt)**" as the Save as type.
   c. Name the file "**grid**" and select "**Save**". This will save the values of our grid box we have a record of the dimensions for future reference.
6. Close **AutoDock Tools**.
7. Open the **Notepad app** and type (or copy and paste) the following:
   a. receptor = PBP2.pdbqt
   b. center_x = -17.869
   c. center_y = 24.64
   d. center_z = -9.923
   e. size_x = 40
   f. size_y = 40
   g. size_z = 40
   h. energy_range = 4
   i. exhaustiveness = 8

8. Go to **File > Save As**.
   a. Select your docking folder.
   b. Select '**Text Documents (*.txt)**" as the Save as type.
   c. Name the file "**conf_vs**" and select "**Save**" and close Notepad.

D. Implementation of the virtual screening

1. Open **File Explorer** and select on the **C:** drive.
   a. Open the **Program Files (x86)** folder.
   b. Open the **Scripps Research Institute** folder.
   c. Open the **Vina** folder.
   d. Select all three files in the **Vina** folder and copy and paste them into your Docking folder. These files are: vina, vina_license, and vina_split.
2. Download the **vina_windows.pl** file from the online Appendix materials and move the file into your Docking folder.
3. Go to the Windows search bar and type "**cmd**" and select "**Command Prompt**" to open the Windows Terminal.
4. Open your Docking folder in Windows **File Explorer**. In the address bar, **right select** and select "**copy address**".
5. Go back to the Command Prompt window. Type "**cd**" and hit the spacebar once. Then paste the address that you just copied and press enter. Note: there should be no space ***before*** cd (Fig. 33.3).

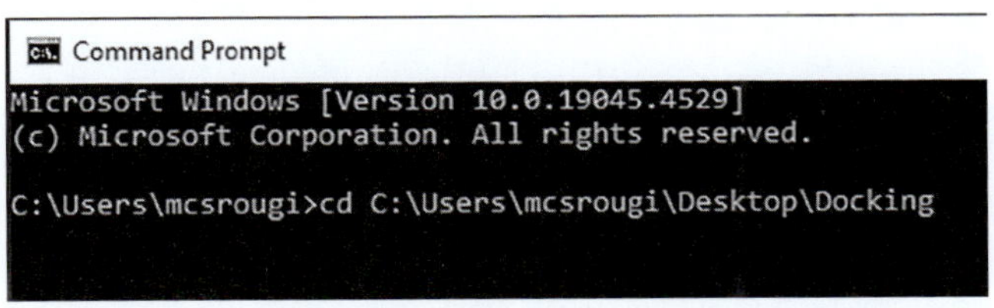

FIGURE 33.3 Command prompt window. *No Permission Required.*

6. Next type "**dir /B > ligand.txt**" and press enter (Fig. 33.4).
   A new file named "**ligand.txt**" should appear in your Docking folder. Open the file with **Notepad** and delete all the files that are not **oxacillin.pdbqt**. Save the file and close it.

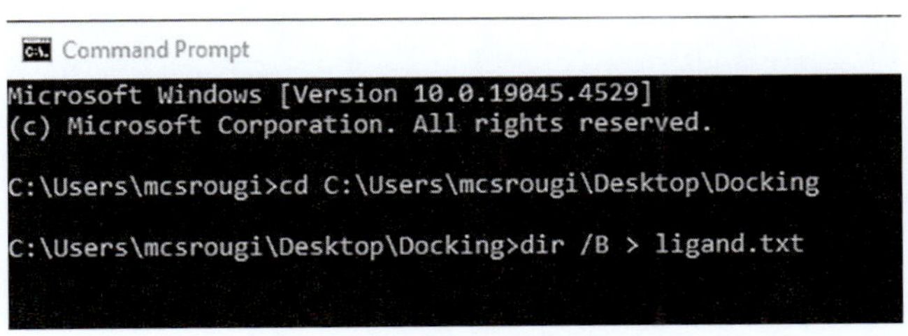

FIGURE 33.4 Command prompt window indicating the directory location for the ligand file. *No Permission Required.*

7. Return to the Command Prompt and type "**perlvina_windows.pl**" and press enter.
8. On the next line, type "**ligand.txt**" and press enter. *Note: If an error pops up, double check your command for typos* (Fig. 33.5).

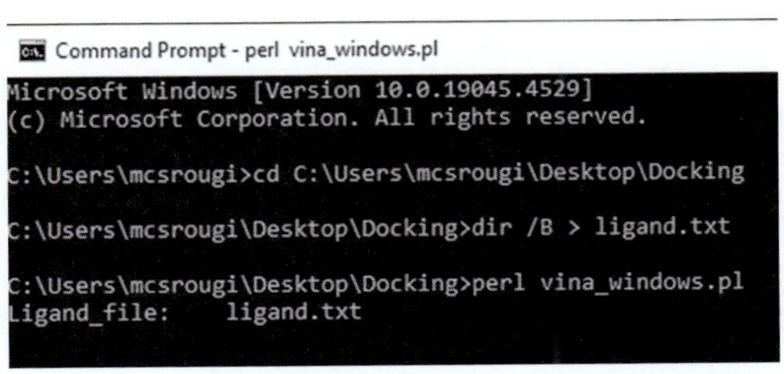

FIGURE 33.5 Command prompt window indicating the code to access the virtual screening program. *No Permission Required.*

9. The virtual screening is now running. After the drug oxacillin has been virtually docked, a table of results (i.e., binding energies) will appear in the Command Prompt window. The results are saved as two files in your virtual screening folder: a log file that contains the table of binding affinities and an output file that contains the molecular conformations of the drug (or compound) in relation to the protein (i.e., PBP2) (Fig. 33.6).

```
Command Prompt

C:\Users\mcsrougi>cd C:\Users\mcsrougi\Desktop\Docking

C:\Users\mcsrougi\Desktop\Docking>dir /B > ligand.txt

C:\Users\mcsrougi\Desktop\Docking>perl vina_windows.pl
Ligand_file:    ligand.txt
oxacillin.pdbqt



oxacillin.pdbqt
#################################################################
# If you used AutoDock Vina in your work, please cite:          #
#                                                               #
# O. Trott, A. J. Olson,                                        #
# AutoDock Vina: improving the speed and accuracy of docking    #
# with a new scoring function, efficient optimization and       #
# multithreading, Journal of Computational Chemistry 31 (2010)  #
# 455-461                                                       #
#                                                               #
# DOI 10.1002/jcc.21334                                         #
#                                                               #
# Please see http://vina.scripps.edu for more information.      #
#################################################################

WARNING: The search space volume > 27000 Angstrom^3 (See FAQ)
Output will be oxacillin_out.pdbqt
Detected 8 CPUs
Reading input ... done.
Setting up the scoring function ... done.
Analyzing the binding site ... done.
Using random seed: 801699584
Performing search ...
0%   10   20   30   40   50   60   70   80   90   100%
|----|----|----|----|----|----|----|----|----|----|
***************************************************
done.
Refining results ... done.

mode |   affinity | dist from best mode
     | (kcal/mol) | rmsd l.b.| rmsd u.b.
-----+------------+----------+----------
   1         -8.1      0.000      0.000
   2         -8.0      9.271     12.378
   3         -7.8      9.262     12.095
   4         -7.3     10.399     13.361
   5         -7.3      3.013      6.580
   6         -7.3     11.346     13.602
   7         -7.2      3.408      6.787
   8         -7.2     11.786     14.633
   9         -7.1     10.963     13.111
Writing output ... done.
```

FIGURE 33.6 Command prompt window indicating the virtual screening output results. *No Permission Required.*

10. The table ranks the different compound conformations and the associated binding affinity for each **pose**. Each pose consists of both the conformation and orientation of molecules relative to the binding site of the macromolecular target.[6] The most energetically favorable pose will be the first one listed in the table and have the lowest binding affinity. The **binding affinity** is the strength by which two (or more) molecules interact or bind. It is reported by the K_d or equilibrium dissociation constant. The smaller the value, the greater the affinity between two molecules.

Root mean square deviation (rmsd) values indicate how two molecular structures differ from one another spatially. It is the measure of the average distance between corresponding atoms of two superimposed structures. Rmsd lower bound (lb) refers to the minimum possible rmsd between two aligned structures. It is the best possible alignment and the smallest possible deviation in the structure alignment. The rmsd upper bound (ub), on the other hand, represents the maximum rmsd based on the alignment. It indicates the maximum deviation between the two models when they are not well aligned. In summary, rmsd lb gives us the best possible alignment between two molecular structures, while rmsd ub gives us the worst possible alignment. The rmsd values help evaluate the accuracy of each structural pose.

Keep the binding affinities table open. We will next use PyMOL to visualize the poses.

Note: *Predicted binding affinities may vary, even within the exact same SAR modeling runs. This is due to a number of factors, some which include the algorithm methodology, which involves random sampling. In addition, the conformational flexibility of the proteins/ligands will affect predicted binding. The program will attempt to find the best fit between the protein and ligand, and the multiple conformations of each can lead to different docking poses and thus scores in each simulation that is run.*

E. Visualizing simulated docking poses using PyMOL

The predicted binding affinities alone do not give us the full picture for how PBP2 and oxacillin interact with one another. You next need to visualize each of these poses with their associated binding affinities using PyMol to evaluate which pose is the best. In particular, you will evaluate the proximity of oxacillin to the PBP2 active site as well as identify any non-covalent interactions that occur within this region to support its binding to the protein. Collectively, this information will enable you to determine which predicted binding pose is the most favorable. For example, you might hypothesize that the pose with the lowest binding affinity (i.e., greater negative value) will be the most favorable. However, if you visualize this pose and find that oxacillin is not near the known PBP2 active site residues and/or it is not in position to form any non-covalent interactions with the active site residues, then you can determine that this particular pose is not the most favorable. In this part of the exercise, you will evaluate and rank each of the nine poses predicted by SAR modeling based on three factors: (1) binding affinity, (2) distance from key active site amino acid residues, and (3) non-covalent interactions that occur between oxacillin and the PBP2 active site region. Record this information in Table 33.1.

1. Open **PyMOL**.
2. Go to **File > Open**.
   a. Choose the "**PBP2.pdbqt**" file from your Docking folder and select "**Open**".
3. Go to **File > Open**.
   a. Choose the "**oxacillin_out.pdbqt**" file from your Docking folder and select "**Open**".

Both the protein and the docking results should now be visible in the PyMOL Viewer window. You can change between the different docking results by using the **left and right arrow keys** on your keyboard. In the next section, we will go over a brief introduction to PyMOL so you can further analyze your docking results.

TABLE 33.1 Binding affinity and intermolecular interactions between oxacillin and PBP2

Image of Docking Pose # (screenshot)	Rank Order (from 1 to 9)	Binding Affinity (kcal/mol)	Identified intermolecular interactions between oxacillin and PBP2 active site
Example from Fig. 33.8	1	-8.1	TYR 437, OH group forms H bond with N atom from isoxazole ring

Brief introduction and general PyMOL functions

Refer to Fig. 33.7.

1. **Command line:** This is where you can type in commands. The command line is generally most useful for experienced users. We will primarily use the mouse instead of the command line.
2. **Object Control Panel:** This contains the list of objects that are currently being visualized. Selecting the object names in the **Object Control Panel** will hide them in **PyMol Viewer**. Selecting the object names again will cause the object to reappear. There are a set of options after each object that can be used to: (1) adjust viewing modes, (2) change colors, (3) add labels, etc. The options and their functions are briefly summarized below:

   A – *Actions*: Rename, duplicate, remove, apply presets (like "ball-and-stick" or "publication"), perform computations.

   S – *Show*: Change the way things appear, e.g., change to stick or cartoon view.

   H – *Hide*: Things that are shown using **S** accumulate and do not automatically replace the last view. **H** is the opposite of **S** and hides unwanted representations.

   L – *Label*: Label atoms, residues, etc.

   C – *Color*: Change the color of atoms and groups.
3. **Mouse Control Legend:** This is the guide for using the mouse. The default is for a three-button mouse. These settings can be changed for different mouse types. The most useful mouse controls for visualizing the protein are listed below.

   Left select: Select (the default is to select residues)

   Hold Left select: Rotate protein

   Ctrl Left select: Move

   Hold Right select: Zoom in and out
4. **Selection details:** This panel will describe what has been selected in the PyMOL viewing window. The default selection is residues. However, other options include chains, molecules, and atoms. For the purposes of this activity, we will use the default option of residues.
5. **Frame navigation:** When the ligand and protein have been virtually docked with multiple poses, this option enables you to navigate through each pose. In this example, nine poses were docked; therefore there are nine separate frames. Each frame visualizes one pose. Use the arrow keys or the scroll bar to navigate through each pose.
6. **PyMOL viewer:** This window allows for the visualization of your protein and compounds. Left selecting on objects in PyMOL Viewer selects them. select on the objects again deselects them. Objects that are selected are listed as (sele) in the Object Control Panel.

Tools for the evaluation of non-covalent interactions between the ligand and protein

In order to determine the **protein-ligand interaction(s)**, it is helpful to view the protein surface. To do this:

1. Select any region on PBP2 with your cursor and press the mouse button. On the **Object Control Panel**, select the **S**(how) **as** button for PBP2 and select "**surface**."

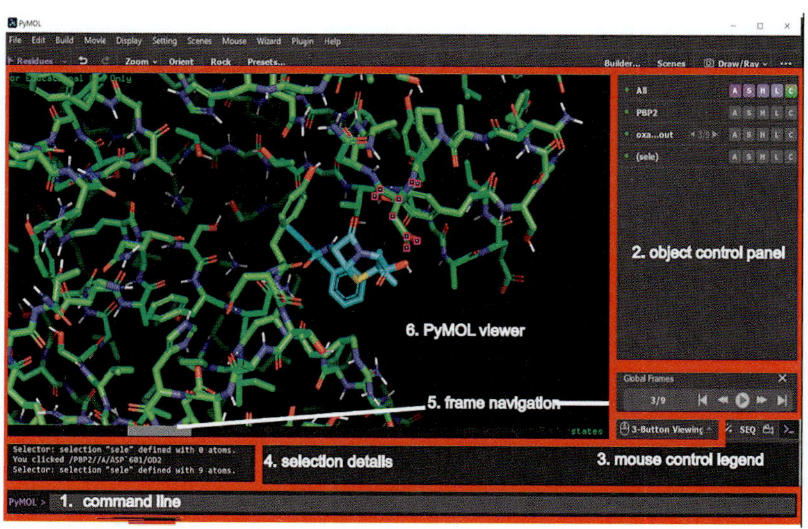

FIGURE 33.7 Annotation of PyMOL functions. *No Permission Required.*

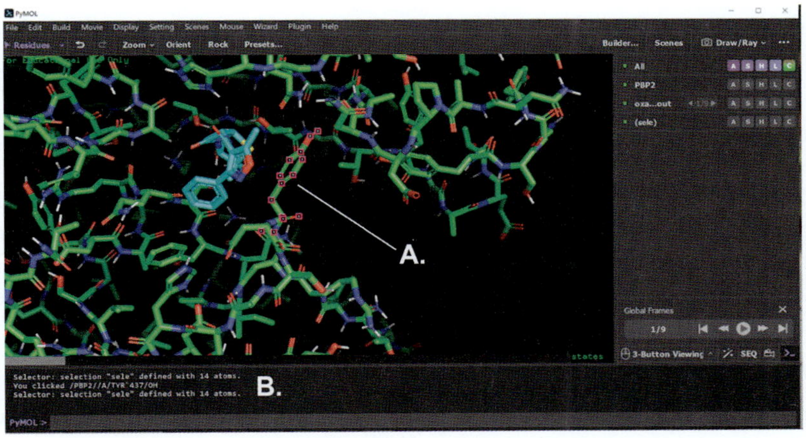

FIGURE 33.8 Screenshot of PyMOL viewer window. Only amino acids in PBP2 that are in close proximity with oxacillin are shown for pose 1. *No Permission Required.*

To visualize individual **amino acid side chains** within the **active site** of PBP2:

1. Select the **S**(how) **as** button for PBP2 in the **Object Control Panel**.
   a. Select **side chain > sticks**.
   b. All of the protein amino acid side chains should now be visible.
2. With your cursor, select oxacillin. On the top of the PyMol window select **zoom >active selection**.
   a. The view window will now be zoomed in on the region of oxacillin-PBP2 putative binding.
3. Next with your cursor, select any region of the PBP2 protein in close proximity to oxacillin
   a. Use the scroll wheel on the mouse or two fingers on a laptop trackpad to reduce the number of visible amino acid chains. Continue to do this until only a few amino acid side chains remain that are in close proximity to oxacillin (Fig. 33.7 and Fig. 33.8).
   b. Using your cursor, select amino acids on PBP2 that may form non-covalent interactions with oxacillin. The residue you select will be denoted with red boxes (A) and identified in the bottom of the view window (B). In this example, pose 1, which has a predicted binding affinity of -8.1 kcal/mol (Fig. 33.8), is being shown. TYR 437 is highlighted as the hydroxyl group has potential to form a non-covalent interaction (i.e., hydrogen bond) with the nitrogen atom in the isoxazole ring of oxacillin, thus supporting the protein-ligand interaction of this binding pose.
   c. Using your cursor again, select any region of black space. Hold your cursor down and move the mouse to twist and turn the pose. This will enable you to examine other perspectives and interactions of the selected pose.
   d. Record **all potential** non-covalent interactions between amino acids in close proximity to oxacillin in Table 33.1. Be sure to specify the functional groups, type of non-covalent interaction, and the distance. **Take an image** of this pose, using the screenshot feature on your computer. An example from Fig. 33.8 has been done for you. **Repeat for each pose.**

TOOLS FOR THE EVALUATION OF LIGAND-ACTIVE SITE DISTANCE

It is also informative to determine the **distance between atoms** in a docking structure. To measure these distances in PyMOL:

1. Go to **Wizard > Measurement**.
   a. This will bring up the **Measurement Wizard** below the **Object Control Panel**.
2. Select the first atom and then select the second atom.
   a. The distance in Ångstroms will appear between the two atoms.
3. To delete the measurement, select "**Delete Last Object**".
4. Selecting "**Done**" will close the Measurement Wizard but leave the measured values in the PyMOL Viewer.

To save your Pymol session go to **File > Save Session As**.

1. Select your Docking folder.
2. Name the file and select "**Save**".

Analyzing the data

1. Complete Table 33.1. All table entries should include a picture (i.e., screenshot) of the visualized results from PyMOL for each of the nine poses entered in the first column.
2. Once you have completed Table 33.1, summarize your findings from your in silico SAR modeling for each docking pose.
3. Next, rank your findings in order of **highest to lowest** with poses predicted to be the most favorable listed at the top (i.e., rank order 1). Remember to take both binding affinity and the orientation of the molecule into account. Be sure in your table to provide notes about important non-covalent interactions that were seen in your modeling output. This should include the amino acid residues in the PBP2 binding site that play a prominent role in those interactions.

DISCUSSION QUESTIONS

1. Compare your results to that of your lab partner or someone else in the class. Do they match? Why?
2. If you perform the computational docking analyses but are unable to validate your findings in the laboratory, how would you explain the discrepancy?

REFERENCES

1. A.G. Fikes, M.C. Srougi, Design and implementation of an accessible and open-sourced in silico drug screening activity for cancer drug discovery. Journal of Chemical Education. 100 (10) (2023), 4125–4130.
2. Oleg Trott, Arthur J. Olson, AutoDock Vina: Improving the speed and accuracy of docking with a new scoring function, efficient optimization, and multithreading. Journal of Computational Chemistry. 31 (2) (2010), 455–461.
3. 2024 The PyMOL Molecular Graphics System, version 3.0 Schrödinger, LLC.
4. S.S. Ambade, V.K. Gupta, R.P. Bhole, P.B. Khedekar, R.V. Chikhale, A review on five and six-membered heterocyclic compounds targeting the penicillin-binding protein 2 (PBP2A) of methicillin-resistant Staphylococcus aureus (MRSA). Molecules. 28 (20) (2023).
5. Milan Dabhi, Rohit Patel, Vidhi Shah, Richa Soni, Meenu Saraf, Rakesh Rawal, Dweipayan Goswami, Penicillin-binding proteins: the master builders and breakers of bacterial cell walls and its interaction with β-lactam antibiotics. Journal of Proteins and Proteomics. 15 (2) (2024), 2[illegible]5–232.
6. P.H.M. Torres, A.C.R. Sodero, P. Jofily, F.P. Silva, Jr, Key topics in molecular docking for drug design. International Journal of Molecular Sciences. 20 (18) (2019).

Appendix A

Equipment

SHARED EQUIPMENT

- −80°C freezer
- −20°C freezer (if the freezer has an automatic defroster, store all enzymes in insulated coolers, such as Stratacoolers, inside the freezer)
- 4°C refrigerator
- tabletop centrifuge
- microcentrifuge
- floor-model shaker incubator or tabletop shaker incubators for 100 mL bacterial cultures
- 37°C incubator for bacterial plates and enzymatic reactions
- stir plate and stir bars
- microwave oven for melting agarose
- sonicator with microimmersion tip (such as the Fisher Sonic Dismembrator 50), optional, for physical lysis only in Lab Session 11, Extraction of Recombinant Protein From *Escherichia coli* Using a Glutathione Affinity Column
- microvolume spectrophotometer (*We highly recommend a microvolume spectrophotometer such as the Nanodrop due to the minimal amount of sample that needs to be used; however, a spectrophotometer with a nucleic acid-compatible cuvette that measures absorbance at 260 and 280 nm could be used if samples are scaled up appropriately.* Alternatively, disposable cuvettes, such as the Eppendorf UVettes hold as little as 50 μL. Samples can be recovered to use after spectrophotometry).
- microtiter plate reader capable of reading absorbance (EGFP fluorescence is optional: excitation: 485 nm; emission: 528 nm) or spectrophotometer(s) capable of reading at 595 nm
- bench-top orbital rotator or platform rocker
- UV transilluminator with gel documentation system (or UV transilluminator in a darkroom with camera)
- PCR thermocycler (a real-time-capable thermal cycler is optional)
- autoclave
- large water bath for cooling media to 60°C and incubating tubes
- precision balance
- ice machine
- each lab group should have the items detailed in the Equipment Checklist in Lab Session 1, Getting Oriented; Practicing With Micropipettes
- For Part V only:
  - temperature controlled (to 37°C) CO_2 incubator for mammalian cells
  - Several BSL-2 Class II Type A2 biological safety cabinets
  - Inverted fluorescence microscope(s)
    - Objectives: 10×, 20×, and preferably 40×
    - camera

Molecular Biology Techniques. DOI: https://doi.org/10.1016/B978-0-443-26714-7.00047-8

Appendix B

Prep List

NOTES TO PREP STAFF

Plasmids and *Escherichia coli* Host Strains

The three plasmids you will need for Parts I–IV are pET-41a (Novagen catalog number 70556-3), pEGFP-N1, and pBIT (request from Dr. Scott Witherow at switherow@ut.edu; include "pBIT or pEGFP-N1 request" in subject heading). The two plasmids you will need for Part V are pLentiCRISPR-E (Addgene plasmid #78852) and pLenti-CRISPR-EGFP-2 (Addgene plasmid #51761).

The two host strains used are the expression host BL21(DE3) (Novagen catalog number 70235-3) and the nonexpression host NovaBlue (Novagen catalog number 70181-3). The catalog numbers given are for chemically competent cells. If you choose to make competent cells, you may purchase regular, noncompetent cells. Alternatively, other similar host strains may be purchased from other companies. Keep in mind that you will need one strain for isolating DNA (such as NovaBlue) and that strain should *not* be specialized for protein expression. You will need a second strain that allows transcription from the T7 promoter; this means that the expression strain must be an λ lysogen [such as BL21(DE3)].

Before the course begins, you will need a freezer stock of the listed *Escherichia coli* host strains (or similar strains) carrying the listed plasmids.

***Escherichia coli* Host Strains (or Similar Strains) Transformed with the Indicated Plasmids**

E. coli Host Strain	Plasmid	Purpose
NovaBlue	pEGFP-N1	Template for PCR
NovaBlue	pET-41a	Plasmid isolation
BL21(DE3)	pET-41a	Negative control for fusion protein expression
BL21(DE3)	pBIT	Positive control for fusion protein expression
Stbl3	pLentiCRISPR-E	Negative control for CRISPR-mediated gene editing of EGFP and vector backbone for cloning designed sgRNAs
Stbl3	pLenti-CRISPR-EGFP-2	Experimental (although validated) for CRISPR-mediated gene editing of EGFP

No Permission Required.

Before the course begins, it is recommended that you purify maxiprep amounts of pEGFP-N1 and pET-41a, pLenti-CRISPR-E, and pLentiCRISPR-EGFP-2. In practice, only small quantities of pEGFP-N1 are needed for PCR template (100 ng per student group), although more will be required if you plan on proceeding through Part V. Sufficient quantities of pET-41a should be prepared by students during their miniprep purification. It is generally good practice to have extra purified plasmid available in case experiments do not work as planned.

Molecular Biology Techniques. DOI: https://doi.org/10.1016/B978-0-443-26714-7.00046-6

Antibiotics

pEGFP-N1, pET-41a, and pBIT each use kanamycin for selection. Make a stock solution of kanamycin at 30 mg/mL in water, filter, sterilize, and store at −20°C. Add kanamycin to media at a dilution of 1:1000 from the stock solution (i.e., 1 mL in 1 L) after autoclaving. It is critical to wait for media to cool to ~60°C before adding the antibiotic. Adding the antibiotic while the medium is too hot can cause inactivation of the antibiotic. It is also important to make sure that the antibiotic being used is effective; bad lots do occur. This can be tested by inoculating a freshly made LB/kan plate with wild-type *E. coli* (lacking a plasmid-containing antibiotic resistance). *Note*: pLentiCRISPR-E and pLentiCRISPR-EGFP-2 both use ampicillin for selection. Once ordered from Addgene, these plasmids arrive already transformed into the Stbl3 strain as an agar stab. Directions for growth are included with shipment.

Experiments in Part V require the use of a penicillin/streptomycin solution, which is purchased as a sterile 100× solution (5000 U/mL) (Gibco catalog number 15070063; 100 mL bottle) as well as a G418 (50 mg/mL) solution (aka geneticin) (Gibco catalog number 10131035).

Aliquots

Restriction enzymes and buffers should be aliquoted and centrifuged prior to individual use. Whenever feasible, make more aliquots than are needed for the class. Students often misplace aliquots or need to repeat a procedure. Extra aliquots come in handy. Additionally, extra volume (typically ~10%) is included in each aliquot to ensure students can pipette the full amount of reagent. Individual aliquots should be labeled either by hand or using laser-printed labels.

Bacterial Waste

Solid

Solid bacterial waste includes tubes, tips, plates, etc., that have been used to grow bacteria or that have come into contact with any living bacteria or biohazardous waste. Recombinant DNA should also be considered biohazardous. The prep staff should provide several containers with appropriate-sized BIOHAZARD bags (Fisher Scientific, cat. #03-411). Replace the bags when a little over half full. All BIOHAZARD bags must be autoclaved before disposing. It is also recommended to provide double containment by using two BIOHAZARD bags in case of leaks or punctures. *Note*: Do not allow regular trash to be placed in BIOHAZARD bags. It can be dangerous to autoclave and puts an additional burden on the prep staff.

For convenience, students should have a small, bench-top biohazard container lined with a BIOHAZARD bag at each station as well as in the cell culture facility. This should be used for pipette tips and other small items that are used at the lab bench. Students should empty their small biohazard waste containers into the large common container at the end of each lab.

Details in the workflow for the disposal and containment of biohazardous waste material may vary from institution to institution. Therefore, please refer to your institution's environmental health and safety office for further information.

Liquid

Prepare a carboy labeled "bacterial graveyard" for liquid bacterial waste only. Treat with bleach (~15%–20% by volume). Wait 10–20 minutes. Bacteria are dead if the liquid has cleared and cell debris has fallen to the bottom of the container. Decontaminated liquid may be poured down the drain.

Autoclaving

Autoclaved materials should be ready at least the day before the material is needed.

Liquid

Fill bottles to only 60% full or less. Caps should be loose to prevent pressure build-up and bursting. Liquids should be sterilized for 20 minutes. Tighten the caps fully only after the solution has cooled to room temperature.

Biohazard Waste

BIOHAZARD bags must be sealed and labeled. If leakage is a problem, they may have to be double bagged or absorbent material added (e.g., vermiculite, bentonite, etc.). Always provide secondary containment for autoclaved items.

Dry Goods

All pipette tips and microfuge tubes should be autoclaved on gravity/dry cycle before use, unless they are purchased as sterile. If performing Part IV or V, use pipette tips and microfuge tubes certified as RNase-free. These come wrapped in plastic and should never be autoclaved.

General Lab Preparation

Designate an area or cart for reagents for student use. Group together materials needed for a particular lab session. Use 5 × 7 in. note cards to label all reagents and include information such as "for Lab 7A, step 2." Include phrases like "Return when finished," "Do not throw away, label with your initials and store at 4°C," or "Discard following use." Be sure to also label each individual aliquot as students will often mix up tubes! Providing clear directions whenever possible will help avoid problems and is especially important to ensure a safe working environment.

Periodically check the BIOHAZARD bags and remove them for autoclaving when they are half full. Make sure that 1 × TBE carboys and 70% ethanol carboys are full.

Students

Students will ask for help and for more reagents, particularly during the first several sessions. Advanced preparation and having additional aliquots on hand will help to avoid frustration.

Supplies and Reagents

Supplies and reagents for each week are listed in this appendix according to lab session. Volumes listed for reagents generally contain an extra ~10% to account for pipetting discrepancies. There are certain supplies and reagents that should be available at all times. These will not necessarily be listed under each lab session, but should be available for students and prep staff for every lab session. These are listed next.

Supplies and Reagents for General Use

- Laboratory safety glasses
- Nonlatex gloves (small, medium, large, and extra-large); powder-free gloves are preferable for working with nitrocellulose membranes
- Sterile pipette tips (all sizes)
- Sterile serological pipettes (all sizes)
- Microcentrifuge tubes
- Sterile deionized water
- 95% Ethanol
- 70% Ethanol (carboy)
- Agarose (can be kept at a general use area containing top-loading balance and microwave)
- Petri dishes (square and round)
- 50 mL Conical tubes
- 15 mL Conical tubes
- Weigh boats
- Centrifuge tubes capable of being centrifuged at 15,000 × *g*
- Sterile toothpicks for inoculating cultures
- LB broth
- LB/kan agar plates (stored at 4°C)
- 1× TBE (carboy)
- GelRed (10,000×; should be stored in foil and protected from light)
- Hand soap
- Kimwipes
- Paper towels

RECIPES FOR GENERAL USE

LB Broth and LB Agar

1. Mix these compounds:

LB Broth Recipe

Compound	Amount per Liter
Tryptone	10 g
NaCl	10 g
Yeast extract	5 g
Deionized water	1000 mL

No Permission Required.

   Alternatively, a premixed powder (such as Fisher Scientific catalog number BP1426-500) can be purchased and mixed directly with deionized water at 25 g/L.
2. Autoclave for 20 minutes at 121°C under a pressure of at least 15 psi.
3. Cool to 60°C and add filter-sterilized kanamycin (final concentration 30 µg/mL; can use 1000× stock solution) and/or filter-sterilized IPTG (final concentration 20 µg/mL; can use 1000× stock solution), where necessary. If performing experiments in Part V where CRISPR vectors are used, add filter-sterilized ampicillin (final concentration 100 ug/mL).

To prepare LB agar, prepare LB broth, then add 15 g/L agar and a stir bar before autoclaving (or use premixed LB agar, such as Fisher Scientific catalog number 1425-500). Autoclave, then place in a 60°C water bath. When the temperature has equilibrated, place on stir plate and add kanamycin, ampicillin, and/or IPTG where necessary. Pour ~20 mL per Petri dish, being cautious not to form bubbles.

IPTG (20 mg/mL)

Dissolve 1.0 g IPTG in 50 mL deionized water. Filter-sterilize and aliquot. This is a 1000× stock. Store at −20°C.

Kanamycin (30 mg/mL)

Dissolve 1.5 g kanamycin in 50 mL deionized water. Filter-sterilize and aliquot. This is a 1000× stock. Store at −20°C.

Ampicillin (100 mg/mL)

Dissolve 5 g of ampicillin in 50 mL deionized water. Filter-sterilize and aliquot. This is a 1000× stock. Store at −20°C.

5× TBE Stock (Tris-Borate-EDTA)

Mix these compounds in to make a 5x TBE stock solution.

Tris-Borate-EDTA (TBE) Recipe

Reagent	Amount
Tris	54 g
Boric acid	27.5 g
EDTA	4.15 g (or 20 mL 0.5 M EDTA, pH 8)
dH_2O	To 1 L

No Permission Required.

Place in carboy for students to prepare 1× stocks for their own use. Prepare 6 L for 20 student groups.

Alternatively, 10× TBE can be ordered from many different suppliers (Fisher Scientific catalog number BP13334 or Bio-Rad catalog number 1610770) and diluted for the students prior to the start of the lab.

GelRed Stock (10,000×)

Each student group should receive 30 μL in tinted microcentrifuge tubes, or wrapped with aluminum foil since GelRed is light-sensitive. Students should save the GelRed (at room temperature) to use in multiple laboratory exercises.

DNA Primers

DNA primers can be ordered from many commercial sources (e.g., Integrated DNA Technologies). They should be ordered desalted, but need no additional modifications. Upon receipt, prepare a stock solution of the primer by dissolving it in sterile dH_2O to a final concentration of 100 pmol/μL (100 μM). Details on DNA primers, the lab session they are used, and the primer sequence are found in.

List of DNA Primers and Their Sequence

Lab Session	Primer Name	Primer Sequence
3	egfpNco	AAACCATGGTGAGCAAGGGCGA
3	egfpNot	AAAGCGGCCGCTTTACTTGTACA
7D	Sequencing primer	CGAACGCCAGCACATGGACAGC
14	egfp14 F	CGACGGCAACTACAAGACC
14	egfp14 R	GTCCTCCTTGAAGTCGATGC
14	23S14 F	GACGGAGAAGGCTATGTTGG
14	23S14 R	GTTGCTTCAGCACCGTAGTG
16	egfp 16F	CCTGAAGTTCATCTGCACCA
16	egfp 16R	TGCTCAGGTAGTGGTTGTCG
16	23S 16F	ACTGCGAATACCGGAGAATG
16	23S 16R	CCTGTTTCCCATCGACTACG
26	User designed	
29	User designed	

No Permission Required.

siRNA Oligos

siRNA oligos can be ordered from many commercial sources (e.g., Thermo Fisher Scientific). It is recommended to order siRNA oligos on a scale of 20 nmole, desalted with a 3′ dTdT overhang. Using both sterile and RNase-free plasticware and precautions, prepare a working siRNA stock concentration of 100 μM using nuclease-free water (Thermo Fisher Scientific catalog number AM9914G; 1.75 mL/tube). Dilute the stock concentration to 10 μM for immediate use or aliquot and store at −20°C.

LAB SESSION 1

Getting Oriented; Practicing With Micropipettes

Note: This exercise will be done by each student, not by each group.

The students will each use these supplies:

- 10 µL Bovine serum albumin (BSA), 10 mg/mL, store at −20°C (Fisher catalog number FERB14 is 20 mg/mL solution, which can be diluted 1:1)
- Nitrocellulose membrane ~2″ × 3″ (1 per group)
- Whatman 3MM paper, ~2″ × 3″ (1 per group)
- Weigh boat
- Square Petri dish, tip box top, or other convenient tray for staining (1 per group)

These are to be shared:

- Ponceau Red (also referred to as Ponceau S) stain (Sigma-Aldrich catalog number P7170), ~10 mL needed per group (*Ponceau Red stain may be saved and reused a few times before being discarded*)

LAB SESSION 2

Purification and Digestion of Plasmid (Vector) DNA

Note to prep staff:

1. Two days before the lab, streak LB/kan plates with the cloning strain of *E. coli* (such as NovaBlue) containing the pET-41a plasmid.
2. The morning of the day before the lab, pick a single colony from the LB/kan plate and inoculate a starter culture of 2–5 mL LB medium containing kanamycin. Incubate for ~8 hours at 37°C with vigorous shaking (~300 rpm).
3. The afternoon of the day before the lab, dilute the starter culture 1:500 into 100 mL selective LB/kan medium and shake at 37°C in a 500 mL flask.

Have available for prelab setup:

- LB broth containing 30 µg/mL kanamycin
- LB agar containing 30 µg/mL kanamycin
- *E. coli* strain NovaBlue (or other K12 strain) containing pET-41a

The students will use these supplies and equipment:

- QIAGEN QIAprep Spin Miniprep Kit (catalog number 27104)

The listed supplies and reagents are included in the QIAGEN QIAprep Spin Miniprep Kit. Each kit contains columns and reagents for 50 plasmid preps. (*Note*: Catalog number 21706 supplies reagents for 250 minipreps.) Be sure to follow the kit's protocol in adding the provided RNase to Buffer P1 and 100% ethanol to Buffer PE.

Aliquot per group:

- 275 µL Buffer P1, store at 4°C
- 275 µL Buffer P2
- 385 µL Buffer N3
- 550 µL Buffer PB
- 800 µL Buffer PE
- 60 µL Buffer EB
- QIAprep spin column

Aliquot per group for the restriction digest:

- 5.5 µL 10× CutSmart restriction enzyme buffer [supplied with purchase of restriction enzymes from New England Biolabs (NEB)], store at −20°C
- 55 µL sterile dH_2O

Students will also need both NcoI-HF (NEB catalog number R3193) and NotI-HF (NEB catalog number R3189) enzymes. For each reaction, students need 10 units of enzyme, which is 0.5 µL of each enzyme for the NEB enzymes. This is a very small amount to aliquot for each group, so we recommend creating a shared aliquot for all student groups and placing on ice in the lab for students to work from.

Students will purify their own pET-41a, but have extra DNA on hand just in case.

LAB SESSION 3

Completion of Vector Preparation and Polymerase Chain Reaction Amplification of *egfp*

The listed supplies and reagents are included in the QIAGEN QIAquick PCR Purification Kit (catalog number 28104). Each kit contains columns and reagents for 50 DNA "clean-ups."

Aliquot the listed amounts for student use:

- 275 µL Buffer PB
- 800 µL Buffer PE
- 55 µL Buffer EB
- 1 QIAquick column and 2 mL collection tube.

Students will use the listed running their agarose gel. Each group needs:

- 30 µL GelRed stock (10,000×) (see recipe in "Supplies and Reagents for General Use"). Students should save for future labs
- 1 mL DNA loading buffer (aka DNA loading dye). This is supplied with the purchase of HF restriction enzymes from NEB as a 6× solution and can also be purchased separately, if necessary (NEB catalog number B7024). Students should save for future labs
- 15 µL NEB 1 kb DNA ladder (NEB catalog number N0552S). Alternatively, NEB catalog number N3232 can be used, but it needs to be prepared with DNA loading buffer before use. See recipe below for preparing 50 ng/µL stock solution with N3232. Students should save for future labs
- Recipe for 50 ng/µL stock solution of 1 kb NEB ladder: 100 µL N3232 + 733 µL dH_2O + 167 µL 6× DNA loading buffer

Students will share these items:

- Agarose (Bio-Rad catalog number 1613101)
- 1× TBE buffer (see recipe in "Supplies and Reagents for General Use")

The students will use the listed supplies for their PCR reaction. Each group needs:

- 50 µL Sterile dH_2O
- 5.5 µL 10× Thermopol reaction buffer (supplied with purchase of Vent polymerase or purchased individually—NEB catalog number B9004)
- 100 ng pEGFP-N1 plasmid DNA (i.e., 2 µL of 50 ng/µL DNA)
- 5.5 µL dNTP mix (available from many suppliers, such as NEB catalog number N0447), store at −20°C
- 1.2 µL egfpNco primer (AAACCATGGTGAGCAAGGGCGA; 100 pmol/µL in sterile dH_2O or 1× TE buffer), store at −20°C
- 1.2 µL egfpNot primer (AAAGCGGCCGCTTTACTTGTACA; 100 pmol/µL in sterile dH_2O or 1× TE buffer), store at −20°C
- 0.55 µL Vent polymerase, store at −20°C. NEB catalog number M0254
- PCR tube

LAB SESSION 4

Preparation of Insert DNA (*egfp*) PCR Product

The students will use the listed supplies. The supplies and reagents are included in the QIAGEN QIAquick PCR Purification Kit (catalog number 28104).

Aliquot per group:

- 550 µL Buffer PB
- 1.75 mL Buffer PE

- 110 µL Buffer EB (also use for blanking the Nanodrop)
- two QIAquick columns and two 2 mL collection tubes

Aliquot per group for the restriction digest:

- 5.5 µL 10× CutSmart restriction enzyme buffer (supplied with purchase of restriction enzymes from NEB), store at −20°C
- 55 µL Sterile dH_2O

Students will also need both NcoI-HF (NEB catalog number R3193) and NotI-HF (NEB catalog number R3189) enzymes. For each reaction, students need 10 units of enzyme, which is 0.5 µL of each enzyme for the NEB enzymes. This is a very small amount to aliquot for each group, so we recommend creating a shared aliquot for all student groups and placing on ice in the lab for students to work from.

Students will share reagents:

- Agarose
- 1× TBE buffer
- GelRed stock (10,000×)

LAB SESSION 5

DNA Ligation and Transformation of *Escherichia coli*

The students will use these supplies. Aliquot per group:

- 2.5 µL 10× Ligase buffer, store at −20°C
- 1.5 µL T4 DNA ligase, store at −20°C

Note: This protocol uses NEB T4 DNA ligase, catalog number M0202L. If using a different brand of T4 DNA ligase, follow the manufacturer's instructions. The incubation time may be significantly longer.

- 175 µL Competent *E. coli* BL21(DE3) cells (available from numerous manufacturers, such as NEB catalog number C2527)
- Alternatively, you can make your own chemically competent *E. coli* cells using commercial transformation kits such as *Mix and Go!* (Zymo Research, catalog number T3001).

Note: Thaw frozen competent cells *immediately* before use on ice. Mix gently before aliquoting. Cells should not sit on ice for more than a few minutes.

- 3 mL µL LB or SOC broth. SOC broth is supplied with the NEB cells and can also be made or purchased separately (NEB catalog number B9020S)
- Three LB/kan plates

Students will share reagents:

- Agarose (if they don't have solidified 1% agarose from previous labs to melt and use)
- 95% Ethanol for plate spreading
- Plate spreader
- Sterile dH_2O
- 1× TBE

Students should have from previous labs (but have extra on hand, just in case):

- Uncut (aka undigested) pET-41a
- ~100 ng Purified digested egfp insert (Lab Session 4 d)
- 1 kb DNA ladder
- 6× DNA loading buffer
- GelRed stock (10,000×)

LAB SESSION 6A

Interim Laboratory Session

Counting Transformants, Replica Plating, and Inoculating Overnight Cultures

- *Note to prep staff*: The afternoon of the day before meeting, streak out pBIT in host strain BL21(DE3) (positive control) and pET-41a (negative control) on several LB/kan plates. Streak in quadrants in order to ensure isolated colonies. Students will use this as their positive and negative controls, respectively. Having multiple plates facilitates sharing in larger classes. Students who did not obtain any transformants on their plates can use extra transformants from their classmates' plates or you may create a plate of "mixed unknowns" by plating both pBIT and pET-41a in host strain BL21(DE3) on the same plate.

The students will need these supplies per group:

- 1 LB/kan plate
- 7 Snap-cap culture tubes, each containing ~5 mL of LB/kan
- Grid sticker (Diversified Biotech PetriSticker 32-square grid, catalog number PSTK-1000)
- Sterile toothpicks or pipette tips
- Sharpies (or other marker pens)
- Biohazard waste container (for toothpicks used to pick colonies)
- Students' transformation plates from Lab Session 5.
- Parafilm (Bemis catalog number PM999)

Note to prep staff:

1. Place students' culture tubes and replica plates in the refrigerator (lids down) until the afternoon prior to the lab session.
2. The afternoon prior to the lab session, place students' replica plates (lids down) in the 37°C incubator and the culture tubes in a shaking incubator at 37°C.

LAB SESSION 6B

Isolation of Miniprep DNA From Potential Transformants

The listed supplies and reagents are included in the QIAGEN QIAprep Spin Miniprep Kit (catalog number 27104). Each kit contains columns and reagents for 50 plasmid preps. (*Note*: Catalog number 21706 supplies reagents for 250 minipreps.) Be sure to follow the kit's protocol in adding the provided RNase to Buffer P1 and 100% ethanol to Buffer PE.

Needed per group:

- 2 mL Buffer P1, store at 4°C
- 2 mL Buffer P2
- 2.8 mL Buffer N3
- 4 mL Buffer PB
- 6 mL Buffer PE
- 400 μL Buffer EB
- Seven QIAprep spin columns

LAB SESSION 6C

Visualization of Green Fluorescent Protein, Part I

Each student group will use:

- 1 LB/kan/IPTG plate
- 1 Grid sticker

Students will share:

- Parafilm

LAB SESSION 7A

Restriction Digestion of DNA

Each student group will use:

- 16 μL 10× CutSmart buffer
- Students will also need 16 units each of both NcoI-HF and NotI-HF enzymes, which is 0.8 μL each. This is a very small amount to aliquot for each group, so we recommend creating a shared aliquot for all student groups and placing it on ice in the lab for students to work from.
- 100 μL Sterile dH_2O

LAB SESSION 7B

Analysis of Miniprep Digestions Using Agarose Gel Electrophoresis

Students will share:

- Agarose (if they don't have solidified 1% agarose from previous labs to melt and use)
- 1× TBE buffer

Students should have from before (but have extra on hand, just in case):

- 1 kb DNA ladder
- 6× DNA loading buffer
- GelRed stock (10,000×)

Note: For this lab, each student group needs either two gel apparatuses with one comb each, or one gel apparatus with two combs, one placed at the top and the other placed at the midpoint.

LAB SESSION 7C

Students will share a UV transilluminator. Be sure students have appropriate safety precautions (eye/skin protection) for UV light exposure.

LAB SESSION 7D

Analysis of DNA Sequence From a Positive Clone, Part I

Note to prep staff:

Dilute the pBIT sequencing primer (CGAACGCCAGCACATGGACAGC) at a stock concentration of 100 μM using sterile dH_2O or 1× TE. Most companies will include already-calculated volumes to resuspend the nucleic acid with to achieve a 100 μM concentration. Create a 10 μM working solution of primer by diluting the 100 μM stock solution 1:10 in sterile dH_2O.

Each student group will need:

- 3 μL of 10 μM pBIT sequencing primer
- 1 PCR tube
- Undigested plasmid DNA from minipreps in Lab Session 6b
- Sterile dH_2O

LAB SESSION 8

Analysis of DNA Sequence From a Positive Clone, Part II

Bioinformatics exercise where students need data from their sequencing reaction from Lab Session 7 d and a computer with software to analyze a chromatogram, i.e., Chromas (http://technelysium.com.au/wp/chromas/).

LAB SESSION 9A

Interim Laboratory Session

Inoculate Culture for SDS-PAGE

The students will use:

- Four snap-cap tubes with 1 mL LB/kan
- Racks for tubes
- Sterile toothpicks or pipette tips
- Sharpie markers
- Master replica plate

LAB SESSION 9B

Expression of Fusion Protein From Positive Clones, SDS–PAGE, and Western Blot: Part I

Note to prep staff:

1. One day before the lab, incubate the students' LB/kan broth cultures overnight at 37°C
2. Three to five hours before the start of the lab, add 2 mL 2× YT/kan/IPTG to each 1 mL student culture and continue to shake at 37°C

Prep staff use:

- 2× YT broth with kan and IPTG (Recipe for 2× YT broth is provided or purchase premixed 2× YT, such as from Growcells MBPE5032)

The students will use the listed supplies. Each group needs:

- Student cultures that have been incubated in 2× YT/kan/IPTG 3–4 hours before the lab
- 0.9 mL B-PER complete bacterial protein extraction reagent which includes DNase and lysozyme (Thermo Fisher Scientific catalog number 89822)
- 1 Any kD SDS polyacrylamide gel, such as Bio-Rad catalog numbers 4569033 (box of 10 gels). These gels work with protein samples in the range 10–200 kD
  - 0.25 mL 6× Laemmli SDS–PAGE sample (loading) buffer (Bioworld catalog number 105700212). Students should save extra 6× SDS–PAGE loading buffer for future labs
  - 10 μL Molecular weight ladder (Precision Plus Protein Kaleidoscope Prestained Protein Standards, catalog number 1610375)
  - Two pieces Whatman 3MM paper (GE Healthcare 3030221), cut to the size of the fiber pads
  - One piece nitrocellulose (Thermo Scientific catalog number 88014), cut to the size of the gels. (Alternatively, Bio-Rad sells precut filter paper/nitrocellulose sandwiches for western blotting; catalog number 1620235)
- Two fiber pads (NOT disposable)
- One glass casserole dish

These are to be shared:

- 1× Tris–glycine–SDS running buffer (follow recipe or there are many options to purchase 10× stock solution, such as Bio-Rad catalog number 1610772 or Bioworld catalog number 214200232)
- Sodium bicarbonate
- One pack of disposable square Petri dishes (Fisher 08-757-11 A)
- One pack of glass tubes or disposable pipettes for use as "rolling pins"

RECIPES

2 × YT Broth

The recipe for 2× YT broth is provided in .

2xYT Recipe

Component	Amount Needed for 20 Groups
Bacto tryptone	16 g
Yeast extract	10 g
NaCl	5 g
Deionized water	To 1 L

No Permission Required.

Autoclave and store at room temperature.

2 × YT/Kan/IPTG

To 100 mL 2× YT, add 100 µL kanamycin (30 mg/mL stock) and 100 µL IPTG (20 mg/mL stock). The final concentration of kanamycin is 30 µg/mL, and the final concentration of IPTG is and 20 µg/mL or 0.08 mM.

Tris–Glycine–SDS Running Buffer

Recipes for 1× and 10x stock solutions are provided in.

Tris-Glycine-SDS Running Buffer Recipe

	Amount		Final Concentration	
	1×	10×	1×	10×
Tris base	6.0 g	30 g	25 mM	0.25 M
Glycine	28.8 g	141 g	190 mM	1.9 M
SDS	2.0 g	10 g	0.1%	1%
Deionized water	To 2000 mL	To 1000 mL		

No Permission Required.

Place carboys of 1× running buffer in the lab. It is a good idea to keep extra 10× stock on hand.

LAB SESSION 10

Expression of Fusion Protein From Positive Clones, SDS–PAGE and Western Blot: Part II

The students will use the listed supplies. Each group needs:

- 10 mL blocking solution [TBS-T plus 5% (w/v) powdered milk], store at 4°C. Nonfat powdered milk can be purchased through scientific vendors (LabScientific catalog number M0841) but is available at most grocery stores. BSA Fraction V (Fisher Scientific catalog number BP1600-100) can be substituted for nonfat powdered milk.
- 4 µL α-EGFP antibody [Living Colors A.v. monoclonal antibody (JL-8) Takara Bio catalog number 632381], store at −20°C, and limit freeze–thaw cycles
- 8 µL goat anti-mouse antibody conjugated to peroxidase (GAMP) (Sigma-Aldrich catalog number A3673), store at 4°C
- One LB/kan plate
- Sterile toothpicks or sterile pipette tips

These are to be shared:

- Ponceau Red (also referred to as Ponceau S) stain (Sigma-Aldrich catalog number P7170), ~10 mL needed per group (*Ponceau Red stain may be saved and reused*)

- One pack of disposable square Petri dishes (Fisher catalog number 08-757-11A)
- At least 250 mL per group of TBS-T (see Table 7 below)
- *Optional:* Microcentrifuge tube cap locks (Fisher catalog number NC9679153)

For detection of the bands on the gel, you will need to do *either* colorimetric detection *or* enhanced chemiluminescence.

Colorimetric detection:

- 10 mL Peroxide stain (made fresh from chloronaphthol stock; see Recipes) (*30% hydrogen peroxide should be less than 1 month old*)

Enhanced chemiluminescence:

- 3.5 mL of Western Peroxide Reagent
- 3.5 mL of Western Luminol/Enhancer Reagent (both peroxide and luminol reagents sold together as Bio-Rad Clarity ECL reagent, catalog number 1705060)

RECIPES

Tris-Buffered Saline (pH 7.4)

Tris-Buffered Saline (TBS) Recipe

	Amount	Final Concentration
Tris base	2.42 g	20 mM
NaCl	8 g	137 mM
HCl (1 M), to pH 7.4	3.8 mL	10%
Deionized water	To 1 L	

No Permission Required.

The recipe for Tris-buffered saline is found in.

Alternatively, 10× TBS (Tris-buffered saline) solution can be bought from a number of sources, such as Fisher Scientific, catalog number BP24711.

TBS-T: Add Tween 20 to TBS to a final concentration of 0.1%. Prepare a large amount for common use.

Blocking solution: TBS-T plus 5% nonfat milk powder (or BSA fraction V).

Chloronaphthol Stock Solution

This is used to make the peroxide stain. Wear gloves as chloronaphthol is a suspected carcinogen. Combine the listed compounds and store at −20°C in a bottle protected from light:

Chloronaphthol Stock Solution Recipe

	Amount	Final Concentration
4-Chloro-1-naphthol	150 mg	(0.3% final concentration)
Ice-cold methanol	50 mL	

No Permission Required.

Peroxide Stain

This should be made fresh immediately before use. Purchase new hydrogen peroxide each semester.

Peroxide Stain Recipe

	Amount Added	Final Concentration
Chloronaphthol stock	17 mL	0.05%
Deionized water	680 mL	
Tris (pH 7.4), 1 M	5 mL	50 mM
NaCl, 1 M	10 mL	100 mM
H_2O_2,[a] 30%	35 µL	0.02%

No Permission Required.

LAB SESSION 11A

Interim Laboratory Session

Inoculate Cultures for Protein Purification

The students will use the listed supplies. Each group needs:

- Sterile toothpicks or pipette tips
- 1 Snap-cap culture tube with 2 mL LB/kan
- Sharpie marker
- Students' replica plates of positive clone (streaked last lab session)
- Rack for culture tubes

LAB SESSION 11B

Extraction of Recombinant Protein From *Escherichia coli* Using a Glutathione Affinity Column

Note to prep staff:

1. The afternoon of the day before the lab: incubate students' cultures (that were stored in the refrigerator after the interim lab) overnight shaking at 37°C. *Also inoculate and incubate E. coli strain BL21(DE3) harboring the pBIT plasmid. This can be used as a backup for purification if students accidentally spill their own cultures.*
2. Three to five hours prior to the start of the lab: subculture students' 2 mL cultures and backup culture into 250 mL flasks containing 100 mL 2× YT/kan/IPTG each. Incubate with shaking at 37°C.

The students will use the listed supplies. Each group needs:

- 2× 50 mL Centrifuge tubes capable of withstanding 10,000 × *g* force
- 50 or 100 mL 2× YT/kan/IPTG (in 250 mL flasks for prep staff to use for subculturing)

For purification using the Thermo Scientific Pierce GST Spin Purification Kit, 0.2 mL (Thermo Scientific Catalog Number 16106), each group will need:

- 1 spin column with collection tube
- 1.8 mL of Equilibration/Wash buffer (provided in kit)
- 700 µL of 1X elution buffer
- 10X elution buffer (100 mM reduced glutathione) is prepared by adding 6 mL of Equilibration/Wash Buffer to the vial of reduced glutathione (184 mg) provided. This stock solution can be stored at −20°C for up to 18 months. Fresh 1X (10 mM reduced glutathione) aliquots of Elution Buffer should be made immediately prior to use.

For purification using gravity chromatography with bulk resin, each group will need:

- Glutathione Sepharose resin: 0.75 mL settled bead volume in 15 mL conical tube (GE Healthcare catalog number 17513201), store at 4°C until use

- 1 Chromatography column and stopcock (Bio-Rad catalog numbers 7311550 and 732-8102)
- 20 mL 1× PBS (phosphate-buffered saline). See recipe or purchase 10× stock (Fisher Scientific BP 3991)
- 3.5 mL 10 mM Reduced glutathione diluted with 1× PBS (Sigma-Aldrich catalog number G4251)

In addition, students will need reagents for cell lysis which can be done two different ways:

- *Sonication*
  - 800 µL 1× PBS with Pefabloc, 4-(2-aminoethyl)benzenesulfonyl fluoride hydrochloride, AEBSF (Sigma-Aldrich catalog number 76307), 0.1 M stock solution (prepare stock solution with sterile, deionized water, and freeze at −20°C)
  - 20 µL Lysozyme (10 mg/mL), store at −20°C (Spectrum Chemical catalog number L3008-1GM)
  - Dry ice (for freeze–thawing)
- *Chemical lysis*
  - B-PER complete bacterial protein extraction reagent which includes DNase and lysozyme. This is the same reagent used in Lab Session 9 (Thermo Fisher Scientific catalog number 89822)
  - Pefabloc, AEBSF (Sigma-Aldrich catalog number 76307), 0.1 M stock solution (prepare stock solution with sterile, deionized water, and freeze at −20°C)

Note: Including the protease inhibitor (Pefabloc) reduces protease activity, preventing degradation of the fusion protein. When omitted, most of the purified product is still full-length, but some degradation does occur.

These are to be shared:

Sterile LB for resuspending bacteria (a few small bottles per lab or one 1.3 mL tube per group)

A refrigerated microcentrifuge or place several microcentrifuges in a refrigerator at 4°C

RECIPES

Phosphate-Buffered Saline, pH 7.4

Phosphate-Buffered Saline (PBS) Recipe

	Amount Added (g)	Final Concentration (mM)
NaCl	8	137
KCl	0.2	2.7
Na_2HPO_4	1.44	8.0
KH_2PO_4	0.24	2.0

No Permission Required.

The recipe for 1x phosphate-buffered saline is found in.

Dissolve the above in 800 mL of deionized water, adjust pH to 7.4 using HCl, and then bring the total volume to 1 L. Sterilize by autoclaving.

Phosphate-Buffered Saline With Pefabloc

Add 1 mL of the 0.1 M Pefabloc stock solution to 50 mL 1× PBS. Mix and keep at 4°C until used.

Phosphate-Buffered Saline With 10 mM Reduced Glutathione (GST Elution Buffer)

Add 154 mg of reduced glutathione to 50 mL 1× PBS. Mix well and use immediately. Make fresh each use. (Alternatively, GST elution buffer can be made from the reduced glutathione and Equilibration/Wash Buffer provided in the GST spin purification kit.)

LAB SESSION 12A

Analysis of Purification Fractions

All students should perform an SDS–PAGE analysis. Students should also quantify purified protein using fluorescence or Bradford assay. Only students continuing to Lab Session 13 (Purification of Total mRNA) need to perform a replica plate.

Students should have from the previous week:

Cell lysate and fractions collected from purification

SDS–PAGE

- 1 Any kD SDS polyacrylamide stain-free gel (see Lab Session 9)
- 5 μL Kaleidoscope prestained molecular weight marker for SDS–PAGE
- 1× Tris–glycine running buffer, recipe as described in Lab Session 9
- 6× SDS–PAGE sample buffer (Bioworld catalog number 105700212)
- 20 mL Gelcode Blue staining reagent (Pierce catalog number 24594), store at 4°C; order fresh every semester (if using regular or homemade SDS–PAGE gels; not used with stain-free gels).
- Square Petri dish or other tray for gel staining

Fluorescence Analysis

- 7 mL 1× PBS, pH 7.4
- 3.2 μL of recombinant EGFP: 1 mg/mL solution in PBS (Fisher Scientific catalog number NC0567714)
- 96-well plate

Bradford Protein Analysis

Quick Start Bradford Protein Assay Kit 4 (Bio-Rad catalog number 5000204)

Test tubes or disposable cuvettes

LAB SESSION 12B

Replica Plating (Only If Performing Lab Session 13)

1 LB/kan plate

LAB SESSION 13A

Interim Laboratory Session

Inoculate Cultures for RNA Purification

The students will use the listed supplies. Each group needs:

- Three empty snap-cap tube
- One snap-cap tube with 3 mL of LB/kan
- Sterile toothpicks
- Tube racks
- Sharpies (or other marker pens)
- Access to their replica plate from Lab Session 12b

Note to prep staff:

1. Place students' empty tubes and inoculated tubes in the 4°C refrigerator until the afternoon of the day before the next lab session
2. In the afternoon of the day before the lab session, place the 3 mL inoculated tubes in the shaking (~250 rpm) incubator at 37°C

LAB SESSION 13B

Total RNA Purification

Note to prep staff:

1. Two hours before the beginning of the lab session, induce each 3 mL culture as described in Lab Session 13b, steps 1–3.
2. Every 1 or 2 groups will need one box of barrier tips in each size: P1000, P200, P20, and P10. These tips are wrapped and certified DNase and RNase-free. They should not be autoclaved. These tips will be used in Lab Sessions 13, 14, 16, 20, and 21b and any other time you might be working with RNA samples.

   Treat benches and pipettes with a 10% bleach solution to hydrolyze contaminating plasmid DNA from previous experiments. Wipe to dry with a paper towel. This treatment aims to hydrolyze DNA that may contaminate subsequent amplifications. It can become problematic, especially when working over multiple semesters with plasmid DNA in the same area as PCR setup. Plasmid DNA contamination is minimized with DNase treatment as written in the protocol, but without clean PCR reaction setup areas in a teaching lab, bleach treatment will also serve to help minimize this unwanted carryover.
3. It is recommended that students wipe down bench tops and micropipettes with RNase*ZAP* (Ambion AM9780) immediately prior to starting Lab Sessions 13, 14, 16, and 21b.

RECIPES

Lactose (20 mg/mL)

Dissolve 0.1 g α-lactose monohydrate in 5 mL deionized water. Filter-sterilize. Store at 4°C until needed. Make fresh each time.

2 × YT/Kan/IPTG

To 100 mL 2× YT, add 100 μL kanamycin (30 mg/mL stock) and 100 μL IPTG (20 mg/mL stock). *The final concentration of IPTG is 20 μg/mL or 0.08 mM.*

2 × YT/Kan/Lac

To 100 mL 2× YT, add 100 μL kanamycin (30 mg/mL stock) and 144 μL lactose (20 mg/mL stock). *The final concentration of lactose is 28.8 μg/mL or 0.08 mM.*

Lysozyme (40 mg/mL)

Dissolve 0.04 g lysozyme in 1 mL deionized water. This is a 100× stock. Make 10 μL aliquots, at least one aliquot per group. Each group will only require 5 μL. Store the aliquots and remaining solution at −20°C.

DNaseI (2.73 Kunitz Units/L)

Add 550 μL RNase-free water provided in QIAGEN DNaseI set to the vial. Cap and invert to mix. Make 35 μL aliquots. Store at −20°C for up to 9 months. These aliquots are for once-off use; do not refreeze after thawing.

The students will use the listed supplies. Each group needs:

10 μL 100× Lysozyme stock

- QIAGEN RNeasy Plant Kit (QIAGEN, catalog number 74904):
  - 2.3 mL Buffer RLT (containing 1% β-mercaptoethanol)
  - 3 QIAshredder spin columns (lilac)
  - 1.7 mL 100% Ethanol
  - 6 RNeasy spin columns (pink)
  - 2.3 mL Buffer RW1
  - 7 mL Buffer RPE
  - 350 μL RNase-free water
- QIAGEN DNaseI set (QIAGEN, catalog number 79254):
  - 35 μL Aliquot of prepared DNaseI

- 35 μL Buffer RDD
- 100 μL RNase-free water

Students will share:

- Extra RNase-free water from QIAGEN RNeasy kit for blanking the Nanodrop

LAB SESSION 14

Analysis of *gst::egfp* mRNA Levels by RT-qPCR: Part I

Note to prep staff:

Instructors should treat micropipettes and lab bench surfaces with a 10% bleach solution prior to performing qPCR. This treatment aims to hydrolyze DNA that may contaminate subsequent amplifications. It can become problematic, especially when working over multiple semesters with plasmid DNA in the same area as qPCR setup. Plasmid DNA contamination is minimized with DNase treatment as written in the protocol, but without clean PCR reaction setup areas in a teaching lab, bleach treatment will also serve to help minimize this unwanted carryover.

The students will use the listed supplies. Each group needs:

- Bio-Rad iScript Select cDNA Synthesis Kit (Bio-Rad, catalog number 1708890):
  - 18 μL random primer mix
  - 4.5 μL iScript reverse transcriptase
  - 35 μL 5× iScript select reaction mix
  - 50 μL Nuclease-free water
- 18 total RNase-free PCR tubes. 6 tubes are needed for reverse transcription, 12 tubes are needed for qPCR (Bio-Rad catalog number TBS-0201)
- 3 optical flat 8-cap strips, for 0.2 mL tubes and plates, ultraclear, 120 (Bio-Rad TCS-0803) *Note*: Please choose the appropriate tubes/plates for your real-time capable thermocycler
- 6 μL aliquot of 10 pmol/μL mix of *egfp* primers:
  - *egfp* 16F: CGACGGCAACTACAAGACC
  - *egfp* 16R: GTCCTCCTTGAAGTCGATGC

Resuspend each original, lyophilized primer in sterile dH_2O to a concentration of 100 pmol/μL. Then combine 50 μL of the forward primer with 50 μL of the reverse primer, plus 400 μL water and vortex before aliquoting for students.

- 6 μL aliquot of 10 pmol/μL stock mix of *23S* primers:
  - *23S* 16F: GACGGAGAAGGCTATGTTGG
  - *23S* 16R: GTTGCTTCAGCACCGTAGTG
- 195 μL Bio-Rad iQ SYBR Green Supermix
- Sterile dH_2O

LAB SESSION 15

Analysis of *gst::egfp* mRNA Levels by RT-qPCR: Part II

Analysis of RT-qPCR data will require the.odm file saved from Lab Session 14. Each station will also need access to a computer. If student computers do not have Bio-Rad MyiQ software (or other instrument-specific software), instructors can complete steps 1–17 before the lab session and have students work from a Microsoft Excel file.

LAB SESSION 16

Analysis of *gst::egfp* mRNA Levels by Semiquantitative RT-PCR: Part I

The students will use the listed supplies. Each group needs:

- Bio-Rad iScript Select cDNA Synthesis Kit (Bio-Rad, catalog number 1708890):
  - 18 μL random primer mix
  - 4.5 μL iScript reverse transcriptase

    - 35 µL 5× iScript select reaction mix
    - 50 µL Nuclease-free water
- 18 total RNase-free PCR tubes: 6 tubes needed for reverse transcription (part I), 12 tubes needed for PCR (part II)
- 3 optical flat 8-cap strips, for 0.2 mL tubes and plates, ultraclear, 120 (Bio-Rad TCS-0803)
- 4 µL of 10 pmol/µL stock mix of *egfp* primers:
    - *egfp* 16F CCTGAAGTTCATCTGCACCA
    - *egfp* 16R TGCTCAGGTAGTGGTTGTCG
- 4 µL of 10 pmol/µL stock mix of *23S* primers:
    - *23S* 16F ACTGCGAATACCGGAGAATG
    - *23S* 16R CCTGTTTCCCATCGACTACG
- 40 µL 10× Taq buffer
- 8 µL dNTP mix (stock of 10 mM each dNTP)
- 3 µL Taq (5 U/µL)
- Sterile dH_2O

LAB SESSION 17

Analysis of *gst::egfp* mRNA Levels by Semiquantitative RT-PCR: Part II

The students will use the listed supplies. Students should have from before:

- 1 kb DNA ladder
- 6× DNA loading buffer
- GelRed (10,000× stock)
- PCR samples from Lab Session 16

Students will share:

- Agarose
- 1× TBE buffer

Analysis of semiquantitative RT-PCR data can be performed using NIH ImageJ software (or other imaging software), if available. Each student should download NIH ImageJ from the National Institutes of Health (https://imagej.net/ij/download.html). Quantification can be performed using Microsoft Excel.

LAB SESSION 18

Culturing Mammalian Cells

Note to prep staff:

1. ~1–2 weeks prior to the start of lab, thaw and culture HEK293 cells in complete medium.
2. All cell culture experiments will require each student group to receive one box of sterile, autoclaved tips in each size: P1000, P200, P20, and P10. These tips will be used for the cell culture experiments delineated in Lab Sessions 18–30. Therefore, they should always be handled sterilely and only used for mammalian cell culture in the biological safety cabinet. Have students label their items for personal use to decrease potential contamination issues.
3. Every biological safety cabinet should have a dedicated spray bottle of 70% EtOH, BIOHAZARDOUS waste liquid disposal system, tube rack, and P1000, P200, P20, and P10 micropipettes as well as a Sharpie marker.
4. Any reagent or plasticware used for Lab Sessions 18–30 should be sterile and handled sterilely.
5. Sterile cell culture reagents, such as medium, 1× PBS, trypsin solution, etc., are aliquoted per lab session for each student group. They should be saved if at all possible to help reduce costs. It is recommended to remind students to only handle their own reagents in a sterile manner and only open them in a sterile environment, as they will use them for multiple lab sessions.

Note: If you wish to purchase stable HEK293 cells expressing GFP (not EGFP), they can be purchased from Cell BioLabs (catalog number AKR-200). These cells can be cultured in D-MEM (high glucose), 10% fetal bovine serum (FBS), 0.1 mM MEM NonEssential Amino Acids (NEAA), 2 mM L-glutamine, 1% Pen-Strep, 10 µg/mL Blasticidin.

RECIPE

Complete Medium

High glucose Dulbecco's Modified Eagle's Medium (DMEM) containing l-glutamine (GlutaMAX-I), sodium pyruvate, and high glucose (4500 mg/L) (Gibco catalog number 10569010; 500 mL bottle)

Take one 500 mL bottle of high glucose DMEM containing l-glutamine and sodium pyruvate and remove 50 mL. Then, add to a final concentration these components:

- 10% Fetal bovine serum (Gibco catalog number 16000044; 500 mL bottle)
- mM MEM nonessential amino acids (100×) (Gibco catalog number 11140050; 100 mL bottle)
- 1% Penicillin-streptomycin (5000 U/mL) (Gibco catalog number 15070063; 100 mL bottle)
- 1500 mg/L Sodium bicarbonate (Gibco catalog number 25080094)

1× PBS, pH 7.4 autoclave to sterilize [alternatively sterile 1× PBS (aka DPBS)] pH 7.4 can be purchased (Gibco catalog number 14190136; 1 L bottle).

Cell Culture Preparation

Thaw HEK293 cells (American Type Culture Collection catalog number CRL-1573). Follow the manufacturer's instructions for culturing, maintaining, and storing frozen stocks. When cells reach ~80%–90% confluency, passage them at a dilution of 1:2. Expand your cells until you have an appropriate amount for each student group to receive a 10 cm^2 or similar cell culture certified vessel at a cellular confluency of 5×10^6 cells for Lab Session 18. You will want to continue to maintain a T75 vessel of parental HEK293 cells throughout Lab Sessions 18–30 as a backup. This will require passaging the cells ~1:5 every 3–4 days

The students will use the listed supplies. Each group needs:

- ≥50% Confluent 10 cm^2 dish of HEK293 cells
- A spray bottle with 70% EtOH
- 50 mL of sterile 1× PBS; saved and used for subsequent lab sessions; store at room temperature
- Sterile tin (tin sterilizing boxes; Fisher Scientific catalog number 03-475-1) containing glass Pasteur pipettes; saved and used for subsequent lab sessions
- 10 mL of 0.25% Trypsin–EDTA solution (Gibco catalog number 25200056; saved and used for subsequent lab sessions), store at 4°C
- 50 mL of complete medium, store at 4°C
- 6-well sterile cell culture plate
- 50 mL sterile conical tube
- Five each of: 5, 10 and 25 mL wrapped serological pipettes

Students will share:

- Hemacytometer
- Cover slips
- Hand counter
- Brightfield microscope

LAB SESSION 19A

Interim Laboratory Session

Transient Transfection of pEGFP-N1 in HEK293 Cells Using FuGENE HD

The students will use the listed supplies. Each group needs:

- 6-well plate of HEK293 cells (from Lab Session 18)
- Sterile pipette tips, P1000, P200, P20, P10
- 5 µL of a 1 µg/µL pEGFP-N1 plasmid, store at room temperature or 4°C
- 5 µL of a 1 µg/µL pcDNA3 plasmid, store at room temperature or 4°C
- 500 µL Opti-MEM (Thermo Fisher, catalog number 31985062), store at 4°C
- 4 Sterile microcentrifuge tubes
- Sharpie marker

Students will share:

- Brightfield microscope
- FuGENE HD transfection reagent (Promega, catalog number E2311)

Note to prep staff: FuGENE HD cannot be aliquoted; it must remain in the original glass vial that it ships in. Therefore, we recommend that the instructor or TA hands the vial to each group as they need it in the biological safety cabinets while the vial remains tightly capped.

LAB SESSION 19B

Analysis of EGFP Expression Using Fluorescence Microscopy

Note to prep staff:

This session uses NIH ImageJ (https://imagej.net/ij/download.html) and we recommend students download this on their personal computers either before or at the start of lab. There are no required reagents to prepare.

LAB SESSION 19C

Creating Stable EGFP-Expressing HEK293 Cells

Note to prep staff:

We recommend that the kill curve should be completed two weeks prior to the start of Lab Session 18 to allow adequate time for selection to take place. Instructors should trypsinize and plate parental HEK293 cells 24–48 hours prior to Lab Session 19c at a cell density of ~3–4 × 10^4 cells/well of a 6-well plate to achieve 20%–30% confluency. Each well should contain a total of 2 mL complete medium.

Performing a Kill Curve

The students will use the listed supplies. Each group needs:

- 6-well plate of parental HEK293 cells at a 20%–30% confluency
- 80 µL of a 50 mg/mL G418 solution (aka geneticin) (Gibco catalog number 10131035), protect from light and store at 4°C
- 50 mL complete medium, store at 4°C
- 6–15 mL conical tubes
- Sterile pipette tips, P1000, P200, P20, P10
- Five each of: 5, 10, and 25 mL wrapped serological pipettes

Students will share:

- Brightfield microscope

Selecting for Stables

Note to prep staff:

Students will use their HEK293 cells transfected from Lab Session 19a. They will use the appropriate dose of G418 selection determined from the kill curve in Lab Session 19c. Once they have verified EGFP expression, it is recommended to expand the cell culture into a T25, and freeze 1 mL aliquots in liquid nitrogen for long-term storage. If you are planning on proceeding to Lab Session 20 and on, continue to maintain one culture of HEK293 EGFP-expressing cells for use in subsequent experiments. The number of cells to maintain will vary depending on class size.

The students will use the listed supplies. Each group needs:

- 80 µL G418; saved and used for subsequent lab sessions
- 50 mL conical tube
- 50 mL complete medium; saved and used for subsequent lab sessions
- Sterile pipette tips, P1000, P200, P20, P10
- Five each: 5, 10, and 25 mL wrapped serological pipettes

LAB SESSION 20

RNAi-Mediated Knockdown of EGFP: Part I

Note to prep staff:

When working with RNA, remember to use RNase-free pipette tips and tubes. Furthermore, wipe down bench tops, the inside of the biological safety cabinet, and micropipettes with RNase*ZAP* (Ambion AM9780).

It is recommended to order siRNA oligos on a scale of 20 nmole, desalted with a 3′ dTdT overhang (Thermo Fisher Scientific). Using both sterile and RNase-free plasticware and precautions, prepare a working siRNA stock concentration of 100 µM using nuclease-free water (Thermo Fisher Scientific catalog number AM9914G; 1.75 mL/tube) Dilute the stock concentration to 10 µM for immediate use.

To resuspend siRNAs:

1. Briefly centrifuge the tube to ensure dried siRNA is at the bottom of the tube
2. Resuspend the siRNA in sterile, nuclease-free water for final concentration of 100 µM. Vortex well to get all the siRNA into solution
3. Make 10 µM working stock solutions using sterile, nuclease-free water
4. Vortex very well
5. Make aliquots of 25 µL per tube and freeze at −20°C until use. Avoid repeated freeze-thaws

LAB SESSION 21A

RNAi-Mediated Knockdown of EGFP: Part II

Counting and Plating EGFP Expressing Mammalian Cells

Note to prep staff:

Two weeks prior, expand HEK293 cells stably expressing EGFP (Lab Session 19) until you have an appropriate amount for each student group to receive a 10 cm^2 or similar cell culture certified vessel at a cellular confluency of 5 × 10 ^{6} cells for Lab Session 21a.

Students should trypsinize and plate their HEK293 cells stably expressing EGFP 24–48 hours prior to Lab Session 21b at a cell density of 3×10^5 cells/well in a 6-well plate. Each well should contain a total of 2 mL complete medium *without antibiotics* (e.g., penicillin/streptomycin and G418).

The students will use the listed supplies. Each group needs:

- ≥80% Confluent 10 cm^2 dish of HEK293 cells stably expressing EGFP protein (created from Lab Session 19)
- 50 mL of sterile 1× PBS, store at room temperature
- Sterile tin of glass Pasteur pipettes
- Sterile pipette tips P1000, P200, and P20
- 10 mL of 0.25% Trypsin–EDTA solution, store at 4°C
- 50 mL of complete medium *without antibiotics*, store at 4°C
- 6-well sterile cell culture plate
- 50 mL sterile conical tube
- Five each of: 5, 10, and 25 mL wrapped serological pipettes

Students will share:

- Hemacytometer
- Cover slips
- Hand counter

LAB SESSION 21B

Interim Laboratory Session

Transient Transfection of siRNAs

It is recommended that students wipe down bench tops, the inside of the biological safety cabinet, and micropipettes with RNase*ZAP* (Ambion AM9780) immediately prior to starting each of the following labs.

The students will use the listed supplies. Each group needs:

- ≥50% confluent 10 cm^2 dish of HEK293 EGFP cells from Lab Session 21a
- One box of barrier tips in each size: P1000, P200, P20, and P10. These tips are sterile, wrapped, and certified DNase and RNase-free. They do not need to be autoclaved
- 50 mL of sterile 1× PBS; saved and used for subsequent lab sessions; store at room temperature
- Sterile tin of glass Pasteur pipettes; saved and used for subsequent lab sessions
- 10 mL of 0.25% Trypsin–EDTA solution; saved and used for subsequent lab sessions; store at 4°C
- 50 mL of complete medium *without any antibiotics*; store at 4°C
- 50 mL sterile and RNase-free conical tube
- 7 sterile and RNase-free microcentrifuge tubes
- 700 µL Opti-MEM, store at 4°C
- Five each of: 5, 10, and 25 mL wrapped serological pipettes.
- 25 µL of each respective experimental and scrambled control siRNA oligo(s) designed, synthesized, and ordered from Lab Session 20 (resuspended in nuclease-free water at a working concentration of 10 µM); store at −20°C until use
- 5 µL of a 10 µM stock of EGFP positive control siRNA oligo (silencer GFP (eGFP) siRNA from Thermo Fisher Scientific catalog number AM4626). This siRNA comes as 50 M stock solution; dilute at a working concentration of 10 µM with nuclease-free water. Store remainder at −20°C

Note to prep staff: FuGENE HD cannot be aliquoted; it must remain in the original glass vial that it ships in. Therefore, we recommend that the instructor or TA hands the vial to each group as they need it in the biological safety cabinets while the vial remains tightly capped.

LAB SESSION 22

RNAi-Mediated Knockdown of EGFP: Part III

Note to prep staff: Seventy-two hours after transfection with siRNA oligonucleotides targeted to EGFP in HEK293 cells (Lab Session 21b), students will acquire and save brightfield and fluorescence images of each well as previously described in Lab Session 19b.

This session uses NIH ImageJ (https://imagej.net/ij/download.html) and we recommend students download this on their personal computers either before or at the start of lab. There are no required reagents to prepare.

LAB SESSION 23A

CRISPR-Mediated Knockout of EGFP: Part I

Note to prep staff: At least 3 days before Lab Session 23, streak out the following cultures on LB/amp plates: pLentiC-RISPR-E and pLenti-CRISPR-EGFP-2. Plates need to be incubated overnight at 37°C. The next morning, plates should be placed at 4°C until the afternoon. One isolated colony from each plate should be inoculated into 100 mL of LB/amp and grown with shaking at ~250 rpm overnight at 37°C. The following morning, a maxiprep kit should be used to purify plasmid DNA from each culture. These plasmids will need to be quantified and shared among student groups.

Only 1.8×10^5 HEK293 cells stably expressing EGFP are needed per group. In order for student groups to gain experience in these techniques, we recommend one flask or dish be cultured ahead of time for each group. The laboratory exercise and prep list details are therefore written assuming each group has their own flask or dish of cells.

The students will use the listed supplies. Aliquot per group:

- 2.5 mL sterile 1× PBS; store at room temperature or 4°C
- 3 mL 0.25% Trypsin–EDTA; store at 4°C
- 8 mL Complete DMEM *without any antibiotics*; store at 4°C
- Sterile pipette tips, P1000, P200, P20, and P10
- Sterile tin containing glass Pasteur pipettes
- Five 5 mL wrapped serological pipettes
- One sterile 48-well plate
- HEK293 cells stably expressing EGFP (see note above)

Students will share:

- Hemacytometer
- Cover slips
- Hand counter
- Brightfield microscope

LAB SESSION 23B

Interim Laboratory Session

Transfection of CRISPR Vectors

The students will use the listed supplies. Each group needs:

1. 120 μL Opti-MEM or serum-free medium; store at 4°C
2. At least 2 μL of 0.85 μg/μL pLentiCRISPR-E purified plasmid (Addgene plasmid #78852); store at 4°C
3. At least 2 μL of 0.85 μg/μL pLentiCRISPR-EGFP-2 purified plasmid (Addgene plasmid #51761); store at 4°C
4. Sterile pipette tips, P1000, P200, P20, P10
5. 48-well plate of EGFP-expressing HEK293 cells (prepared from Lab Session 23a)
6. 3 Sterile microcentrifuge tubes

Students will share:

1. FuGENE HD reagent
2. Brightfield microscope

Note to prep staff: FuGENE HD cannot be aliquoted; it must remain in the original glass vial that it ships in. Therefore, we recommend that the instructor or TA hands the vial to each group as they need it in the biological safety cabinets while the vial remains tightly capped.

LAB SESSION 23C

Interim Laboratory Session

Expansion of Transfected Cells

Note to prep staff: If it is not feasible to meet with the class for this interim lab, the cells must still be passaged to 6-well plates at this time.

The students will use the listed supplies. Each group needs:

- 400 μL 1× PBS; store at room temperature or 4°C
- 400 μL 0.25% Trypsin–EDTA solution (Gibco catalog number 25200056); store at 4°C
- 8 mL Complete DMEM containing G418 (dose of G418 determined from kill curve in Lab Session 19c); store at 4°C
- Sterile pipette tips, P1000, P200, P20, P10
- Sterile tin containing glass Pasteur pipettes
- 3–5 mL Wrapped serological pipettes
- 1 Sterile 6-well plate

LAB SESSION 24

CRISPR-Mediated Knockout of EGFP: Part II

Note to prep staff:

This session uses NIH ImageJ (https://imagej.net/ij/download.html) and we recommend students download this on their personal computers either before or at the start of lab. There are no required reagents to prepare.

LAB SESSION 25

Advanced CRISPR: Part I

sgRNA design and analysis

Note to prep staff:

This session uses Benchling software (https://www.benchling.com). Students will create a free account or can use one the instructor creates. The steps in Section A can be performed in class or outside of class as homework. Primers designed here are for use in Lab Session 26. Therefore, we recommend ordering them at least one week before Lab Session 26 is performed.

Primer design

Note to prep staff:

Section B also involves primer design in preparation for Lab Session 29. Primers designed in Section A as well as Section B should be ordered together at one time. Standard, desalted oligos (usually the most inexpensive synthesis) are sufficient. If not already resuspended, dilute each oligo to 100 µM in sterile dH_2O or TE. Most companies will include already-calculated volumes to resuspend with to achieve that concentration.

It is key that the instructor coordinates BOTH oligos for CRISPR guide design (how many genes, how many unique regions) but also PCR primer design. You do not want to unnecessarily order the same or even overlapping primers. One primer pair may work for assessing multiple sgRNAs, depending on their target sequences. Students are instructed to save their designed oligos in this lab session. It is helpful to review these oligos either within the Benchling notebook or on a master spreadsheet to avoid redundant choices in primers.

The laboratory exercise and prep list details are written assuming each group is performing annealed oligo cloning with one pair of oligos to create one guide. If students are attempting more than one guide per group, reagents will need to be scaled up accordingly.

Students will ideally have access to two water baths: one preset to 55°C for the BsmBI-v2 restriction enzyme digestion and one preset to 37°C for the CIP incubation. Factor in additional time if only one water bath is used, as temperature will need adjusted between steps.

Restriction enzyme digestion and dephosphorylation of the CRISPR vector

The students will use the listed supplies. Aliquot per group:

- At least 2 µL of 5 µg/µL pLentiCRISPR-E, store at −20°C
- 7 µL 10× NEB buffer r3.1 (shipped with BsmBI-v2 below), store at −20°C
- 4 µL NEB BsmBI-v2 (10 U/µL) catalog number R0739, store at −20°C
- 1 µL NEB Quick CIP Calf Intestinal Phosphatase (5 U/µL) catalog number M0525, store at −20°C

Agarose gel electrophoresis and gel extraction of digested CRISPR vector backbone

The listed supplies and reagents are included in the QIAGEN QIAquick Gel Extraction Kit (QIAGEN, catalog number 28704). Each kit contains columns and reagents for 50 gel extractions. Be sure to follow the kit's protocol in adding 100% ethanol to Buffer PE.

- 600 µL Buffer QG, store at room temperature
- 1 QIAquick spin column, store at room temperature
- 1–2 mL collection tube, store at room temperature
- 1 mL Buffer PE, store at room temperature
- 40 µL Buffer EB, store at room temperature
- 300 µL isopropanol, store at room temperature

Students will share:

- Agarose
- 1× TBE buffer (see recipe in "Supplies and Reagents for General Use")
- New razor blades for band excision
- 3 M sodium acetate pH 5.0
- NEB Quick-Load 1 kb Extend DNA Ladder (New England BioLabs, catalog number N3239S), store at −20°C

- For agarose gels using GelRed, dilute the DNA ladder to 12.5 ng/μL and load 150 ng per lane in a total volume of 12 μL. To dilute to 12.5 ng/μL: 30 μL Quick-Load 1 kb Extend DNA ladder + 70 μL of d*I* H_2O + 20 μL of 6× DNA loading dye

Students should have from before (but have extra on hand, just in case):

- GelRed stock (10,000×)
- 6× DNA loading buffer (dye)

LAB SESSION 26

Advanced CRISPR: Part II

Note to prep staff:

Predesigned oligos will be needed in this session. They were to be designed and ordered at least one week prior to this session. If they have not arrived, students cannot proceed with annealed oligo cloning. It is recommended that prep staff (rather than the students) resuspend the oligos in sterile *d*H_2O or TE to a final concentration of 100 μM and store at −20°C.

The laboratory exercise and prep list details are written assuming each group is performing annealed oligo cloning with one pair of oligos to create one guide. If students are attempting more than one guide per group, reagents will need scaled up accordingly.

The students will use the listed supplies. Aliquot per group:

- Students' designed oligos; store at −20°C
  - 2 μL of oligo 1 (100 μM)
  - 2 μL of oligo 2 (100 μM)
- 8 μL NEB 10× T4 DNA ligase buffer; store at −20°C
- 1 μL NEB T4 PNK (polynucleotide kinase, catalog number M0201S); store at −20°C
- 50 ng of uncut pLenti-CRISPR-E; store at −20°C
- 3 μL NEB T4 DNA ligase (catalog number M0202L); store at −20°C

Note: This protocol uses NEB T4 DNA ligase, catalog number M0202L. If using a different brand of T4 DNA ligase, follow the manufacturer's instructions. The incubation time may be significantly longer.

Four 50 μL aliquots of Invitrogen One-Shot Stbl3 chemically competent *E. coli* (Thermo Fisher Scientific catalog number C737303)

Note: These competent cells are unique from those used in Lab Session 5. The Stbl3 strain minimizes recombination events that may occur in lentiviral plasmids. Thaw frozen competent cells immediately before use. Mix gently before aliquoting. Cells should not sit on ice for more than a few minutes. Cells should not sit at room temperature for any period of time.

Money-Saving Tip: Stbl3 cells shipped from the manufacturer are in 50 μL aliquots. You may choose to make 25 μL aliquots for students to increase the number of transformations possible.

- 1.2 mL SOC or LB prewarmed to room temperature
- 4 LB/amp agar plates

Students will share:

- 95% ethanol for plate spreading
- Plate spreader
- Sterile *d*H_2O

Students should have from a previous lab:

- BsmBI digested, CIP-treated, gel-purified pLentiCRISPR-E from Lab Session 25

LAB SESSION 27A

Advanced CRISPR: Part III

Interim Laboratory Session

Counting Transformants and Inoculating Cultures for Miniprep DNA

The students will use:

- 1 LB/amp agar plate with a grid sticker (Diversified Biotech PetriSticker 32-square grid, catalog number PSTK-1000)
- 3 Culture tubes containing 2 mL LB/amp

LAB SESSION 27B

Advanced CRISPR: Part III

The students will use:

- QIAGEN QIAprep Spin Miniprep Kit (catalog number 27104)

The listed supplies and reagents are included in the QIAGEN QIAprep Spin Miniprep Kit. Each kit contains columns and reagents for 50 plasmid preps. Be sure to follow the kit's protocol in adding the provided RNase to Buffer P1 and 100% ethanol to Buffer PE.

Aliquot per group:

- 1 mL Buffer P1, store at 4°C
- 1 mL Buffer P2, store at room temperature
- 1.2 mL Buffer N3, store at room temperature
- 3 QIAprep spin columns, store at room temperature
- 3 Collection tubes, store at room temperature
- 2.5 mL Buffer PE, store at room temperature
- 200 μL Buffer EB, store at room temperature
- 3 PCR tubes for shipping samples for sequencing

LAB SESSION 28A

Advanced CRISPR: Part IV

Note to prep staff:

Bioinformatics exercise where students need data from their sequencing reactions from Lab Session 27 and a computer with Chromas (http://technelysium.com.au/wp/chromas/) software for analysis.

Only 3.6×10^5 of EGFP-expressing HEK293 cells are needed per group. In order for student groups to gain experience in these techniques, we recommend one flask or dish of HEK293 cells stably expressing EGFP be cultured ahead of time for each group. The laboratory exercise and prep list details are therefore written assuming each group has their own flask or dish of cells.

The students will use the listed supplies. Aliquot per group:

- 2.5 mL 1× PBS, store at room temperature or 4°C
- 3 mL 0.25% Trypsin–EDTA, store at 4°C
- 8 mL complete DMEM *without any antibiotics*, store at 4°C
- Sterile pipette tips, P1000, P200, P20, P10
- Sterile tin containing glass Pasteur pipettes
- Five 5 mL wrapped serological pipettes
- One sterile 48-well plate
- HEK293 cells stably expressing EGFP (see note above)

Students will share:

- Hemacytometer
- Cover slips
- Hand counter

LAB SESSION 28B

Interim Laboratory Session

Transfection of CRISPR Vectors

The students will use the listed supplies. Aliquot per group:

- 150 µL Opti-MEM or serum-free medium, store at 4°C
- At least 0.28 µg of pLentiCRISPR-E, store at 4°C
- At least 0.28 µg of pLentiCRISPR-EGFP-2, store at 4°C

Students will share:

- FuGENE HD reagent

Note to prep staff: FuGENE HD cannot be aliquoted; it must remain in the original glass vial that it ships in. Therefore, we recommend that the instructor or TA hands the vial to each group as they need it in the biological safety cabinets while the vial remains tightly capped.

Students should have from a previous lab:

- At least 2 µg of pLenti-CRISPR-sgRNA miniprepped DNA from Lab Session 27b

LAB SESSION 29

Advanced CRISPR: Part V

Note to prep staff:

The reagent volumes listed here and in the protocol assume that four lysates are processed per group (eliminating the Opti-MEM only and pLentiCRISPR-E lysates transfected in Lab Session 28).

Biohazardous liquid will be removed from pelleted cells outside of the cell culture lab; therefore it is required to have waste containers available containing 100% bleach to 10% of the total liquid volume.

The students will use the listed supplies and reagents. In the biological safety cabinet, aliquot per group for use at the bench. *Note*: these aliquots will no longer be used in a sterile environment and *should not be reused in cell cultures*.

- 1 mL 1× PBS, store at room temperature
- 1 mL 0.25% Trypsin–EDTA, solution (Gibco catalog number 25200056), store at 4°C.
- 1 mL Complete DMEM, store at 4°C
- 2 µL Each forward and reverse oligo designed and ordered in Lab Session 25, already resuspended in sterile dH_2O or 1 × TE to a final concentration of 100 µM.
- 9 PCR tubes

The listed reagents are found in the Invitrogen GeneArt Genomic Cleavage Detection Kit (Invitrogen catalog number LSA24372). This kit is enough for 20 reactions (lysates). If you are using a different indel detection assay, please refer to the manufacturer's instructions.

- 220 µL Cell lysis buffer, store at −20°C
- 10 µL Protein degrader, store at −20°C
- 1.5 µL Control template and primers, store at −20°C
- 150 µL AmpliTaq Gold, store at −20°C

Students will share:

- Agarose
- 1× TBE

Students should have from before (but have extra on hand, just in case):

- GelRed stock (10,000×)

Note to prep staff:

The protocol calls for each group to pour three separate 2% agarose gels. This is based on a typical "mini" gel with an 8-well comb because each group will need to run the listed samples in Lab Session 30:

Each group will need six lanes for the gel used to verify PCR product:
1 DNA ladder
1 Positive control reaction
1 Lane for each lysate (pLentiCRISPR-EGFP, pLenti-CRISPR-sgRNA×3 transfection conditions)

Each group will need an additional 11 lanes for the gel used to visualize cleavage:
1 DNA ladder
2 Lanes for the control reaction
2 Lanes for each lysate (8 total)

The gels are to be poured in Lab Session 29 wrapped in cellophane and saved at 4°C for use in Lab Session 30. If you cannot accommodate this number of gel trays and combs being used simultaneously in your classroom laboratory, you may choose to prepare the gels ahead of time. You may also use "midi" size gels to accommodate more samples. Keep in mind that when they are being run in Lab Session 30, you will need enough tanks and power supplies to accommodate up to two mini gels simultaneously run per group.

Time-Saving Tip: You can order premade 2% agarose gels from a number of manufacturers.

LAB SESSION 30

Advanced CRISPR: Part VI

Note to prep staff:

The reagent volumes listed here and in the protocol assume that four lysates are processed per group (eliminating the Opti-MEM only and pLentiCRISPR-E lysates transfected in Lab Session 28).

The listed reagents are found in the GeneArt Genomic Cleavage Detection Kit (Invitrogen catalog number LSA24372). If you are using a different indel detection assay, please refer to the manufacturer's instructions.

- 12 μL 10× Detection reaction buffer, store at −20°C
- 8 μL Detection enzyme, store at −20°C
- 60 μL Water, store at room temperature

Students will use the listed reagents and supplies. Each group needs:

- 10 PCR tubes
- NEB 50 bp DNA ladder (catalog number N3236), dilute to a 15 ng/μL working stock solution, see recipe, store at −20°C
- Recipe for 15 ng/μL stock solution of 1 kb NEB ladder: 30 μL N3236 + 803 μL dH_2O + 167 μL 6× DNA loading buffer

Students will share:

- 1× TBE
- If gel analysis will not occur during this lab, students are instructed to add 1 μL of 100 μM EDTA to each reaction to stop digestion and store samples at –20°C. The EDTA solution is not included in the GeneArt kit

Students should have from before (but have extra on hand, just in case):

- 6× DNA loading buffer
- 2% agarose gels saved from Lab Session 29

LAB SESSION 31

Gibson Assembly In Silico Cloning

This is an in silico molecular cloning exercise using Gibson assembly cloning. Students will need a computer, Internet access, and a free account to Benchling (https://www.benchling.com). It is recommended that students register for a Benchling account prior to beginning the lab session.

LAB SESSION 32

Transcriptomic Analysis

This is a data analysis exercise where students need a computer with Microsoft Excel and Internet access.

LAB SESSION 33

Structure-Activity Relationship Modeling

This is a virtual modeling exercise that uses four open-source software programs: (1) AutoDockTools, (2) Vina, (3) Strawberry Perl, and (4) PyMol. Students will require a computer and access to the Internet. Specific download instructions are found in Appendix E. It is recommended that students download these software programs prior to beginning the lab session. It is important to note that AutoDockTools (part of the MGLTools suite) is *not compatible* with Mac OS.

Appendix C

Preparation of Competent *Escherichia coli* Cells

INTRODUCTION

Escherichia coli can be made competent for DNA uptake by treatment with divalent cations, and they can be stored in this state indefinitely at −80°C. Many parts of the procedure for making competent cells are critical for obtaining cells with a high transformation efficiency. The cells must be harvested in log phase, and, once chilled, they must not warm to room temperature or they will lose competence. For optimal transformation efficiencies, the cells should be aliquoted in a cold room. In the absence of a cold room, cells can be aliquoted on ice with extreme care to prevent cells from warming.

Because the bacterial expression vector we are using, pET-41a, has a T7 promoter that drives transcription of the gene of interest, the *E. coli* host strain must be a λDE3 lysogen so that it will have a T7 RNA polymerase to bind the promoter for expression. A suitable strain is BL21(DE3).

PROTOCOL

Preparation of Chemically Competent Cells by Calcium Chloride Treatment

Remember that you are not using antibiotics in this procedure. Use sterile technique and make sure the centrifuge bottles are sterile. The following steps are modified from Chung et al.[1]

1. Start an overnight culture of *E. coli* by using a sterile inoculating loop to scrape cells from a single colony into 2 mL of Luria-Bertani (LB) in a polypropylene snap-cap tube. (*Note*: Antibiotics are not added.) Incubate at 37°C overnight in a shaking incubator.
2. The next morning, start 100 mL cultures of *E. coli* BL21(DE3) with the fresh overnight culture as inoculum. Based on the number of students in your class, inoculate several 1 L flasks containing 100 mL of LB broth with 0.1 mL of overnight culture. You will need to start approximately one 100 mL culture per 10 lab groups.
3. Incubate the flask at 37°C with shaking. Grow the bacteria to log phase, indicated by an OD_{600} value between 0.3 and 0.5. It will take 2.5–4 hours for the culture to grow to the proper stage for making competent cells. Use sterile technique (including sterile pipettes) to withdraw samples from the culture to measure their OD_{600} using a spectrophotometer. *Once cultures appear cloudy, they are in log phase growth and their* OD_{600} *will increase rapidly.*
4. When the proper optical density is reached, pour the culture into two prechilled sterile large centrifuge tubes. Balance the tubes and then place on ice for 15 minutes. From now on, it is important that the cells *never* warm to room temperature.
5. Harvest the cells by centrifugation at 6000 rpm (5000 × *g*) for 10 minutes at 4°C. Make sure the rotor is prechilled.
6. Pour off the supernatant and, keeping the cells on ice, remove excess LB medium with a sterile Pasteur pipette or a 1 mL micropipette. Resuspend the cells gently by pipetting up and down in 1 mL of ice-cold sterile transformation and storage solution (TSS; LB broth with 10% (w/v) PEG 3350 or 8000, 5% (v/v) DMSO, and 50 mM $MgCl_2$, at a final pH of 6.5).

Molecular Biology Techniques. DOI: https://doi.org/10.1016/B978-0-443-26714-7.00037-5

7. Transfer the contents of one tube to the other to consolidate your cells. Be very gentle; the cells are now fragile. TSS contains 50 mM $MgCl_2$; other procedures use 50 mM $CaCl_2$. The dimethylsulfoxide (DMSO) in TSS is necessary if you intend to freeze some cells. Because there is little loss in transformation efficiency, it is always useful to freeze extra cells.
8. Pipette 50 μL aliquots quickly into cold microcentrifuge tubes on ice. These cells can be used immediately or frozen at −80°C. You should keep one tube on ice for the test transformation and freeze the rest. You will need three 50 μL aliquots of frozen competent cells per lab group for Lab Session 5.
9. To flash freeze the competent cells, take your ice bucket to a dry ice/ethanol bath near a −80°C freezer. Using a pre-cooled microcentrifuge tube holder, submerge the tubes. Retrieve the tubes into a labeled cardboard or Styrofoam box and immediately place at −80°C.

Transformation Control

Transformation efficiency is measured by determining the number of colony-forming units (cfu)/μg DNA. Only super-coiled DNA should be used (linear DNA does not transform efficiently). In this section you will use 10 ng of uncut pET-41a DNA to confirm the competence of the cells.

1. Label the tops of the two prechilled sterile microcentrifuge tubes with a designation for the following treatments: TE, pET-41a.
2. Aliquot 20 μL of competent cells into each tube. Remember to keep the cells on ice at all times.
3. Add to the appropriate tube:
   a. 2 μL TE.
   b. 2 μL of a 5 ng/μL pET-41a DNA. Mix gently and leave on ice for 5 minutes.
4. Heat shock the cells for 30 seconds at 42°C.
5. Incubate tubes on ice for 2 minutes.
6. Add 80 μL of liquid medium (either LB or SOC); gently invert the tubes to mix and incubate with shaking at 37°C, 225 rpm for 45 minutes.
7. Plate the entire volume (100 μL) on a selective medium (LB/kan).
8. Incubate the plates at 37°C overnight.
9. Observe plates. TE control should have no colonies; pET-41a control should have numerous colonies or a lawn of growth.
10. Calculate the transformation efficiency of your cells by counting the number of cells on your pET-41a control plate. If there are too many cells on the entire plate to count, you can count a fraction of the plate (1/2 or 1/8) and multiply the number of cells in that fraction by the appropriate number to estimate the total number of colonies on the entire plate.

$$\frac{\text{cfu}}{\mu\text{g}} = \text{number of colonies on plate}/0.010\ \mu\text{g DNA}$$

Preparation of Chemically Competent Cells Using Commercial Kits

Commercial kits are also available that streamline competent cell preparation, such as Zymo Research *Mix and Go!* transformation kit (catalog number T3001). As with any procedure, it is important to always perform a transformation control as described above.

REFERENCE

1. C.T. Chung, S.L. Niemela, R.H. Miller, One-step preparation of competent Escherichia coli: transformation and storage of bacterial cells in the same solution, Proceedings of the National Academy of Sciences of the United States of America. 00278424. 86 (7) (1989) 2172–2175. https://doi.org/10.1073/pnas.86.7.2172.

Appendix D

Pre-Lab Questions

LAB SESSION 1

1. What is your major(s)/department (grad students: who is your principle investigator, if chosen)?
2. What are you hoping to learn in this class?
3. What are your career goals?
4. Have you had previous experience with molecular biology/biochemistry? If so, what (relevant coursework/research experience)?

LAB SESSION 2

1. What are the volume ranges (max. and min. amount to be measured) of the following micropipettes?
   a. P10
   b. P20
   c. P200
   d. P1000
2. What is the volume (in microliters) that would be pipetted using the following settings?

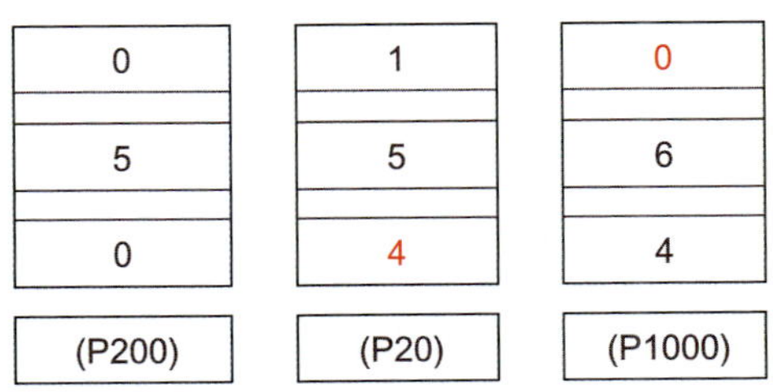

a. __________ b. __________ c.__________

No Permission Required

3. What method will we use to separate plasmid from chromosomal DNA?
   a. Alkaline lysis.
   b. Silica adsorption.
   c. NanoDrop/spectrophotometry.
   d. Restriction digestion.
4. An $A_{260/280}$ of________ indicates optimal purity of double-stranded DNA.
5. Why will you use two different restriction enzymes to cut the vector pET-41a?
   a. In case one of the enzymes fails to cut.
   b. Because the two enzymes have compatible cohesive ends.

Molecular Biology Techniques. DOI: https://doi.org/10.1016/B978-0-443-26714-7.00042-9

c. Because cutting the vector with two enzymes that leave incompatible ends and then cutting the insert with the same two enzymes will force the insert into the correct orientation when cloning.
d. a and b.

LAB SESSION 3

1. What DNA serves as the template to amplify *egfp* in today's PCR?
   a. pET-41a.
   b. egfpNco.
   c. egfpNot.
   d. pEGFP-N1.
   e. b and c.
2. What is the basis for selecting the annealing temperature to use in a specific PCR reaction?
   a. The length and GC content of the primers.
   b. The length and GC content of the desired PCR product.
   c. The restriction sites engineered into the primers.
   d. The ideal reaction temperature of the polymerase.
3. What is the purpose of running our digested vector through the spin column?
   a. To remove any live *E. coli* that might be mixed in the sample.
   b. To remove restriction enzymes from the digest.
   c. To remove unwanted salts prior to ligation.
   d. a, b, and c.
   e. b and c.
4. In DNA agarose gel electrophoresis, which side of the apparatus should your wells be closer to?
   a. The black side (cathode).
   b. The red side (anode).
5. Because a 73 base pair fragment is removed from pET-41a during digestion, the uncut vector should always run slower through the agarose gel than the cut/digested vector.
   a. True.
   b. False.

LAB SESSION 4

1. Approximately what size band do you expect to see on your PCR gel?
   a. 700 kb
   b. 5 kb
   c. 6 kb
   d. 1 kb
   e. 700 bp
2. If your Nanodrop reading was 75 ng/μL, what volume of PCR product would you need to add to your restriction digest?
   a. 0.15 μL
   b. 6.67 μL
   c. 150 μL
   d. 6670 μL
3. What is a purpose of taking a Nanodrop reading of your PCR product?
   a. To determine concentration of DNA.
   b. To determine the volume of DNA.
   c. To determine the fluorescence intensity of egfp.
   d. To determine whether you have non-specific PCR products.
4. If your clean, digested *egfp* PCR product has a concentration of 7 ng/μL, what volume would you need to use to have 21 ng for the following week's ligation (use the equation given at the end of the lab session)?
5. What should you do if you have an aliquot of enzyme and cannot pipette the entire volume out of the tube?
   a. Try to centrifuge down the liquid for 5 seconds.
   b. Vortex.

c. Complain to your TA as a first resort.
d. Cry and go home.

LAB SESSION 5

1. Linear DNA will not be replicated in *E. coli* even if it is taken up into the cell.
   a. True.
   b. False.
2. Ligase buffer does not need to be kept on ice when not in use since it is active at room temperature.
   a. True.
   b. False.
3. For today's experiment, what is the vector-to-insert ratio that you will be using?
   a. 1:1
   b. 5:1
   c. 1:5
   d. 1:10
4. What control will you include to ensure that your cells are competent?
   a. Uncut pET-41a.
   b. Digested vector with insert.
   c. Digested vector without insert.
   d. None of the above.
5. Gently mix competent cells by flicking the tube instead of vortexing them before use in transformation.
   a. True.
   b. False.

LAB SESSION 6

1. While performing a miniprep, you need to vortex your samples well to ensure complete mixing after the addition of Buffer P2.
   a. True.
   b. False.
2. The miniprep protocol you'll be performing uses alkaline lysis to purify the plasmid DNA.
   a. True.
   b. False.
3. The purpose of the minipreps in today's lab is to purify DNA______.
   a. to see if it glows green when subjected to UV light.
   b. to use in restriction digestion reactions.
   c. to send out for DNA sequencing.
   d. b and c.
4. In the PCR screen, how are the primers designed?
   a. Both are designed to bind to opposite ends of the insert DNA.
   b. Both are designed to bind to vector DNA, on opposite sides of the insert.
   c. Both are designed to bind to vector DNA, on the same side of the insert.
   d. One is designed to bind insert DNA and one is designed to bind adjacent vector DNA.
5. If you transferred a colony while replica plating to an LB/kan plate instead of an LB/kan/IPTG plate, what would you expect to see next week when visualizing the cells?
   a. Bacteria would not grow on the plate.
   b. Bacteria would grow on the plate, but they would not have the *egfp* gene present and therefore there would be no EGFP protein expression.
   c. Bacteria containing the *egfp* gene would grow, but EGFP protein would not be expressed.
   d. There would be bacteria growing with high levels of EGFP expression.

LAB SESSION 7

1. What is the purpose of preparing a master mix?
   a. Minimizes error inherent in pipetting smaller volumes.
   b. Reduces variability between samples.
   c. Saves time.
   d. All of the above.
2. In your PCR screen, what would you expect to see when you analyze your PCR product on a gel if you had a negative clone (a clone with no insert)?
   a. No band on the gel.
   b. One band about 750 bp (0.75 kb)
   c. One band about 1200 bp (1.5 kb)
   d. Two band.
3. In your PCR screen, what would you expect to see when you analyze your PCR product on a gel if you had a positive clone (a clone with egfp insert)?
   a. No band on the gel.
   b. One band about 750 bp (0.75 kb)
   c. One band about 1200 bp (1.5 kb)
   d. Two bands.
4. In your restriction digestion analysis, what would you expect to see when you analyze your digests on a gel if you have a negative clone?
   a. No band on the gel.
   b. One band about 750 bp (0.75 kb)
   c. One band about 6000 bp (6.0 kb)
   d. Two bands, at approximate sizes of 750 and 6000 bp
5. Sequencing primers should be designed to bind where?
   a. ~60 nucleotides upstream of the beginning of the region to be sequenced.
   b. ~5 nucleotides upstream of the beginning of the region to be sequenced.
   c. ~5 nucleotides downstream of the beginning of the region to be sequenced.
   d. As long as the primer binds within the region to be sequenced, it doesn't matter exactly where it is designed.

LAB SESSION 8

1. Each peak in the chromatogram corresponds to:
   a. A fluorescent deoxynucleotide triphosphate (dNTP) which has been released from the DNA fragment resulting in the termination of synthesis.
   b. A fluorescent deoxynucleotide triphosphate (dNTP) which has been incorporated into the DNA fragment resulting in the termination of synthesis.
   c. A fluorescent dideoxynucleotide triphosphate (ddNTP) which has been released from the DNA fragment resulting in the termination of synthesis.
   d. A fluorescent dideoxynucleotide triphosphate (ddNTP) which has been incorporated into the DNA fragment resulting in the termination of synthesis.
2. A typical sequencing read will typically yield_____ nucleotides of unambiguous sequence.
   a. 80–100.
   b. 300–500.
   c. 800–1000.
   d. 3000–5000.
3. Where do you expect to see multiple "Ns" within your sequencing read?
   a. In the beginning of your sequence (first 25 nucleotides).
   b. In the middle of your sequence.
   c. In the end of your sequence (last 25 nucleotides).
   d. a and b.
   e. a and c.

4. Which tool will you use to look for similarity between your sequencing data and the expected pBIT sequence?
   a. BLAST.
   b. Sanger sequencing.
   c. Chromas Lite.
   d. Genbank.
5. If there is a mutation within your *egfp* sequence, the mutation is most likely introduced by:
   a. The enzyme during PCR.
   b. Bacteria during the transformation step.
   c. Student error in one of the previous steps.
   d. The bacteria incorrectly copying the DNA during replication.

LAB SESSION 9

1. In SDS–PAGE, what chemical is used to ensure that all protein molecules are coated with a negative charge?
   a. IPTG.
   b. β-Mercaptoethanol.
   c. SDS.
   d. X-gal.
2. In SDS–PAGE, what chemical is used to ensure that protein disulfide bonds are broken?
   a. IPTG.
   b. β-Mercaptoethanol.
   c. SDS.
   d. X-gal.
3. What is the advantage of running a discontinuous protein gel rather than a continuous gel?
   a. Proteins can be separated by size alone, rather than being dependent on size, shape and charge.
   b. For better resolution of the protein bands.
   c. The gel runs faster.
   d. a and b.
   e. None of the above.
4. What is the advantage of treating protein samples with SDS and β-mercaptoethanol?
   a. Proteins can be separated by size alone, rather than being dependent on size, shape and charge.
   b. For better resolution of the protein bands.
   c. The gel runs faster.
   d. a and b.
   e. None of the above.
5. The non-polymerized forms of acrylamide (powder and liquid) are far more dangerous than polymerized (solidified) polyacrylamide.
   a. True.
   b. False.

LAB SESSION 10

1. What is the purpose of staining with Ponceau Red?
   a. To visualize the molecular weight markers.
   b. To confirm the expression of the specific GST::EGFP fusion protein.
   c. To block non-specific sites on the membrane prior to adding primary antibody for the western blot.
   d. To confirm that proteins, in general, were transferred from your gel to the nitrocellulose.
2. When performing the Ponceau stain, you may see a band of approximately 25 kD in your negative control that is not present in your positive control or your other positive clones. What does this band likely represent?
   a. Degraded GST::EGFP fusion protein.
   b. Light-chain component of the antibody used for staining.
   c. GST protein alone (no associated EGFP).
   d. LacI.

3. At the completion of the western blot, what will you see on the membrane?
   a. The molecular weight markers.
   b. Only the specific GST::EGFP fusion protein.
   c. All of the proteins expressed by *E. coli* in the lanes where you loaded cell lysates.
   d. a and b.
   e. a and c.
4. Approximately what size will the GST::EGFP fusion protein be?
   a. 25 kDa
   b. 30 kDa
   c. 60 kDa
   d. 6653 kDa
5. To make a 1:5000 dilution of an antibody, what volume of antibody would be added to 10 mL of a TBS-T solution?
   a. 2 μL
   b. 5 μL
   c. 20 μL
   d. 50 μL

LAB SESSION 11

1. What is the purpose of the GST portion of your fusion protein?
   a. It makes the protein fluoresce green.
   b. It is responsible for the IPTG induction of the protein.
   c. It allows for the affinity purification of the protein using a glutathione affinity column.
   d. It allows the fusion protein to be cleaved into two fragments.
2. Where did the *gst* gene come from?
   a. It was already engineered into the pET-41a expression vector.
   b. Fluorescent jellyfish.
   c. We cloned it into the pET-41a expression vector after gel-purifying it.
   d. Both b and c.
3. Which fraction from an affinity chromatography purification experiment has the highest total amount of the protein of interest?
   a. Cell lysate.
   b. Flow-through (unbound).
   c. Wash.
   d. Eluate.
4. Elution of your fusion protein from the column occurs by adding reduced glutathione. How does the glutathione do this?
   a. Competing for binding sites on the column, causing the fusion protein to be knocked off.
   b. Binding to the GST portion of the fusion protein, causing the fusion protein to be unable to bind column.
   c. Binding to the EGFP portion of the fusion protein, causing the fusion protein to be unable to bind column.
5. What does the enzyme lysozyme do?
   a. Breaks down the cell wall in bacteria.
   b. Degrades DNA to facilitate protein purification.
   c. Degrades RNA to facilitate protein purification.
   d. Both b and c.
   e. All of the above.

LAB SESSION 12

1. We need to use an EGFP-specific antibody in order to assess if we have successfully purified our fusion protein from contaminants.
   a. True.
   b. False.

2. Additional bands in eluate fractions, i.e. other than the band representing the GST::EGFP fusion protein, could have resulted from:
   a. Contaminants present in your sample (i.e., column not washed well).
   b. Degradation of the fusion protein.
   c. Fusion protein's inability to bind to column.
   d. Both a and b.
   e. All of the above.
3. Assume you have three elution fractions, each with a volume of 500 μL. The concentration (and total amount) of GST::EGFP in the elution fractions is as follows:
   Elution 1: 0.1 μg/mL (0.05 μg)
   Elution 2: 0.4 μg/mL (0.2 μg)
   Elution 3: 0.5 μg/mL (0.25 μg)
   If the elutions were combined together, how much total GST::EGFP would be present?
   a. 0.5 μg
   b. 1.0 μg
   c. 1.5 μg
   d. 500 μg
   e. 750 μg
4. GelCode Blue reagent will stain which of the following?
   a. Only the molecular weight ladder.
   b. Only the GST::EGFP fusion protein.
   c. All proteins except the GST::EGFP fusion protein.
   d. All proteins present on the gel.
5. Which fraction would you expect to have the lowest total protein concentration, as measured by a Bradford protein assay?
   a. Cell lysate.
   b. Last flow through fraction.
   c. The most fluorescent elution fraction.

LAB SESSION 13

1. List two reasons why homogenization of *E. coli* is performed.
2. Why are you using a second RNeasy spin-column after the DNase step?
   a. To purify additional RNA you missed with the first column.
   b. To inactivate the DNase.
   c. To remove salts.
   d. Both a and c.
   e. Both b and c.
3. What nucleic acid will you purify from *E. coli*?
   a. Genomic DNA (gDNA).
   b. Complementary DNA (cDNA).
   c. Messenger RNA (mRNA).
   d. Total RNA.
4. What is the extinction coefficient of RNA?
5. If you read the absorbance of your samples and the A_{260}/A_{280} ratio is 1.5, what is the most likely contaminant?

LAB SESSION 14

1. What type of enzyme will be used to convert RNA to cDNA?
   a. Reverse transcriptase.
   b. RNA polymerase.
   c. Ligase.
   d. Oligo-dT.

2. In qPCR, what is used to quantify how much DNA is amplified?
   a. Spectrophotometry.
   b. Gel electrophoresis.
   c. Fluorescence.
   d. Luminescence.
3. Why is a no RT reaction prepared?
   a. To help quantify the level of the reference gene.
   b. To control for protein contamination.
   c. To control for DNA contamination.
   d. All of the above.
4. What are the two required qualities of a reference gene?
5. How many master mixes will you be preparing for the reverse transcription step?
   a. One.
   b. Two.
   c. Three.
   d. Four.

LAB SESSION 15

1. What do you call the cycle at which an amplification plot crosses the threshold?
   a. Exponential cycle.
   b. Threshold cycle.
   c. Baseline cycle.
   d. Plateau cycle.
2. Assuming a threshold was properly set; if amplicon A has a *CT* value of 22 and amplicon B has a *CT* value of 23:
   a. There was approximately twice as much amplicon B as amplicon A in the sample.
   b. There was approximately twice as much amplicon A as amplicon B in the sample.
   c. There was only slightly more amplicon B than amplicon A in the sample.
   d. There was only slightly more amplicon A than amplicon B in the sample.
3. In this lab session, what will serve as your calibrator?
   a. *egfp*.
   b. *23S*.
   c. No induction.
   d. IPTG induction.
4. In this lab session, what will serve as your reference?
   a. *egfp*.
   b. *23S*.
   c. No induction.
   d. IPTG induction.
5. What is one concern of setting your threshold too low?
   a. You could be analyzing amplification in the plateau phase.
   b. You could be analyzing amplification in the exponential phase.
   c. You could be analyzing background signals.
   d. Both a and b.

LAB SESSION 16

1. Which of the following are properties of both semi-quantitative PCR AND qPCR?
   a. They are "end-point" methods.
   b. They require a gel electrophoresis step.
   c. They can detect large (10-fold) changes in gene expression.
   d. a, b, and c.

2. How many master mixes will you be preparing for the reverse transcription step?
   a. One.
   b. Two.
   c. Three.
   d. Four.
3. How will you control for the tendency of smaller amplicons to be more efficiently amplified than larger amplicons?
   a. Primer pairs for both amplicons will be mixed in the same tube.
   b. Primer lengths will all be identical.
   c. Amplicon lengths will all be identical.
   d. Extension times will all be identical.
4. How will you avoid analyzing levels in the plateau phase of amplification?
   a. The cycle numbers corresponding to the exponential phase were previously determined.
   b. Aliquots of each sample will be removed during the early cycles of the program.
   c. Reagents will be at concentrations high enough to never become limiting.
   d. The thermal cycler program will run for only 1.5 hours, ensuring the plateau is not reached.
5. Why is *23S* rRNA being amplified?
   a. It serves as a loading control.
   b. Its level is predicted to change with IPTG and/or lactose induction.
   c. Both a and b.
   d. None of the above.

LAB SESSION 17

1. What type of nucleic acid are you visualizing on the agarose gel?
   a. Genomic DNA.
   b. Complementary DNA.
   c. Messenger RNA.
   d. Total RNA.
2. If the "+RT" reactions worked as expected, how many bands will be visible in each of those lanes of the gel?
   a. Zero.
   b. One.
   c. Two.
   d. Three.
3. If the "−RT" reactions worked as expected, how many bands will be visible in each of those lanes of the gel?
   a. Zero.
   b. One.
   c. Two.
   d. Three.
4. In the +RT lanes, if the *23S* amplicons vary in intensity from lane to lane, what is/are the most likely reason(s)?
   a. There was a gel loading error.
   b. Varying amounts of RNA were reverse transcribed.
   c. The different inducers tested induced *23S* gene expression to varying degrees.
   d. Both a and b.
   e. Both a and c.
5. In the +RT lanes, if the *23S* amplicons do not vary in intensity from lane to lane, and the *egfp* amplicons do vary, what can you conclude?
   a. There was a gel loading error.
   b. Varying amounts of RNA were reverse transcribed.
   c. The different inducers tested induced *egfp* gene expression to varying degrees.
   d. Both a and b.
   e. None of the above.

LAB SESSION 18

1. A transformed cell line will only grow for a finite (limited) amount of time.
   a. True.
   b. False.
2. Confluency can be defined as:
   a. Cell appearance.
   b. A particular cell type.
   c. The percentage of surface area occupied by the cells.
   d. Whether a cell is adherent or floating.
3. Signs of unhealthy or contaminated adherent cells include (choose all that apply):
   a. Cloudy medium.
   b. Detachment.
   c. Cell shrinkage.
   d. Elongated cell body.
4. What is Trypsin-EDTA solution and how does it work?
5. If the complete medium on your cells turns yellow, what could that be an indication of?
   a. Contamination.
   b. Overly confluent cells.
   c. Acidic medium.
   d. All of the above.

LAB SESSION 19

1. Which of the following method(s) can be used to introduce nucleic acids into mammalian cells?
   a. Chemical.
   b. Physical.
   c. Cationic lipid.
   d. All of the above.
   e. None of the above.
2. Cationic lipids are amphiphilic molecules with negatively charged head groups that interact with positively charged nucleic acids to help facilitate their entry through a cell's lipid bilayer.
   a. True.
   b. False.
3. Stable transfection of a nucleic acid in a mammalian cell:
   a. Requires foreign DNA to be integrated into the genome.
   b. Enables the DNA to be expressed for a short period of time.
   c. Enables the DNA to be expressed for an indefinite period of time.
   d. Does not require foreign DNA to be integrated into the genome.
   e. Both a and c.
   f. Both b and d.
4. What is the excitation/emission spectra for EGFP? Why are those properties important to know before using a fluorescence read-out like microscopy?
5. Photobleaching is
   a. When a fluorophore reacts with molecular oxygen.
   b. Caused by continuous exposure to excitation light.
   c. When a fixed or live sample gradually loses fluorescence.
   d. All of the above.

LAB SESSION 20

1. Which of the following small non-coding RNAs is 100% complementary to its targeted mRNA and results in its degradation?
   a. piRNAs.
   b. miRNAs.
   c. siRNAs.
   d. rasiRNAs.

2. Which RNAi associated-enzyme functions as a RNAse II ribonuclease?
   a. AGO.
   b. RISC.
   c. Exportin 5.
   d. Dicer.
3. List three controls for siRNA experiments.
4. What is the purpose of using BLAST to when you design putative siRNA effector molecules?
5. Which of the following could be considered an off-target effect of siRNA-induced silencing?
   a. Activation of the interferon response.
   b. Knockdown of the target mRNA.
   c. siRNA binding to a class of similar mRNAs.
   d. Short-term knockdown.
   e. Both a and c.
   f. Both b and d.

LAB SESSION 21

1. Why is it important to titrate the concentration of siRNA in your experiments?
   a. To ensure knockdown of your target mRNA is achieved.
   b. To prevent knockdown of non-targeted mRNAs.
   c. To prevent cellular toxicity.
   d. Both a and b.
   e. All of the above.
2. Calculate the volume of 50 mM siRNA needed to achieve a final concentration of 20 nM in 3 mL of medium.
3. Why does it take a few days for knockdown to be observed after transfection with siRNA oligos?
4. You transfect your HEK293 EGFP expressing cells with siRNA selective for *egfp*. What are the methods you could employ to verify the effectiveness of your siRNA?
   a. Fluorescence microscopy.
   b. SDS–PAGE followed by western blotting.
   c. RT-qPCR.
   d. ELISA.
   e. All of the above.
5. Why is knockdown using RNAi usually only 90% effective at its maximum?
   a. RNAi targets the transcript not the gene.
   b. RNAi targets the protein not the gene.
   c. RNAi targets ribosome function.
   d. RNAi targets the gene not the transcript.
   e. RNAi targets the protein not the transcript.

LAB SESSION 22

1. NIH Image J is a program available from the National Institutes of Health and can be used to perform a variety of functions including gel electrophoresis and fluorescence quantification.
   a. True.
   b. False.
2. What is corrected total cell fluorescence?
3. Would you expect ALL HEK293 EGFP expressing cells transfected with the positive control EGFP siRNA to exhibit fluorescence knockdown?
   a. Yes, this is a verified siRNA shown to selectively and potently knockdown EGFP.
   b. Yes, this siRNA can enter all the cells.
   c. No, there are very few (if any) siRNAs that achieve complete knockdown.
   d. No, even with selective siRNAs transfection efficiencies can vary from cell line to cell line.
   e. Both a and b.
   f. Both c and d.

4. Calculate the standard deviation from the following set of numbers:
   a. 208295230.
   b. 160458435.
   c. 154125432.
   d. 468391207.
   e. 100498270.
5. Some cells in your experimental RNAi-treated samples may still exhibit EGFP-mediated fluorescence. Since it doesn't fit your hypothesis for how RNAi-directed to *egfp* mRNA should work, you should not include those cells in your computational analysis for total fluorescence.
   a. True.
   b. False.

LAB SESSION 23

1. CRISPR can be used to accomplish all of the following:
   a. Knockout a gene.
   b. Knockin a gene.
   c. Edit the sequence of a gene.
   d. All of the above.
2. What type of enzyme is Cas9?
   a. DNA polymerase.
   b. DNA ligase.
   c. An endonuclease.
   d. An exonuclease.
3. What type of molecule complexes with Cas9?
   a. ssDNA.
   b. dsDNA.
   c. ssRNA.
   d. dsRNA.
4. The NHEJ pathway requires a homologous donor DNA sequence to repair a DSB.
   a. True.
   b. False.
5. Why does the number of cells per well plated in Laboratory Session 23 differ from the protocol in Laboratory Session 18?
   a. A different type of multiwell plate is used.
   b. A different cell line is used.
   c. A different number of days will pass before transfection.
   d. All of the above.

LAB SESSION 24

1. Which well(s) is(are) considered a negative control in this experiment?
   a. Opti-MEM only.
   b. pLentiCRISPR-E.
   c. pLentiCRISPR-EGFP.
   d. Both a and b.
2. Which tool on the NIH Image J toolbar will be used to outline each analyzed cell?
3. What will serve as a background reading for fluorescence?
   a. The brightest cellular region you can find.
   b. The dimmest cellular region you can find.
   c. A randomly selected non-cellular region.
   d. A region the same size as your analyzed cell with the same approximate fluorescence.
4. Your measurements will account for cell size.
   a. True.
   b. False.

5. What will you use to calculate averages and standard deviations?
    a. Microsoft Excel.
    b. NIH Image J.
    c. You will calculate by hand.
    d. Any of the above.

LAB SESSION 25

1. Which tool will be used to design sgRNAs?
    a. NIH Image J.
    b. Nucleotide BLAST.
    c. Benchling.
    d. Microsoft Excel.
2. This number is a score that predicts the on-target activity of a particular sgRNA:
    a. Activity Score.
    b. Off-target Score.
    c. *P*-value.
    d. T_m.
3. What type of nucleic acids will be ordered in this session for use in annealed oligo cloning?
    a. Two dsDNAs.
    b. One dsDNA.
    c. Two ssDNAs.
    d. One ssDNA.
4. Which enzyme will be used to dephosphorylate pLentiCRISPR-E?
    a. BsmBI.
    b. NcoI.
    c. PNK.
    d. CIP.
5. What is the approximate size of the DNA band that you will excise from the agarose gel?
    a. 13 kb
    b. 2 kb
    c. 700 bp
    d. 20 bp

LAB SESSION 26

1. Which enzyme will be used to phosphorylate the oligos?
    a. BsmBI.
    b. NcoI.
    c. PNK.
    d. CIP.
2. The prepared vector is also phosphorylated.
    a. True.
    b. False.
3. How will you achieve annealing between the oligos?
    a. You will use T4 DNA ligase.
    b. You will slowly increase the temperature of the reaction.
    c. You will slowly decrease the temperature of the reaction.
    d. You will transform the oligos into *E. coli*.
4. Why are you preparing a transformation control with only cut pLentiCRISPR-E?
    a. You want to visualize any transformants that arise from remaining uncut vector.
    b. You want to visualize how viable your cells are.
    c. You want to visualize any transformants that arise from vector recircularizing.
    d. You want to visualize any transformants that arise from taking up only insert.

5. Why do the agar plates used in this session contain ampicillin (amp)?
   a. The oligos encode ampicillin resistance.
   b. pLentiCRISPR-E encodes ampicillin resistance.
   c. Ampicillin prevents unwanted contaminants.
   d. Ampicillin can be used in place of kanamycin.

LAB SESSION 27

1. Which transformation plate(s) should have few if any colonies?
   a. Uncut vector only.
   b. Cut vector only.
   c. Cut vector only with ligase.
   d. Both b and c.
2. How many sequencing primers will be used to perform Sanger sequencing on the isolated plasmids?
   a. One.
   b. Two.
   c. Three.
   d. Four.
3. What purpose does Buffer P1 serve in the miniprep protocol?
   a. It resuspends the cells.
   b. It lyses the cells.
   c. It neutralizes the lysate.
   d. It precipitates gDNA.
4. What purpose does Buffer P2 serve in the miniprep protocol?
   a. It resuspends the cells.
   b. It lyses the cells.
   c. It neutralizes the lysate.
   d. It precipitates gDNA.
5. Quality Sanger sequencing results are independent of plasmid DNA purity.
   a. True.
   b. False.

LAB SESSION 28

1. What tool(s) will you use to analyze your sequencing results?
   a. Chromas.
   b. NCBI Nucleotide BLAST.
   c. Both of the above.
   d. Neither of the above.
2. What is the approximate length of your insert that you should find in the sequencing results?
   a. 13 kb
   b. 2 kb
   c. 700 bp
   d. 20 bp
3. Why should "CACCG" be found immediately upstream of your sgRNA sequence?
   a. This represents the guide RNA scaffold.
   b. This represents the BsmBI recognition site.
   c. This represents sequence complementary to the overhang in your vector created by BsmBI digestion.
   d. This represents the PAM site.
4. Using a hemacytometer, you count 137 cells total among 4 sets of 16 squares. What is the cell suspension's concentration?
   a. 3.43×10^5 cells/mL
   b. 1.37×10^4 cells/mL
   c. 8.56×10^4 cells/mL
   d. 6.85×10^3 cells/mL

5. What is the purpose of PBS (phosphate buffered saline)?
   a. It will chemically detach cells from the flask.
   b. It will supply growth factors to the cells.
   c. It will wash away residual medium.
   d. It will prevent contamination in your cell cultures.

LAB SESSION 29

1. What is the definition of transfection efficiency?
   a. The percentage of cells that are viable.
   b. The percentage of cells that have taken up an introduced nucleic acid.
   c. The percentage of cells that show CRISPR-Cas9 induced DSBs.
   d. The percentage of cells that have been detached from the flask.
2. What cellular material will be used as a template in this PCR?
   a. Plasmid DNA.
   b. Genomic DNA.
   c. Complementary DNA.
   d. All of the above.
3. What will the PCR attempt to amplify in this session?
   a. Your sgRNA sequence target sequence of ~20 bp.
   b. A region of your gene of interest ~400–500 bp containing your sgRNA target sequence.
   c. The Cas9 sequence.
   d. The PAM sequence of 3 bp
4. The DNA agarose gels you pour will be used in the next session for:
   a. Confirming today's PCR produced one specific band of the correct size.
   b. Detecting indels in a proportion of your cells.
   c. Visualizing EGFP levels.
   d. Both a and b.
5. How many cell lysates will you prepare in this lab session?
   a. two.
   b. four.
   c. eight.
   d. none of the above.

LAB SESSION 30

1. Denaturation of PCR products creates what type of molecules?
   a. ssRNAs.
   b. dsRNAs.
   c. ssDNAs.
   d. dsDNAs.
2. In addition to two possible cleaved bands on the agarose gel, what other band will you identify?
   a. Stuffer fragment.
   b. Daughter band.
   c. Parental band.
   d. All of the above.
3. The mismatch cleavage assay will serve as a primary screen.
   a. True.
   b. False.
4. Why are you analyzing PCR products on an agarose gel before proceeding to denaturing and reannealing reactions?
5. What sizes will cleavage products from the kit's provided control be?
   a. 20 and 40 bp
   b. 400 and 500 bp
   c. 2 and 13 kb
   d. 225 and 291 bp

LAB SESSION 31

1. Which type of cloning method relies on DNA recombination and requires the enzyme integrase?
   a. Gibson assembly cloning.
   b. Restriction enzyme cloning.
   c. Gateway cloning.
   d. TOPO cloning.
2. What of the following is an advantage to using molecular cloning simulation tools?
   a. They allow researchers to bypass experimental validation of results.
   b. They provide a cost-effective way to design and test cloning experiments before performing them in the lab.
   c. They replace the need for laboratory skills in molecular biotechnology.
   d. They offer detailed, real-time tracking of biomolecules in a cell.
3. For this lab, we will be simulating the cloning of PBP2. What is the primary function of PBP2 in bacterial cells?
   a. It facilitates the transport of nutrients across the cell membrane.
   b. It serves as a regulatory protein in gene transcription.
   c. It is involved in cell wall synthesis and plays a role in antibiotic resistance.
   d. It functions as an energy source for cellular processes.
4. In the context of gene sequences, what does the CDS stand for?
   a. Chromosomal DNA sequence.
   b. Coding DNA sequence.
   c. Central DNA sequence.
   d. Cellular Development Site.
5. Which of the following best describes the Gibson Assembly method used in molecular cloning?
   a. It uses restriction enzymes to cut DNA at specific sequences for fragment assembly.
   b. It relies on homologous recombination to assemble DNA fragments in living cells.
   c. It involves the use of overlapping DNA fragments, exonuclease, DNA polymerase, and ligase to seamlessly assemble multiple DNA fragments in a single reaction.
   d. It assembles DNA fragments by ligating blunt ends without requiring homologous overlap.

LAB SESSION 32

1. True or False: RNA-seq and RT-qPCR are both techniques that measure mRNA levels.
2. Which process converts RNA to complementary DNA (cDNA)?
   a. replication.
   b. transcription.
   c. reverse transcription.
   d. translation.
3. What is the correct order of steps in an RNA-seq workflow?
   a. RNA purification, cDNA library synthesis, reference genome alignment, RNA sequencing, quantification.
   b. RNA purification, reference genome alignment, cDNA library synthesis, RNA sequencing, quantification.
   c. RNA purification, cDNA library synthesis, RNA sequencing, reference genome alignment, quantification.
   d. RNA purification, RNA sequencing, cDNA library synthesis, reference genome alignment, quantification.
4. What website will you visit to access existing RNA-seq data?
5. If a gene is more highly expressed in a treated sampled compared to an untreated sample, will the log2 fold change be a negative or positive number?

LAB SESSION 33

1. Why is structure-activity relationship (SAR) modeling primarily used for in scientific research?
   a. Predicting the physical structure of a new protein based on DNA sequence data.
   b. Understanding the relationship between the chemical structure of a compound and its biological activity to design more effective drugs.
   c. Determining the stability of DNA under different environmental conditions.
   d. Measuring the metabolic rate of cells in response to drug treatments.

2. How will structure-activity relationship (SAR) molecular docking be used in this lab session?
   a. To analyze the predicted binding affinity between oxacillin and PBP2.
   b. To analyze the predicted binding affinity between EGFP and PBP2.
   c. To design molecules that will bind to the active site of PBP2.
   d. To determine the enzymatic efficiency of PBP2.
3. Why might predicted binding affinities vary when using different Structure-Activity Relationship (SAR) modeling programs?
   a. Each SAR program may use different algorithms, conformational flexibility of the proteins and or ligands, as well as other assumptions, which can lead to variations in predictions.
   b. SAR modeling programs are designed to generate random predictions for diverse compounds.
   c. Binding affinities depend solely on the experimental data, and SAR programs have no effect on the predictions.
   d. SAR programs are standardized and should always yield identical results for binding affinities.
4. What is the best definition of "binding affinity" in the context of molecular biology?
   a. The energy required to create a new molecular bond between two compounds.
   b. The strength of the interaction between a molecule, such as a drug, and its target, often represented by the stability or tightness of their binding.
   c. The number of binding sites available on a protein for a specific molecule.
   d. The ability of a molecule to be absorbed by cells in the human body.
5. Binding affinity can be affected by the number and type of non-covalent intermolecular interactions between a drug (i.e. ligand) and its enzyme target. Which of the following are non-covalent intermolecular interactions? Select all that apply.
   a. Hydrogen bonding.
   b. C-C bonding.
   c. Ionic interactions.
   d. Van der Waals forces.
   e. Disulfide bonding.

Appendix E

Software Download Instructions for Lab Session 33: Structure-Activity Relationship Modeling

INTRODUCTION

Prior to beginning Lab Session 33, there are several programs that need to be downloaded. It is recommended that instructors have students download software on their personal computers during class time to ensure successful acquisition of the software. Alternatively, if your laboratory has dedicated laptops to download the software ahead of time on these devices for classroom use.

The instructions listed below are for Windows or Linux operating systems only. MGLTools version 1.5.7 is **currently not compatible with Mac OS**.

Note, if you plan on adapting Lab Session 33 to perform molecular docking with novel ligands, you will require a program to make the ligand files. ChemDraw is a program that allows users to create novel ligand files; however, it requires a subscription license. If your organization does not have a license, ChemSketch is a free alternative. It is available at https://www.acdlabs.com/resources/free-chemistry-software-apps/chemsketch-freeware/ (accessed 1 Dec 2024).

The download instructions detailed below were adapted with permission from "Design and Implementation of an Accessible and Open-Sourced *In Silico* Drug Screening Activity for Cancer Drug Discovery," by Fikes and Srougi, *J. Chem. Educ*. 2023, 100, 10, 4125–4130.[1] Copyright 2023 American Chemical Society.

SOFTWARE DOWNLOAD INSTRUCTIONS

There are a total of four programs required for molecular docking: (1) AutoDock Vina, (2), MGLTools, (3) PyMOL, and (4) Strawberry Perl.

I. AutoDock Vina

The AutoDock Vina download can be found at: https://github.com/ccsb-scripps/AutoDock-Vina/releases (accessed 1 Dec 2024).

1. Download v1.2.5 for Windows
2. Open the downloaded file
    a. A warning might appear about running an unrecognized app. Select More Info and then select Run Anyway.

II. MGLTools

The download for MGLTools can be found at: https://ccsb.scripps.edu/mgltools/downloads/ (accessed 1 Dec 2024).

1. Download the version for Windows (latest version is 1.5.7)
2. Open the downloaded file
3. The MGLTools Setup (InstallJammer Wizard) dialog box will appear. Select Next.

Molecular Biology Techniques. DOI: https://doi.org/10.1016/B978-0-443-26714-7.00035-1

4. Read and accept the terms of the license agree and select yes
5. The next window will be the Choose Destination Location window
   a. The default location should be C:\Program Files (x86)\MGLTools-1.5.7
   b. If this is not the default location, then select Browse and choose C:\Program Files (x86)\MGLTools-1.5.7
   c. Select Next
6. A dialog box about commercial usage and citing software will appear. Read the dialog box and select OK
7. Select Finish
8. The Python Molecular Viewer might open immediately, but you can close it.

III. PyMOL

The download for PyMOL for educational purposes can be found at: https://pymol.org/edu/?q=educational (accessed 1 Dec 2024),

1. Fill out the form and select Continue
2. Read the statements on the next page, type in the characters shown, and select "I Agree"
3. The download credentials should be sent to your email. Select the link in your email and enter the given username and password.
4. Download the PyMOL 2.0 License File
5. Select the Installers: PyMOL Download Page link
6. Select the Windows EXE Installer icon to download PyMOL
7. Open the PyMOL Installer
   a. The PyMOL Setup will recommend closing all other applications before continuing with the installation
8. Select Next
9. Read the terms of agreement and then select "I Agree"
10. Choose to either install PyMOL for All Users (requires admin privileges) and select Next
11. A dialog box will appear asking if you want to allow this app to make changes to your device. Select Yes.
12. Choose the Install location
    a. The default location in the C:\ drive should be fine. If you need to install PyMOL in a different location, select Browse to choose the desired location
    b. Select Next
13. The Advanced Installation Options window will appear. Select the following options:
    a. Register PyMOL as the system Python 3.7
    b. Clear the package cache upon completion
14. Select Install
15. Once the progress bar is complete, select Next
16. Select Finish

IV. Strawberry Perl

The Download for Strawberry Perl can be found at: https://strawberryperl.com/ (accessed 1 Dec 2024).

1. Download the appropriate version (32 bit or 64 bit) for your computer.
   a. If you are not sure what type of processor your computer has, go to Start > Settings > System > About
   b. The information should be listed under System Type
2. Open the downloaded file
3. The Setup Wizard should appear; select Next
4. Check the box next to I accept the terms in the License Agreement and select Next
5. Choose the Destination Folder
   a. Choose the default location in the C:\ drive. If you need to install Strawberry Perl in a different location, select Change to choose the desired location
   b. Select Next
6. Select Install
   a. A dialog box may appear asking if you want to allow this app from an unknown publisher to make changes to your device. Select Yes.

7. Select Finish
8. A README file will appear with more information about Strawberry Perl. You do not need to read this and may close the file. However, if you want to learn more about Strawberry Perl, you can read this and visit the given URLs for more information.

REFERENCE

1. Fikes Audrey G, Srougi Melissa C, Design and implementation of an accessible and open-sourced in silico drug screening activity for cancer drug discovery, Journal of Chemical Education 100 (10) (2023) 4125–4130. https://doi.org/10.1021/acs.jchemed.3c00307 0021-9584.

Index

Note: Page numbers followed by "*f*" and "*t*" refer to figures and tables, respectively.